Two[illegible]

Hematopoietic Stem Cell Biology

Stem Cell Biology and Regenerative Medicine

Series Editor
Kursad Turksen, Ph.D.
kturksen@ohri.ca

For other titles published in this series, go to
http://www.springer.com/series/7896

Motonari Kondo
Editor

Hematopoietic Stem Cell Biology

Humana Press

Editor
Motonari Kondo
Duke University Medical Center
Durham, NC
USA
motonari.kondo@duke.edu

ISBN 978-1-60327-346-6 e-ISBN 978-1-60327-347-3
DOI 10.1007/978-1-60327-347-3
Springer New York Dordrecht Heidelberg London

Library of Congress Control Number: 2009936792

Printed on acid-free paper

Humana Press is part of Springer Science+Business Media (www.springer.com)

Preface

In the summer of 1988, my developmental biology professor announced to the class that hematopoietic stem cells (HSCs) had finally been purified. Somehow, I never forgot the professor's words. When I started working in Dr. Irv Weissman's laboratory at Stanford as a postdoctoral fellow, I realized that the findings mentioned by the professor were from Weissman's laboratory and had been published in a 1988 edition of the journal *Science*. It has been over 20 years since the publication of that seminal paper, and since then tremendous advances in understanding the biology and maturation of HSCs, namely the process of hematopoiesis, which includes lymphocyte development, have been made. These discoveries were made possible in part by advancements in technology. For example, recent availability of user friendly fluorescence activated cell sorting (FACS) machines and monoclonal antibodies with a variety of fluorescent labels has allowed more scientists to sort and analyze rare populations in the bone marrow, such as HSCs.

All classes of hematopoietic cells are derived from HSCs. Stem cell biology draws enormous attention not only from scientists, but also from ordinary people because of the tremendous potential for development of new therapeutic application to diseases that currently lack any type of effective therapy. Thus, this type of "regenerative medicine" is a relatively new and attractive field in both basic science and clinical medicine. The potential of regenerative medicine has been proven with bone marrow transplantation, which is a common therapy used to treat patients with leukemia or congenital immunodeficiency. Discovery and identification of somatic stem cells, such as neuronal stem cells and liver stem cells, has lead to the idea that somatic stem cells may also be used in regenerative medicine. However, for clinical application, the barrier we must overcome is the source of the somatic stem cells used. Embryonic stem (ES) cells, which can be maintained and expanded in culture, can give rise to all types of cells in the body. However, since human embryos are necessary for the establishment of human ES cells, it is ethically challenging to obtain human ES cells. Since HLA matching is a critical factor for success in organ transplantation, enormous numbers of ES cells need to be established if ES cells are to be used for therapies. Therefore, it may not be realistic to use ES cells for clinical application. An alternative approach is to establish pluripotent stem cells with the nuclear transfer technique. Although nuclear transfer can be used to establish custom-made stem cells, it is again not realistic to use this technique for clinical application because nuclear transfer is a very difficult technique. In this sense, identification of inducible pluripotent stem (iPS) cells from mature cells by introduction of a set of transcription factors (approximately four genes) is striking. Establishment of iPS cells is accomplished by transfection of the set of genes into mature cells using methods already common to most laboratories. Although there are various issues that have to be overcome, we have finally reached the starting point where we can begin to apply stem cells to clinical settings.

Hematopoiesis research is the leading field in stem cell biology because experimental systems have already been established and HSCs can be purified relatively easily by FACS sorting. However, a lot of issues remain to be resolved. For example, how self-renewal potential is maintained in HSCs is still not understood. To understand this issue, we need to examine HSCs themselves as well as the bone marrow microenvironment, namely the HSC niche. In Chapter 1, Drs. Ema, Kobayashi, and Nakauchi provide an overview of the regulation of self-renewal and asymmetric cell division of HSCs. In Chapter 2, Drs. Iwasaki and Suda review recent advances in understanding the HSC niche. Drs. Tan, Kim, Wagers, and Mayack provide more insight by comparing HSCs to other somatic stem cells in Chapter 3.

Investigation of hematopoiesis and lymphopoiesis is a very important subject in hematology and immunology. Since HSCs localize at different organs during the development of animals, it is necessary to characterize HSCs from animals of different ages. Drs. Ezine, Gautreau, Parcelier, and Canque review characteristics of HSCs in animals at different maturational stages in Chapter 4. The role of transcription factors in hematopoiesis from HSCs is reviewed by Drs. Moriguchi, Suzuki, Engel, and Yamamoto in Chapter 5. Drs. Ikuta and Tani-ichi provide an overview of the role of cytokines in lymphopoiesis from HSCs by focusing on IL-7 function in γδT cell development in Chapter 6. Among hematopoietic cells, dendritic cells are unique because they are derived from either lymphoid or myeloid progenitors. The nature of dendritic cells is precisely described in Chapter 7 by Drs. Merad and Manz. Finally, recent advances in the understanding of leukemia stem cells are reviewed by Drs. Dao and Jamieson in Chapter 8.

These authors are experts and frontrunners in their fields. I sincerely appreciate each author's time and contribution to each chapter.

Motonari Kondo, M.D., Ph.D.

Acknowledgments

The preparation of this book required the efforts of many people. I especially thank the chapter authors for their contribution. I also thank Anne Lai and Eva Chung for critical reading of the text. I also wish to acknowledge the efforts of the editorial and production staffs at Springer.

Contents

Contributors

Bruno Canque
UMR CNRS 7151-Institut Universitaire d'Hématologie (IUH), Centre Hayem,
1 avenue Claude Vellefaux, 75475 Paris cedex 10, France

Kim-Hien T. Dao
University of California at San Diego School of Medicine, Moores Cancer Center,
3855 Health Sciences Road, Mail Code 0820 La Jolla, CA 92093, USA

Hideo Ema
Division of Stem Cell Therapy, Center for Stem Cell and Regenerative Medicine,
Institute of Medical Science, University of Tokyo, 4-6-1 Shirokanedai, Minato-ku,
Tokyo 108-8639, Japan

James Douglas Engel
Cell and Developmental Biology, University of Michigan Medical School,
Ann Arbor, MI 48109-2220, USA

Sophie Ezine
Inserm U591-Institut Necker, Université Paris Descartes, 156 rue de Vaugirard,
75730 Paris Cedex 15, France

Laetitia Gautreau
Inserm U591-Institut Necker, Université Paris Descartes, 156 rue de Vaugirard,
75730 Paris Cedex 15, France

Koichi Ikuta
Laboratory of Biological Protection, Department of Biological Responses,
Institute for Virus Research, Kyoto University, Kyoto 606-8507, Japan

Hiroko Iwasaki
Department of Cell Differentiation, The Sakaguchi Laboratory of Developmental
Biology, Keio University School of Medicine, 35 Shinanomachi, Shinjuku,
Tokyo 160-8582, Japan

Catriona H. M. Jamieson
University of California at San Diego School of Medicine, Moores Cancer Center,
3855 Health Sciences Road, Mail Code 0820 La Jolla, CA 92093, USA

Francis S. Kim
Section on Developmental and Stem Cell Biology, Joslin Diabetes Center, One Joslin Place, Boston, MA 02215; Department of Stem Cell & Regenerative Biology, Harvard University, and Harvard Stem Cell Institute, 42 Church Street, Cambridge, MA 02138, USA

Toshihiro Kobayashi
Division of Stem Cell Therapy, Center for Stem Cell and Regenerative Medicine, Institute of Medical Science, University of Tokyo, 4-6-1 Shirokanedai, Minato-ku, Tokyo 108-8639, Japan

Markus G. Manz
Institute for Research in Biomedicine (IRB), Via Vincenzo Vela 6, 6500 Bellinzona, Switzerland

Shane R. Mayack
Section on Developmental and Stem Cell Biology, Joslin Diabetes Center, One Joslin Place, Boston, MA 02215; Department of Stem Cell & Regenerative Biology, Harvard University, and Harvard Stem Cell Institute, 42 Church Street, Cambridge, MA 02138, USA

Miriam Merad
Department of Gene and Cell medicine and the Department of Medicine, Mount Sinai School of Medicine, 1425 Madison Ave, New York, NY 10029, USA

Takashi Moriguchi
Department of Medical Biochemistry, Tohoku University Graduate School of Medicine, 2-1 Seiryo-cho, Aoba-ku, Sendai 980-8575, Japan

Hiromitsu Nakauchi
Division of Stem Cell Therapy, Center for Stem Cell and Regenerative Medicine, Institute of Medical Science, University of Tokyo, 4-6-1 Shirokanedai, Minato-ku, Tokyo 108-8639, Japan

Aude Parcelier
UMR CNRS 7151-Institut Universitaire d'Hématologie (IUH), Centre Hayem, 1 avenue Claude Vellefaux, 75475 Paris cedex 10, France

Toshio Suda
Center for Integrated Medical Research, The Sakaguchi Laboratory of Developmental Biology,Keio University School of Medicine, 35 Shinanomachi, Shinjuku, Tokyo 160-8582, Japan

Mikiko Suzuki
Department of Medical Biochemistry, Tohoku University Graduate School of Medicine, 2-1 Seiryo-cho, Aoba-ku, Sendai 980-8575, Japan

Kah Yong Tan
Section on Developmental and Stem Cell Biology, Joslin Diabetes Center, One Joslin Place, Boston, MA 02215; Department of Stem Cell & Regenerative Biology, Harvard University, and Harvard Stem Cell Institute, 42 Church Street, Cambridge, MA 02138; Department of Molecular and Cellular Biology, Harvard University, 7 Divinity Avenue, Cambridge, MA 02138, USA

Shizue Tani-ichi
Laboratory of Biological Protection, Department of Biological Responses, Institute for Virus Research, Kyoto University, Kyoto 606-8507, Japan

Amy J. Wagers
Section on Developmental and Stem Cell Biology, Joslin Diabetes Center, One Joslin Place, Boston, MA 02215; Department of Stem Cell & Regenerative Biology, Harvard University, and Harvard Stem Cell Institute, 42 Church Street, Cambridge, MA 02138, USA

Masayuki Yamamoto
Department of Medical Biochemistry, Tohoku University Graduate School of Medicine, 2-1 Seiryo-cho, Aoba-ku, Sendai 980-8575, Japan

Principles of Hematopoietic Stem Cell Biology

Hideo Ema, Toshihiro Kobayashi, and Hiromitsu Nakauchi

Abstract Hematopoietic stem cells (HSCs) are able clonally and persistently to produce all blood cells while maintaining the HSC pool size, permitting sustained hematopoiesis throughout life. HSCs have served as an excellent model in stem cell biology, and HSC transplantation has served as a prototype for stem cell therapy in regenerative medicine. The research field of stem cell biology and regenerative medicine has been conceptually and practically established through the study of HSCs. After many years of work, we have finally begun to understand self-renewal and multilineage differentiation in HSCs at the molecular level. This chapter briefly introduces how particular molecules play roles in HSC function and regulation. The principles of HSC biology, however, are the focus of this review, which provides an overview of basic concepts in stem cell biology with reference to some historic work.

Introduction

Hematopoiesis is the tightly regulated process of blood cell formation. Short-lived mature blood cells are replenished on a daily basis by hematopoietic stem cells (HSCs) through large numbers of HSC progeny such as multipotent progenitors and lineage-committed progenitors. HSCs are the first-identified and the best-characterized stem cells, providing a first model for the stem cell system applicable to many other stem cells. Figure 1 shows the hematopoietic system, which is organized hierarchically with a series of cell populations arranged in rank from stem

H. Ema (✉), T. Kobayashi, and H. Nakauchi
Division of Stem Cell Therapy, Center for Stem Cell and Regenerative Medicine, Institute of Medical Science, University of Tokyo, 4-6-1 Shirokanedai, Minato-ku, Tokyo, 108-8639, Japan
e-mail: hema@ims.u-tokyo.ac.jp

M. Kondo (ed.), *Hematopoietic Stem Cell Biology,* Stem Cell Biology and Regenerative Medicine, DOI 10.1007/978-1-60327-347-3_1,

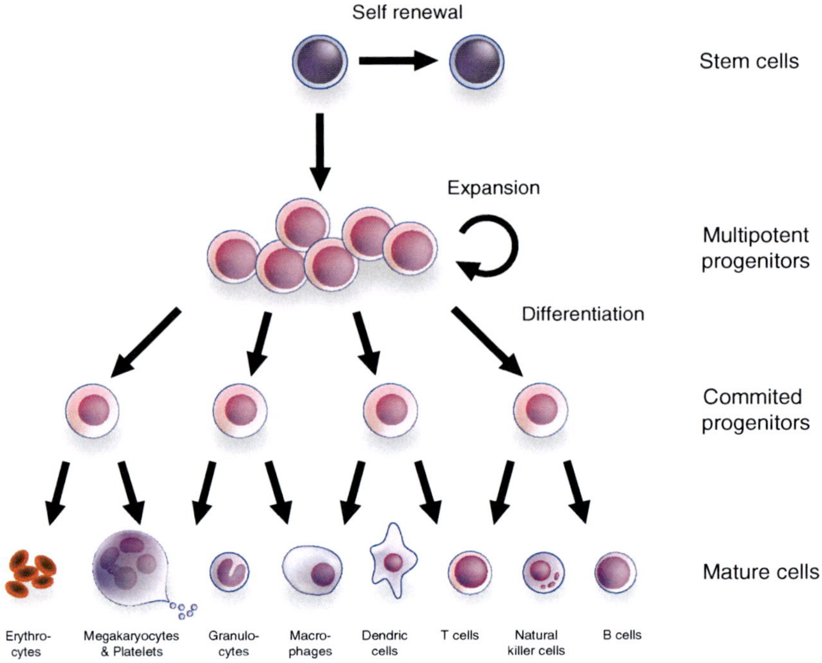

Fig. 1 The hematopoietic system. The hematopoietic hierarchy is illustrated. In concept, one stem cell asymmetrically gives rise to a stem cell to maintain its line and to a daughter cell destined for multilineage differentiation. Multipotent progenitor cells rather rapidly undergo symmetric self-renewal a limited number of times, leading to their amplification in a short interval. The entire hematopoietic system is established and maintained by a hematopoietic stem cell (HSC)

cells to mature blood cells. Most importantly, one stem cell is sufficient for establishment of the entire system.

Many basic concepts in stem cell biology have been created based on the study of HSCs. An asymmetric division model, a stochastic model (Till et al. 1964), a clonal succession model (Kay 1965), and a stem cell niche model (Schofield 1978) are good examples of such work. Some of these important concepts are revisited in this review. Although much progress has recently been made in this field, our understanding of the molecular mechanisms underlying the regulation of HSCs has just begun. This review presents the current understanding of the regulation of HSCs at the molecular level and highlights molecules controlling self-renewal. How do individual HSCs make a decision to self-renew or not self-renew in each division? How many times, and through what pathways, can they repeat this event? This review discusses why these two questions are still the main theme in the study of self-renewal.

The Definition and Entity of HSCs

Definition of Stem Cells

Stem cells are defined as cells with both self-renewal and differentiation potentials. This definition was initially framed by J. E. Till and E. A. McCulloch, based on their discovery and characterization of colony-forming units in the spleen (CFU-S) (Till et al. 1964; Till and McCulloch 1961). When mouse bone marrow or spleen cells are injected into lethally irradiated mice, visible colonies develop in the spleen of the recipient mice during a relatively narrow window of time after transplantation (e.g., 8–12 days). These colonies consist mainly of myeloid cells such as neutrophils, macrophages, erythroblasts, and megakaryocytes. The clonal nature of CFU-S was proven using chromosomal markers (Abramson et al. 1977). The sizes and cell compositions of colonies change with time after transplantation, indicating that CFU-S are heterogeneous. It is now known that CFU-S belong to a certain class of progenitors distinct from HSCs. The multipotent progenitor population is highly enriched in day 12 CFU-S (Yang et al. 2005), and the erythroid/megakaryocytic progenitor population is highly enriched in day 8 CFU-S (Pronk et al. 2007). Nonetheless, most of the concepts established, based on detailed analyses of CFU-S, have been proven to apply to stem cells, an instance of how concepts or models can remain important throughout the history of research in a particular field.

To self-renew is a process of producing one or two daughter cells identical to their parental cell via cell division. More precisely, it copies the entire genome, including epigenetic modifications. In practice, an act of self-renewal is a cell division that produces one or two daughter cells functionally equivalent to the parental cell. Both self-renewal potential and the full range of differentiation potentials ("multipotency") are maintained throughout self-renewing division. Differentiation is the process by which cells become more specialized, functional cells. Stem cells least likely differentiate without cell division. In this regard, to differentiate is a process of epigenetic change via cell division. In practice, an act of differentiation is a cell division that produces one or two daughter cells in which self-renewal potential or some part of multipotency is lost.

HSCs give rise to daughter cells that progressively restrict their differentiation potentials during their differentiation. Lineage commitment is the event that allows cells to differentiate further along one or more particular lineage, but not along other lineages. It can occur in cells at all differentiation stages, from HSC daughter cells to bipotent progenitor cells. In general, cells cannot cancel lineage commitment or reverse differentiation sequences. It is hoped that these events—self-renewal, differentiation, and commitment—will be more clearly and simply defined in molecular terms in the near future.

The Functional Heterogeneity of HSCs

The HSC population is composed of heterogeneous stem cells. HSCs are able both to differentiate into all possible blood lineages and to self-renew. Since this is basically a qualitative definition, if there is any difference among HSCs, this difference must be quantitative. It seems more difficult to quantitate differentiation potential than to quantitate self-renewal potential. The amount of repopulating activity in HSCs has been expressed in terms of the repopulating unit (RU) (Harrison et al. 1993). One RU is defined as the amount of repopulating activity per 10^5 bone marrow cells. The number of RUs in a certain number of test donor cells is calculated based on data from analysis of the peripheral blood of recipient mice in competitive repopulation (see the section "In vivo Assays" and Fig. 2 for details). The formula for RU calculation is: RU = ([% chimerism of test donor-derived cells]×[No. of competitor cells])/([% chimerism of competitor-derived cells] × 10^5). When RUs for individual HSCs were measured by single-cell competitive repopulation, variation in RU per HSC was detected among adult mouse HSCs. Since a positive correlation between RU and the number of secondary HSCs (the number of HSCs generated in a primary recipient mouse, as estimated by secondary transplantation) was found on analysis of single HSCs and their individually cloned descendants (e.g., on a clonal basis), it was suggested that HSCs are a heterogeneous population in regard to self-renewal potential (Ema et al. 2005). Retroviral tracking analysis (Jordan and Lemischka 1990) and single-cell transplantation (Dykstra et al. 2007) have demonstrated that, after transplantation, the in vivo kinetics of individual HSCs differ. How levels of self-renewal potential relate to particular patterns of in vivo kinetics is uncertain. Nonetheless, these data show that mouse HSCs are functionally heterogeneous. Similar heterogeneity has also been suggested for human HSCs (Guenechea et al. 2001).

When HSCs or pre-HSCs emerge in a developing embryo, they must constitute a small and homogeneous population. As these cells divide, the level of self-renewal potential might progressively fall. As a result, HSCs can become diverse in self-renewal capacity. In the generation-age hypothesis (Rosendaal et al. 1979), the previous division history of each stem cell is one determinant of the functional organization of the HSC population. Stem cells that have generated many stem cells (old stem cells) are hypothesized to form blood before stem cells that have generated few stem cells (new stem cells). After a given number of generations, two progenitor cells arise during division, and a stem cell thereby is lost. While we are not sure that order exists in the use of HSCs, we find this hypothesis attractive (Ema and Nakauchi 2003). In the strict sense, HSCs may not be able to regenerate HSCs that are exactly the same as themselves. Elucidation of the molecular mechanisms underlying self-renewal potential will address this issue.

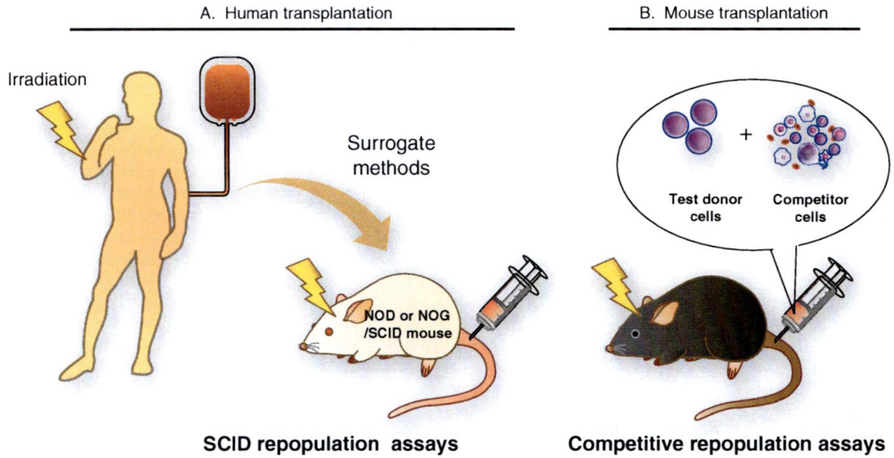

Fig. 2 Bone marrow transplantation and stem cell assays. **(A)** Bone marrow or cord blood transplantation in man. SCID repopulating cells can be detected by xenotransplantation. The nonobese diabetes/severe combined immunodeficiency (NOD/SCID) or NOD with common gamma receptor deficiency (NOG)/SCID mice are currently used as recipient animals after sublethal irradiation. **(B)** Mouse transplantation models are basically similar to human transplantation in regard to procedures. In competitive repopulation, a certain number of test donor cells (among which hematopoietic stem cells are expected) and a standard number of unfractionated bone marrow cells are cotransplanted into each group of lethally irradiated mice

In vivo and In vitro Assays for HSCs

In vivo Assays

Patients with hematological disorders such as leukemia and lymphoma have been successfully treated with bone marrow or cord blood transplantation. Conditioning regimens such as chemotherapy and total body irradiation are used to eliminate dysfunctional or malignant hematopoietic cells and to suppress the immune system before transplantation. When bone marrow cells are intravenously injected into myeloablated patients, the entire hematopoietic systems of the recipients are gradually but durably reconstituted with donor bone marrow cells if immunological reaction and infection are under control. Long-term reconstitution is solely due to engrafted HSCs. Experimental bone marrow transplantation in animal models is basically similar to therapeutic bone marrow transplantation in humans (Fig. 2).

In vivo, long-term, multilineage reconstitution is the gold standard for detection of HSCs. In a mouse model, test donor cells are retrospectively inferred to have contained one or more HSCs when their transplantation into lethally irradiated mice results in long-term multilineage reconstitution. Competitive repopulation is a modified version of transplantation that ensures the survival of lethally irradiated recipient mice and permits the quantitative evaluation of repopulation levels. Competitor cells, which are usually normal bone marrow cells, are cotransplanted in most instances (Fig. 2). In experimental transplantation, test donor cells and competitor cells are distinguished by certain genetic markers. The markers CD45.1 (Ly5.1) and CD45.2 (Ly5.2) are commonly used. CD45.1 C57BL/6 mice are congenic to CD45.2 C57BL/6 mice. CD45 is the leukocyte common antigen expressed on all leukocytes and their progenitor cells, including adult HSCs, but not on red blood cells, platelets, and their committed progenitors. A minimum criterion is that myeloid, B-lymphoid, and T-lymphoid populations need to be demonstrably reconstituted with test donor–derived cells for 3–4 months to show that test donor cells contained HSCs. When these three lineages are demonstrably reconstituted, erythroid and megakaryocytic lineages will almost certainly have been reconstituted with progeny of the same test donor cells. However, a better system, one that permits simultaneous analysis of leukocytes of all lineages, red blood cells, and platelets, should be used when it is available. Among these three lineages, the myeloid lineage is more important than the lymphoid lineages. In rare instances, the myeloid lineage is demonstrably reconstituted 4 months after transplantation, but the lymphoid lineages are not. Lymphoid lineage reconstitution might become detectable later on in such cases.

When multilineage reconstitution takes place 4 months after transplantation, this is conceivably due to self-renewal and differentiation potentials in HSCs. To confirm self-renewal potential, however, secondary transplantation needs to be performed. When long-term multilineage reconstitution takes place again in secondary recipient mice, this is certainly due to self-renewal and differentiation potentials in HSCs. We have recently realized that some HSCs show little reconstitution activity 4 months after primary transplantation, but show significant reconstitution activity after secondary transplantation. To detect all HSC activities, serial transplantation is mandatory. In the strict sense, the criteria of HSCs may have to depend on data from serial transplantation.

We hold that only transplantation of HSCs enables long-term reconstitution of granulocytes, and that unbalanced reconstitution between myeloid and lymphoid lineages, rarely observed, may be associated with HSC aging (Sudo et al. 2000). Because lymphocytes have a long lifespan compared with granulocytes and other myeloid lineage cells, when lymphoid lineages are reconstituted 4 months after transplantation but the myeloid lineage is not, it most likely indicates that test donor cells included lymphoid progenitors, but not HSCs.

Needless to say, we are unable to assay human HSCs experimentally in patients. Nonobese diabetes/severe combined immunodeficiency (NOD/SCID) or NOD/SCID mice crossed with common gamma receptor knockout (NOG/SCID) mice are currently used as a surrogate xenotransplantation assay (Larochelle et al. 1996). This assay is useful to detect human multilineage repopulating cells, but most recipient mice hardly survive beyond a year. We are not certain that such observation periods are long enough to detect self-renewal and multilineage differentiation potential in human HSCs. Primates can be used as an alternative animal model. For instance, monkey HSCs, instead of human HSCs, can be studied. But, who will be able to monitor those animals for up to 50 years of life after bone marrow transplantation to identify monkey HSCs? The only practical way to identify human HSCs is to find human cells equivalent to mouse HSCs in every possible respect. For instance, HSCs might be identified by expression of a combination of functional molecules, such as $p57^{Kip2}$, the c-Kit and c-Mpl receptors, and the ABC transporter G2.

In vitro Assays

In vitro colony assays (Fig. 3) have already become classic, and they remain one of the most important assays in the study of hematopoiesis. This assay was invented in the 1960s by two research groups, one in Israel (Pluznik and Sachs 1965) and the other in Australia (Bradley and Metcalf 1966). Hematopoietic progenitor cells can grow in semisolid media such as agar and methylcellulose, finally forming colonies just like bacteria. Unlike bacteria, hematopoietic progenitor cells need to be supplemented with growth factors to proliferate in culture. Also unlike bacteria, a variety of cell types are derived from a single cell. This assay thus can be called a clonal differentiation assay. Don Metcalf, one of the pioneers in use of this assay, describes colonies as three-dimensional populations of cells of wonderfully variable shapes and sizes that look like galaxies as approached by a fast-moving spaceship (Metcalf 2000). Their beauty has continued to fascinate him for over 40 years! The significance of this assay system has been appreciated, for the most part, after discovery of a number of colony-stimulating factors. These factors, interleukins among them, are collectively called cytokines. To date at least 20 hematopoietic

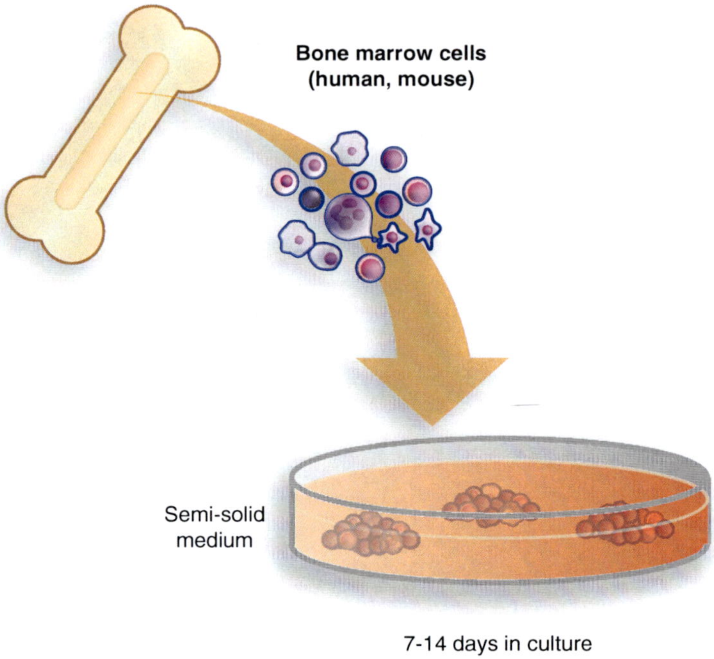

Fig. 3 In vitro colony assays. Cells from hematopoietic tissues are plated in semisolid medium with a combination of cytokines, such as stem cell factor (SCF), thrombopoietin (TPO), interleukin-3 (IL-3), and erythropoietin (EPO) at minimum. One progenitor gives rise to a colony composed of a variety of cell types in 7–14 days

cytokines are known. Had colonies grown in vitro been unattractive to Metcalf and others, this assay might not have been used by so many researchers for so long.

When 10,000 mouse bone marrow cells are plated in methylcellulose culture containing a combination of cytokines minimally required for colony formation, such as stem cell factor (SCF), thrombopoietin (TPO), interleukin-3 (IL-3), and erythropoietin (EPO), a hundred or more colonies are formed. Most colonies are derived from progenitors but not from HSCs; HSCs are barely present in this number of bone marrow cells. Whether HSCs could form in vitro colonies was not known until 2004 (Takano et al. 2004). Using this assay with highly purified cells, testing can show which cytokines directly act on HSCs. Unfortunately, purification levels have not reached 100%. Researchers continue to try to increase the degree of HSC purification, as discussed below. SCF, TPO, IL-3, IL-6, IL-11, and others are the cytokines that affect HSCs. These cytokines were earlier recognized as early acting cytokines, using blast-colony assays in which blast colonies were formed by bone marrow or spleen cells from mice treated with the cytotoxic drug 5-fluorouracil. HSCs are barely present in these blast colonies. But most of what was found in blast-colony assays has also been found to apply to HSCs. A series of blast-colony studies systematically performed by Makio Ogawa (Ogawa 1993) has been found to be of great value by many HSC researchers.

In vitro colony assays are useful for analysis of HSCs and progenitor cells. Myeloid differentiation potential can be examined at the clonal level for newly isolated populations. Using knockout mice, stages at which differentiation block occurs can be analyzed. If early hematopoietic differentiation is blocked in knockout mice, it becomes impossible to detect HSC activity properly. The primary importance in HSC studies is to determine, using in vitro colony assays, whether myeloid differentiation potential in HSCs is normal or not; but many published studies neglect this simple test.

Whether human HSCs can form colonies in vitro is not clear. $CD34^+$ cells from human bone marrow and cord blood contain a subset of HSCs (Kato et al. 2001). These cells unlikely express CD38. Most $CD34^+CD38^-$ cells do not form colonies in methylcellulose culture, even in the presence of a combination of many cytokines. In humans, as in mice, $CD34^-$ cells are considered more immature than $CD34^+$ cells (Bhatia et al. 1998). $CD34^-$ cells do not form colonies at all. This is possibly because culture conditions are not optimal for human HSCs. Human HSCs may be able to form a colony only in contact with stromal cells.

The Purification of Mouse HSCs

High Degrees of Mouse HSC Purification by Flow Cytometry

The frequency of HSCs among bone marrow cells in adult C57BL/6 mice has been estimated to be on the order of 1 in 10^4–10^5 cells. To study HSCs, we must prospectively isolate them from other bone marrow cells. Visser et al. were the first to attempt to enrich HSCs by flow cytometry (Visser et al. 1984). This approach was further refined by Spangrude et al., who succeeded in higher purification of adult mouse HSCs (Spangrude et al. 1988). The Thy-1^{low}, Sca-1^+, and lineage marker–negative fraction was most enriched in adult bone marrow HSCs at that time. Because the Thy-1 marker is only applicable to Thy-1 congenic mice (C57BL/Ka-Thy-1.1), it has been replaced by c-Kit receptor in many laboratories. Single-cell transplantation was very difficult with this Thy-1^{low}, Sca-1^+, and Lin^- population (Smith et al. 1991). Introduction of anti-CD34 antibody significantly increased the level of HSC enrichment. Unexpectedly, HSCs (Osawa et al., 1996), were found in the $CD34^{low/negative}$ fraction of bone marrow cells. This has defined cells with the $CD34^{-/low}$, c-Kit^+, Sca-1^+, Lin^- ($CD34^-$KSL) phenotype as highly enriched in HSCs (Osawa et al., 1996), thus enabling single HSC transplantation and clonal characterization (Ema et al., 2006).

On average, 25% of recipient mice show long-term multilineage reconstitution after transplantation of single $CD34^-$KSL cells from 8- to 10-week-old C57BL/6 mice (Ema et al. 2005). Recently Slam-family receptors have been used for HSC purification. $CD150^+$, $CD48^-$, $CD41^-$ cells are also highly enriched in mouse HSCs (Kiel et al. 2005; Kim et al. 2006). Most of these cells express c-Kit and Sca-1, but not lineage markers. On the other hand, most $CD34^-$KSL cells express CD150. $CD34^-$KSL cells and $CD150^+CD48^-CD41^-$ cells mostly overlap, but the precise relationship of these two phenotypes remains to be determined. The side population (SP) phenotype detected by flow cytometry analysis of cells stained with the supravital dye Hoechst 33342 has also been useful for HSC purification (Goodell et al. 1996). However, it should be kept in mind that not all HSCs show the SP phenotype at any one time (Morita et al. 2006), although most HSCs express the adenosine triphosphate–binding cassette transporter Bcrp1 whose function is responsible for the SP phenotype (Zhou et al. 2002).

In our experience, data vary significantly among individual single-cell transplantation experiments. This might result from a number of reasons, but is perhaps due to variations among individual donor mice, the number of HSCs included in competitor cells, and conditions of competitor cells and recipient mice. To control these conditions appears difficult. Nevertheless, the criteria for long-term multilineage repopulating cells must be unified. The source and number of competitor cells, with the definition of long-term and multilineage reconstitution, are the most important conditions. From both logical and practical points of view, we propose that 8- to 12-week-old male or female C57BL/6 mice be used as HSC donors and recipients, that 2×10^5 normal and unmanipulated bone marrow cells be used as competitor cells, that long term be defined as at least 4 months, and that 1% or more peripheral blood chimerism with detectable test donor cell-derived myeloid, B-lymphoid, and T-lymphoid cells be defined as multilineage reconstitution. This percentage is obtained by ([% test donor cells] × 100)/(% test donor-derived cells + % competitor-derived cells). In some laboratories, 0.1–0.3% chimerism is adopted as defining multilineage reconstitution, which is useful for certain purposes. We, however, believe that if one wishes to compare purification levels between different protocols, the above criteria should be agreed on and shared.

Toward the Ultimate Purification of HSCs

If only 25% of cells in recipient mice show long-term multilineage engraftment after transplantation of single $CD34^-$KSL cells, what are the remaining 75% of cells? Every step from the purification and transplantation processes through the in vivo repopulating process must be carefully reviewed, taking into account cell damage by laser or high-pressure sorting, technical error upon injection of a cell mixture, seeding (or homing) failure, and self-renewal failure by chance. Some find it difficult to believe that every $CD34^-$KSL cell successfully homes to a bone marrow niche and starts self-renewing division. Most $CD34^-$KSL cells exhibit myelopoietic activity on in vitro assay. Their in vitro proliferation capacity varies greatly. Multipotent progenitor cells with variably limited self-renewal potentials may be similar to HSCs in respect to their surface markers. These two types of cells, HSCs and multipotent progenitor cells, are rather artificially separated based on the outcome of transplantation assays, but they could be cells along a continuous gradient.

Nevertheless, we have attempted to detect repopulating cells among $CD34^-KSL$ cells by increasing the sensitivity of competitive repopulation. When compromised bone marrow cells were used as competitor cells, 54% and 26% of mice transplanted with single $CD34^-KSL$ cells showed, respectively, long-term and short-term reconstitution. If technical error is $\leq 10\%$, approximately 90% of cells in this particular experiment proved to be repopulating cells. We still do not know what the remaining 10% of cells were. Nonetheless, HSC seeding efficiency should be very high. Such experiments must be repeated and extended.

We have recently realized that not many antibodies that recognize cell surface markers can be used to separate the $CD34^-KSL$ cell population further into positive and negative subpopulations. Some markers, such as integrin, can further separate cells in this population but do not always lead to further enrichment of HSCs. So far our screening of antibodies that can distinguish subclasses of $CD34^-KSL$ cells has not identified any whose use significantly increases the engraftment rate. Although most cells in the SP tip reportedly have stem cell activity (Matsuzaki et al. 2004), this finding has been difficult to reproduce. Is it possible that $CD34^-$ KSL cells cannot be further functionally separated by surface markers? We shall persevere in this approach to selection of HSCs with high self-renewal activity.

The Fate of HSCs

Fate Determination Units

Francois Jacob said that "The dream of every cell is to become two cells." That is also the dream of HSCs. The task of HSCs is to divide as many times as they can, but most HSCs are asleep, or hibernating, at any given time. HSCs may dream of becoming daughter cells while they are sleeping. Historically, the concept of the G_0 phase in the cell cycle emerged from the study of HSCs (Lajtha 1979). HSCs in the G_0 phase, presumably residing in their niches, seem to maintain their pool size and, in effect, be protected from a variety of pressures, such as proliferative stress and oxidative stress, under physiological and pathological conditions.

Knockout mouse studies have convincingly shown that the lack of a negative regulator in the cell cycle leads to the reduction of the stem cell pool. The Forkhead O (FoxO) subfamily of transcription factors is present in the nucleus of hibernating HSCs (Yamazaki et al. 2006). By activation of the PI3K/Akt signal transduction pathway, FoxO proteins are phosphorylated and exported from the nucleus to the cytoplasm. As a result, HSCs enter the cell cycle. FoxO1, FoxO3, and FoxO4-triple knockout HSCs are defective in long-term reconstitution (Tothova et al. 2007). These results suggest that a return to the G_0 phase after cycling is necessary for HSCs to reconstitute successfully and durably the hematopoietic system of lethally irradiated mice. We do not yet know how the G_0 to G_1 phase transition or its reverse is controlled in HSCs. FoxO proteins might play a role in the HSC G_1/G_0 transition or the maintenance of HSC hibernation. Cyclin C in combination with Cdk3 has been reported to stimulate pRb phosphorylation, which occurs when cells exit from the G_0 phase (Ren and Rollins 2004). It would be interesting to know whether the cyclin C/Ddk3 complex plays a similar role in HSCs.

It is generally accepted that most HSCs in the adult mouse bone marrow enter the cell cycle at least once per month. By calculation, over 10,000 HSCs exist in the bone marrow of an adult mouse. It is important to know how individual cell cycle status is orchestrated among such a large number of HSCs. To imagine that this is controlled only by the local niches is difficult.

Figure 4 sets out possible explanations of many aspects of HSC behavior. Once HSCs enter the cell cycle, fate choices exist. Till et al. provisionally called HSC fate either "birth" or "death" (Till et al. 1964). Certainly, HSCs have a choice of dying by apoptosis. But Till et al. called differentiation "death" since many cells die at the end of terminal differentiation. They assumed that HSC fate is controlled by varying birth and death probabilities. Their concept is still vivid. Our hypothesis is that HSCs undergo either self-renewal or differentiation at each cell division. If this is the case, one cell cycle could be the minimum unit for fate determination in HSCs.

Three Types of Cell Division

Figure 5 illustrates three manners of stem cell division. When two daughter cells are stem cells, this division is termed symmetric self-renewal. When one daughter cell is a stem cell and the other is a progenitor cell, this division is termed asymmetric

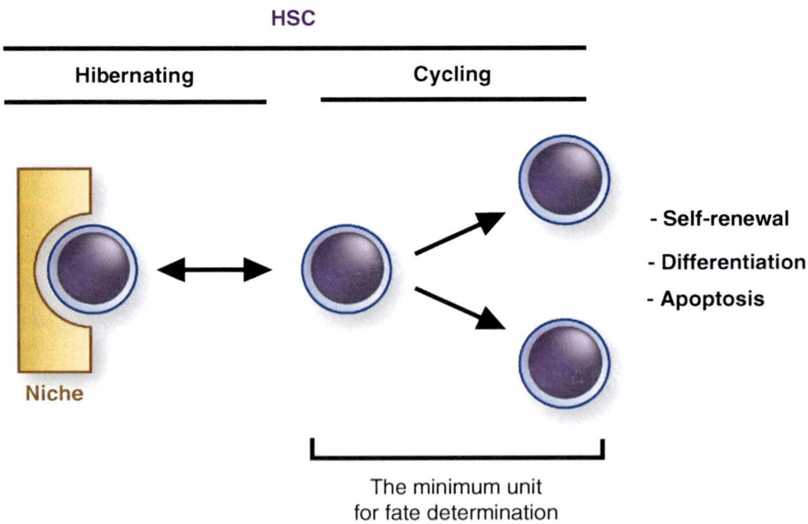

Fig. 4 Fate determination in hematopoietic stem cells (HSCs). Under physiological conditions, most HSCs are likely hibernating in particular niches. HSCs regularly enter the cell cycle by unknown mechanisms. HSCs take one of three choices—self-renewal, differentiation, or apoptosis—via cell division

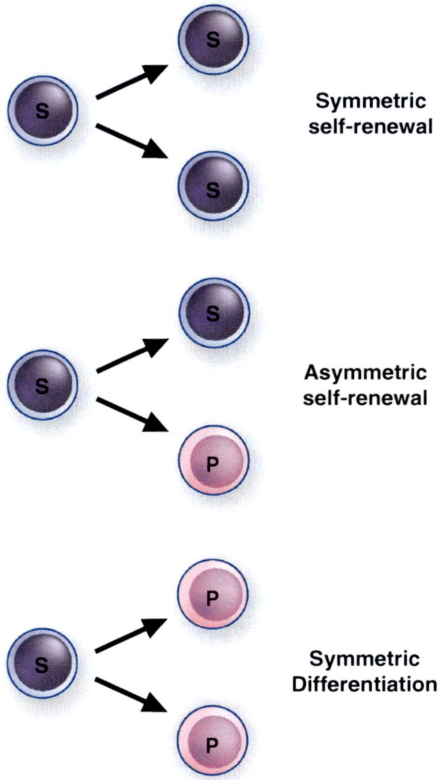

Fig. 5 Manners of stem cell division. Three types of cell division in hematopoietic stem cells are shown

self-renewal. When both daughter cells are progenitor cells, this division is termed symmetric differentiation. HSCs are detectable only by in vivo functional assays. Their surface markers have been characterized extensively, but no specific marker associated with HSC function has yet been identified. Thus we cannot be absolutely sure which cells in situ, as in the bone marrow, are HSCs. At this moment, it is impossible to study each HSC division in vivo. We therefore have attempted to recapitulate the three types of HSC division in culture.

Micromanipulation techniques for so-called paired daughter cells (Suda et al. 1984) have been adapted to single-cell transplantation (Fig. 6). After single HSCs divide once, each daughter cell is separated from the other and is transplanted into a lethally irradiated mouse, together with rescue cells. Recipient mice are analyzed 12 weeks or more after transplantation to evaluate whether transplanted single daughter cells were HSCs. We have so far observed only asymmetric self-renewal, but not symmetric self-renewal, in HSCs from wild-type mice when division was

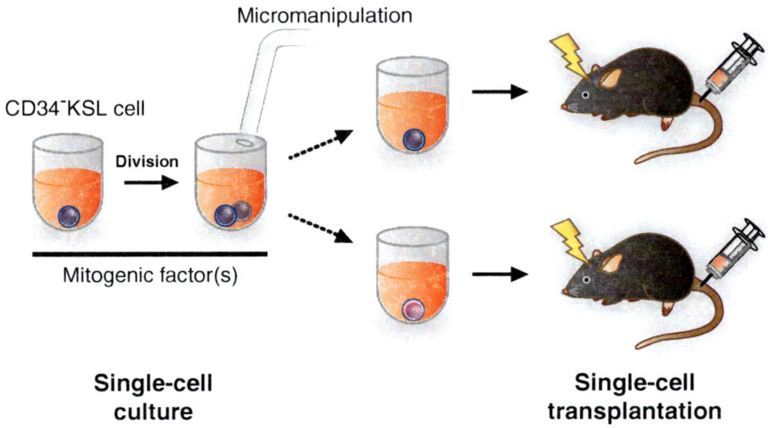

Fig. 6 Paired daughter cell experiments. Single-cell cultures are performed in a 96-well microtiter plate. Single hematopoietic stem cells (HSCs) divide once in the presence of cytokines (e.g., SCF + TPO). Following division, each daughter cell is separated by micromanipulation and is individually injected into a lethally irradiated mouse together with competitor cells. Long-term multilineage repopulation by individual daughter cells is evaluated 3–4 months or more after transplantation

induced in serum-free medium by a combination of minimal cytokines such as SCF and TPO (Ema et al. 2000; Seita et al. 2007). This is why ex vivo expansion and maintenance of HSCs have not been easily achieved. However, we could observe symmetric self-renewal divisions in Lnk-deficient HSC. Lnk seems negatively to control the probability of self-renewal downstream of TPO/c-Mpl signaling in HSCs (Seita et al. 2007). Because of this, Lnk-deficient HSCs have a higher chance of self-renewal divisions even in vitro.

Whether the fate of HSCs is determined after only one division has not yet been formally tested. Whether daughter cells of HSCs are stem cells can be known only retrospectively. This raises the possibility that their fate is actually determined after subsequent divisions. We are not sure that this alternative mode is useful for further study because the latter model is more complicated than the former. For example, the outcome of a first asymmetric self-renewal division might be the outcome of the combination of a first division, resulting in symmetric self-renewal division with, in one daughter cell, a second division, resulting in asymmetric or symmetric self-renewal, and, in the other daughter cell, a second division, resulting in symmetric differentiation.

We are not entirely sure that the data from paired daughter cell experiments faithfully reflect in vivo events, but the experimental system should serve in searches for molecular mechanisms underlying fate determination in HSCs.

Molecular Basis for Self-Renewal

How can we address self-renewal at the molecular level? Two aspects of self-renewal should be kept in mind. First, self-renewal is a type of cell division, and whether to self-renew is chosen by HSCs under the influence of various extrinsic factors. Second, although the self-renewal capacity of HSCs is extensive, it is not infinite. Cytokine networks and niche elements play crucial roles in induction and inhibition of HSC division. Subsequent intracellular signal transduction and transcriptional activation play major roles in HSC fate determination.

We have begun to identify the molecules that regulate the self-renewal division of HSCs. Basic cell cycle machinery has to function all the time for all dividing hematopoietic cells. The genome must be protected from DNA damage or mutation. DNA repair (Nijnik et al. 2007) and maintenance of sufficient telomere length (Allsopp et al. 2003a) are necessary to support a number of self-renewal divisions in HSCs (Rossi et al. 2007). These protective mechanisms should be held in common with most cells, but abnormalities in these mechanisms are more readily detected in HSCs than in HSC progeny. The argument that slowly dividing cells with a long lifespan have increased chances to suffer genomic DNA damage and telomere shortening than do rapidly dividing cells with a short lifespan makes no sense. If HSCs have a problem in their genome, their progeny have the same problem. Hematopoiesis is more severely disturbed by defects in HSCs than by defects in progenitor cells. Hibernation should be a necessary behavior in HSCs. There is no evidence that mechanisms controlling hibernation in HSCs differ from those in memory B- and T-cells. At the moment, $p57^{Kip2}$, which may function in HSCs but not in lymphocytes, is the only possible exception (Yamazaki et al. 2006). All molecular mechanisms regulating or supporting self-renewal in HSCs are closely related to one another, but in studies they must be individually addressed. Otherwise, we may not be able to understand which mechanism is specific to HSCs.

Extrinsic Control of HSCs

In the adult bone marrow, HSCs seem to lie in contact with osteoblasts and vascular endothelial cells. Accordingly, osteoblastic niches (Calvi et al. 2003; Zhang et al. 2003) and vascular niches (Kiel et al. 2005) have been proposed to regulate HSCs. The niche has two hypothesized roles. One role is to induce or support self-renewal in HSCs. The other is to induce and maintain hibernation in HSCs. It is not easy to imagine that both events are controlled by the same niche. This could be explained by an interesting assumption, which is that the osteoblastic niche controls hibernating HSCs and the vascular niche controls cycling HSCs (Arai et al. 2005). So far no published data support this distinction.

To control HSCs in number, the niche might provide specific molecules required for HSCs to self-renew. Reya et al. reported that Wnt3a is one such candidate molecule (Reya et al. 2003; Willert et al. 2003). A constitutively active form of β-catenin reportedly enhances self-renewal in HSCs and the canonical Wnt/β-catenin pathway reportedly is activated in self-renewing HSCs (Reya et al. 2003). Unfortunately, none of these claims has been supported by subsequent studies. We have not observed Wnt3a to act as a mitogen for HSCs in the presence of SCF nor seen that Wnt3a can promote self-renewal in HSCs (Ema, unpublished data, 2007). Two groups reported instead that conditional expression of a constitutively active form of β-catenin in HSCs blocks differentiation in HSCs, leading to loss of their repopulating activity (Kirstetter et al. 2006; Scheller et al. 2006). Other studies using a conditional knockout strategy showed that reconstitution activity in β-catenin$^{-/-}$ HSCs is unaffected (Cobas et al. 2004). Concern that γ-catenin might compensate for absence of its closely related homologue, β-catenin, in the canonical pathway was allayed when two independent groups convincingly showed that γ-catenin and β-catenin double knockout mice have HSCs with normal repopulating activity (Jeannet et al. 2008; Koch et al. 2008). Later, Reya et al. claimed again that repopulating activity in β-catenin$^{-/-}$ HSCs is reduced (Zhao et al. 2007), leaving observers confused. Whether any Wnt protein is provided to HSCs in vivo after transplantation remains uncertain. If there is no Wnt signal, it must be difficult to see any difference between wild-type and β-catenin$^{-/-}$ HSCs after transplantation. We believe that other potential self-renewal factors must be sought rather than to expect too much from the positive effect of one or two particular members of the Wnt family on self-renewal in HSCs (Trowbridge et al. 2006).

Nevertheless, HSCs must be susceptible to external mitogenic stimuli. Cytokines should play a major role in HSC division. The other role of the niche is to keep HSCs in hibernation, as discussed above. Several lines of evidence clearly show that the G_0 phase in HSCs is necessary for maintenance of the HSC pool. This is rather easily understood if we believe that continuously cycling HSCs are soon exhausted because their number of cell divisions is limited. The question again is how their hibernation is controlled. Some cytokines negatively regulate HSCs, for instance TGFβ1 (Yamazaki et al. 2008) and angiopoietin 1 (Arai et al. 2004). But there is little evidence that these play a role in vivo. Members of the TGFβ superfamily signal through a complex consisting of type I and type II serine/threonine kinase receptors. Upon ligand binding, type II receptors recruit and transphosphorylate type I receptors, and then activate Smad transcription factors. The TGFβ signal is processed through Smad2 and Smad3; the BMP signal is processed through Smad 1, Smad5, and Smad8. Receptor-activated Smads of both pathways form heterooligomers with the common mediator Smad4. Studies of conditional Smad4 knockout mice showed that maintenance of the HSC pool depends on activation of Smad4 (Karlsson et al. 2007). Redundancy among negative regulators should exist. However, it can be reasonably assumed that at least certain members of the TGFβ superfamily are involved in the regulation of HSC hibernation.

With respect to asymmetric division, the niche may play a role in cellular polarization in HSCs under conditions of self-renewal or hibernation. The study of germline stem cells (GSCs) in the fruit fly, *Drosophila melanogaster*, has provided great insights into interactions between stem cells and their niches. In the ovary, two or three GSCs reside in a niche comprising cap cells and other cells. When a GSC gives rise to two daughter cells, one daughter cell stays as a stem cell in the niche, given decapentaplegic (Dpp, a homolog of human bone morphogenetic proteins 2 and 4) signaling. The other daughter cell moves out of the niche and begins a program of differentiation (Xie and Spradling 2000). In the testis, GSCs are maintained in a niche composed of hub cells. GSCs in the testis also undergo asymmetric division. One daughter cell remains in the niche as a stem cell, and the other, displaced from the niche, begins to differentiate (Yamashita et al. 2003). Activation of Jak-Stat and E-cadherin/Armadillo (β-catenin) signaling is reportedly required for self-renewal (Yamashita et al. 2003; Tulina and Matunis 2001). In the central nervous system of *Drosophila*, a neuroblast asymmetrically divides into a neuroblast and a ganglion mother cell (GMC) (Wodarz and Huttner 2003; Roegiers and Jan 2004). A GMC in turn gives rise to a neuron and a glial cell. The orientation of the mitotic spindle is perpendicular to the plane of the overlying neuroepithelium. Cell fate determinants are known to be segregated exclusively to one daughter cell or another. For example, Numb, Prospero, Partner of Numb, and Miranda are localized to the basal pole of parental neuroblasts during mitosis and are segregated into GMCs. In contrast, Bazooka, Par-6, and PKC as a complex are localized to the apical pole of parental neuroblasts and are segregated into neuroblasts. Numb acts as an inhibitor in Notch signaling, but it is unclear how Numb and other proteins regulate the fate of neuroblasts.

In all these stem and progenitor cells, polarization of cells upon mitosis seems critical for asymmetrical fate specification, with fate determined in a fixed manner. The number of cell divisions that differentiated daughter cells can undertake is limited. Unlike these stem cells, polarity has not been recognized in HSC division. Even if polarity exists, it might arise in the absence of the niche. The probabilities of asymmetrical self-renewal in HSCs may change under certain conditions, and daughter cells of HSCs can undergo numerous divisions. It is possible that the regulatory mechanisms of asymmetrical division in HSCs fundamentally differ from those in other stem cells. To address this issue, we need to identify HSC-specific niches and the niche-associated molecules that regulate HSCs.

Intrinsic Control of HSCs

Changes in intracellular signal transduction and transcriptional activity remarkably affect the outcome of division in HSCs. In the context of self-renewal, outcomes would be modulation of self-renewal probability and modulation of numbers of divisions remaining to HSCs. No particular self-renewal signal transduction pathway has been identified. Self-renewal may instead result from activation and inactivation of multiple signal transduction pathways.

Lnk is an adaptor protein that negatively regulates multiple signal transduction pathways. HSCs are increased in number and self-renewal potential in $Lnk^{-/-}$ mice (Ema et al. 2005). This has turned out to be due to hypersensitivity to TPO, with increased probability of symmetric self-renewal in response to TPO or to TPO and SCF (Seita et al. 2007). How Lnk negatively regulates self-renewal events and self-renewal potential in HSCs is extremely interesting, but mostly remains to be clarified.

Pten is a phosphatase that negatively regulates phosphatidylinositol-3-OH kinase (PI3K)-Akt signaling. Conditional knockout studies showed that Pten$^{-/-}$ HSCs are continuously driven into cell cycle, resulting in poor maintenance of the HSC pool after transplantation into lethally irradiated mice (Yilmaz et al. 2006). The mammalian target of rapamycin (mTOR) is activated downstream of the PI3K-Akt pathway in Pten$^{-/-}$ HSCs. Interestingly, the drug rapamycin inhibits mTOR activity, leading to restoration of the HSC pool in mice transplanted with Pten$^{-/-}$ HSCs. A consequence of Pten knockout apart from HSC impairment is that T-lymphoblastic leukemia develops from a progenitor compartment in Pten$^{-/-}$ mice, accompanied by additional chromosomal translocation (Guo et al. 2008).

Constitutive activation of Jak2/Stat5 signal transduction at HSC level leads to myeloproliferative diseases. Point mutations that result in a valine to phenylalanine substitution at amino acid 617 of JAK2 kinase (JAK-V617F) are frequently found in patients with polycythemia vera, essential thrombocytopenia, and idiopathic myelofibrosis (Baxter et al. 2005; James et al. 2005; Kralovics et al. 2005). These diseases have recently been created in mouse models (Tiedt et al. 2008). Interestingly, differences in JAK-V617F mutant dosage seem to lead to different disease phenotypes. Expression of constitutively active forms of Stat5 in mouse HSCs results in aggressive myeloproliferative diseases (Schwaller et al. 2000; Kato et al. 2005). These studies together suggest that Jak2/Stat5 signaling is involved in the control of HSC self-renewal.

Transcription factors that play important roles in self-renewal must meet at least two conditions. First, they must be expressed in HSCs. Second, when they are deleted, self-renewal activity in HSCs must be reduced or lost. Runx1 (Okuda et al. 1996), Scl (Tal1) (Porcher et al. 1996; Robb et al. 1996), and Gata2 (Tsai et al. 1994) are considered essential for the development of HSCs because mouse embryos deficient for one of these transcription factors die due to lack of definitive hematopoiesis. Definitive hematopoiesis in the fetal liver is initiated mainly by progenitor cells derived from the yolk sac rather than from HSCs, so that mice lacking one of the above-named factors do not die for want of HSCs. HSCs are responsible for definitive hematopoiesis in later life, but HSCs do not have to contribute to hematopoiesis in murine embryos or fetuses. For instance, after HSCs emerge in E12.5 fetal liver, it takes 1 week or more for them to differentiate into mature neutrophils. Most of their mature progeny are used after birth. Mice deficient in Runx1, Scl, or Gata2 die mostly due to anemia and/or bleeding (impaired definitive hematopoiesis) (Okuda et al. 1996; Porcher et al. 1996; Robb et al. 1996; Tsai et al. 1994), but there is little evidence that generation of HSCs is impaired in such mice. Interestingly, the function of adult HSCs is not significantly impaired by conditional deletion of Runx1 (Ichikawa et al. 2004) or Scl (Mikkola et al. 2003; Hall et al. 2003). The number of HSCs is even increased in Runx1$^{-/-}$ adult mice (Ichikawa et al. 2008). So far there is no evidence that certain molecules essential for generation of HSCs are also required for self-renewal in HSCs. In analysis of mice deficient in a certain molecule, if fetal liver hematopoiesis occurs with more or less normal hematopoietic differentiation, if transplantation of fetal liver cells deficient for that molecule and adult bone marrow harboring a conditional deletion of that molecule into lethally irradiated mice results in long-term reconstitution failure, and if no homing or migration defect in HSCs exists, that molecule would be a good candidate for close association with self-renewal in HSCs.

The Hox family consists of an evolutionally highly conserved group of genes that encode DNA-biding transcriptional factors. Overexpression of Hox family genes in HSCs leads to enhancement of self-renewal potential. HoxB4 overexpression in particular can induce self-renewal in vitro (Sauvageau et al. 1995) as well as in vivo (Antonchuk et al. 2002) without leukemic transformation. When Nucleoporin 98 (NUP98)-HoxB4 or -HoxA10 fusion protein is overexpressed in HSCs, HSC self-renewal is privileged (Ohta et al. 2007). Overexpression of HoxB3, HoxB6, HoxB8, HoxA9, or HoxA10 affected HSCs similarly, but resulted in impaired differentiation along various lineages (Argiropoulos and Humphries 2007). Leukemia developed, with a long latency in some cases. The frequency of leukemia was increased by overexpression of NUP98-HoxA9 fusion protein. NUP98-Hox fusion proteins seem to collaborate with Meis1 to accelerate the onset of leukemia (Argiropoulos and Humphries 2007). Why Hox members other than HoxB4 cause leukemic transformation is a curious point. HoxB4 is not essential for HSC development and maintenance (Brun et al. 2004). HoxA9 plays a crucial role in the maintenance of self-renewal in HSCs (Lawrence et al. 2005; Magnusson et al. 2007a). A conditional transgenic mouse for HOXA10 has been reported (Magnusson et al. 2007b). Interestingly, a moderate but not high level of HOXA10 expression enhanced self-renewal in HSCs and its expression at high levels blocked differentiation along erythroid/megakaryocytic lines. A dose effect of each Hox gene is suggested to be important for the probability of self-renewal in HSCs (McKinney-Freeman et al. 2008).

Gfi1 is a transcriptional repressor expressed in HSCs. Repopulating activity is severely reduced in Gfi1$^{-/-}$ HSCs (Zeng et al. 2004; Hock et al. 2004). The proportion of cycling HSCs is increased in Gfi1$^{-/-}$ mice. It is possible that Gfi1 negatively regulates the G_0/G_1 transition in HSCs. Moz is a zinc finger protein with histone acetyltransferase (HAT) activity. Transplantation of Moz$^{-/-}$ fetal liver cells resulted in poor long-term repopulation (Thomas et al. 2006; Katsumoto et al. 2006). Evi1 is a transcription factor of the SET/PR domain family. Transplantation of Evi1$^{-/-}$ fetal liver cells resulted in poor long-term repopulation (Goyama et al. 2008). The zinc finger protein Zfx has been reported to be a shared transcriptional regulator essential for self-renewal in both embryonic stem cells (ESCs) and HSCs. Since ESCs differ much more from HSCs in functions and gene expression profiles than previously thought, Zfx might have a unique role among many other transcription factors.

More examples of signal molecules and transcription factors related to self-renewal in HSCs certainly exist. The important question is what their final common target molecules are. We need to know what directly controls self-renewal events in HSCs and what determines the number of divisions that HSCs are allowed to undergo.

Basic Machineries Supporting Self-Renewal

The Polycomb group (PcG) proteins have been a focus of attention in stem cell biology (Lessard and Sauvageau 2003; Park et al. 2003; Iwama et al. 2004; Ohta et al. 2002; Molofsky et al. 2003). PcG and trithorax group (trxG) genes are highly conserved throughout evolution. PcG and trxG proteins, respectively, control transcriptionally repressed and activated states of multiple gene loci, including *Hox* loci (Orlando 2003). It has been shown that repopulating activity in fetal liver HSCs is significantly reduced in mice deficient for a member of the PcG proteins, whether Bmi1, Mph1 (Rae28), or Mel18 (Lessard and Sauvageau 2003; Park et al. 2003; Iwama et al. 2004; Ohta et al. 2002). Ring1B seems to play a somewhat different role in adult HSCs, as shown by analysis of Ring1B conditional knockout mice (Cales et al. 2008). These studies indicate that PcG proteins play crucial roles in the maintenance of self-renewal capacity in HSCs. Interestingly, all these molecules are part of a multiprotein complex, called Polycomb repressive complex (PRC)-1. On the other hand, PRC2, containing Eed, Ezh1, Ezh2, and YY1, is essential for early embryonic development (Ohta et al. 2002). Overexpression of Ezh2 in HSCs has been shown to enhance HSC repopulating activity (Kamminga et al. 2006).

Bmi1 was the first molecule to draw our attention to epigenetic control in HSCs. *Bmi1* was initially identified as an oncogene by retroviral insertional mutagenesis in Eμ-*myc* transgenic mice (van Lohuizen et al. 1991; Haupt et al. 1991). Analysis of Bmi1 knockout mice has shown that null mutant mice are smaller than wild-type mice, displaying progressive hypoplasia of bone marrow, thymus, and spleen, with cerebellar atrophy and skeletal deformity (van der Lugt et al. 1994). The *Ink4a* locus, which encodes both $p16^{Ink4a}$ and $p19^{Arf}$ under the control of individual promoters, has been identified as a downstream target of Bmi1 (Jacobs et al. 1999). In the absence of Bmi1, the expression of $p16^{Ink4a}$ and $p19^{Arf}$ reportedly is upregulated in mouse embryonic fibroblasts. As a result, progression of these fibroblasts into the S phase of the cell cycle is impaired, mimicking premature senescence. Conversely, overexpression of Bmi1 suppresses the expression of $p16^{Ink4a}$ and $p19^{Arf}$ and allows fibroblast immortalization.

Marked reductions in repopulating activity in Bmi1-deficient HSCs can be completely reversed by the deletion of both $p16^{Ink4a}$ and $p19^{Arf}$ (Oguro et al. 2006). $p16^{Ink4a}$ inhibits cell cycle progression by inhibiting cyclin D–dependent kinase and prevents phosphorylation of the tumor suppressor Rb, whereas $p19^{Arf}$ prevents degradation and inactivation of the tumor suppressor Trp53 by binding to Mdm2 (Jacobs et al. 1999). Whether $Ink4a^{-/-}Arf^{-/-}$ HSCs can self-renew more so than wild-type HSCs is an interesting point. Stepanova and Sorrentino have attempted to answer this question by analyzing mice deficient in both genes (Stepanova and Sorrentino 2005). Only a modest increase in self-renewal was observed for $Ink4a^{-/-}Arf^{-/-}$ HSCs, but HSC numbers increased approximately tenfold when Ink4a, Arf, and Trp53 were deleted (Akala et al. 2008). Interestingly, numbers of mature blood cells and myeloid progenitors were within normal ranges in the triple mutant mice.

It is possible that the probability of self-renewal in HSCs is under the control of these three cell cycle regulators.

Both self-renewal and differentiation in HSCs occur via division. The cell cycle machinery should function normally, which is true not only for HSCs but also for all dividing cells. Mitogenic stimulation initiates cell cycle progression driven by heterodimeric kinases made of a regulatory subunit, cyclin and a catalytic subunit, cyclin-dependent kinase (Cdk) (Malumbres et al. 2004). Two classes of cyclin-Cdk complexes operate during G1 phase progression. D-type cyclins (D1, D2, D3) complexed with Cdk4 or Cdk6, or E-type cyclins (E1 and E2) complexed with Cdk2, phosphorylate pRb, p107, and p130, which are members of the retinoblastoma (Rb) protein family. After Rb is hyperphosphorylated by the cyclin E/Cdk2 complex, the association of Rb with various E2F family members is disrupted, allowing initiation of the E2F transcriptional program. The E2F6 transcription factor acts as a dominant negative inhibitor through competition with other members of the E2F family. Interestingly, E2F6 can be a component of PRC1 (Trimarchi et al. 2001). It has been reported that cyclin $D1^{-/-}$, $D2^{-/-}$, $D3^{-/-}$ triple knockout mouse embryos die of anemia and heart abnormality by E16.5 (Kozar et al. 2004). The proliferation of HSCs and of progenitor cells is disturbed in these mutant mice, suggesting that HSCs depend on D-type cyclins for their division. It has been shown that mouse embryos deficient in both Cdk4 and Cdk6 display normal organogenesis (deficiency of these proteins is likely compensated for by Cdk2), but die late in gestation due to impaired erythropoiesis (Malumbres et al. 2004). It remains unclear whether these Cdks are essential in HSCs because long-term reconstitution using fetal liver cells from this mutant mouse has not been shown.

Cdk inhibitors (CKIs) are assigned to one of two families based on their structures and Cdk targets. One is the Ink4 family, which consists of $p16^{In4a}/p19^{Arf}$, $p15^{In4b}$, $p18^{In4c}$, and $p19^{In4d}$. The members of this family competitively inhibit D-type cyclin binding to Cdk4 and Cdk6. The other is the Cip/Kip family, which consists of $p21^{Cip1}$, $p27^{Kip1}$, and $p57^{Kip2}$. The members of this second family are potent inhibitors of cyclin E- and A-dependent Cdk2 (Sherr and Roberts 1999).

Cheng et al. have examined HSCs from mice deficient in one of these CKIs in a series of studies. In the absence of $p21^{Cip1/Waf1}$, the number of HSCs increases via an increase in the proportion of cycling HSCs, which are easily exhausted by serial transplantation (Cheng et al. 2000a). Curiously, these phenotypes have not been seen in mice of the C57BL/6 strain (van Os et al. 2007). In the absence of $p18^{Ink4c}$, the number of HSCs markedly increases without premature exhaustion after transplantation (Yuan et al. 2004). Of interest is that in the absence of $p27^{Kip1}$, numbers of HSCs did not change, but the pool size of progenitor cells enlarged under proliferative stress such as serial transplantation (Cheng et al. 2000b). Neither how $p21^{Cip1/Waf1}$, but not $p18^{Ink4c}$, is linked to the proliferative capacity of HSCs, nor why $p27^{Kip1}$ plays a primary role in progenitor cells, but not in HSCs, is known. Nonetheless, these studies suggest that each CKI has a distinct role in regulating the cell cycle at different differentiation stages.

Telomeres are the genetic structures that protect chromosome ends from degradation, recombination, and fusion. The mouse telomere is relatively long, ranging from 20 to 100 kb, while it is relatively short in man, ranging from 3 to 12 kb. Human telomerase is a ribonucleoprotein complex consisting of the RNA and catalytic components (hTR and hTERT). Most somatic cells, including stem cells, show progressive telomere shortening with division because they have little telomerase activity. Average telomere length thus can indicate division history, as shown for human fibroblasts (Allsopp et al. 1992). Self-renewal potential is diminished in HSCs when the telomere becomes too short (Allsopp et al. 2003b). However, self-renewal potential in HSCs decreases with division even when sufficient telomere length is maintained. Overexpression of telomerase in HSCs can prevent telomere shortening but cannot prevent a decrease in self-renewal potential (Allsopp et al. 2003a). In human hematopoietic cells, telomerase overexpression did not prevent proliferation-associated telomere shortening (Wang et al. 2005a). These data show that a decline of self-renewal capacity in HSCs associated with proliferative stress is independent of telomerase activity and telomere length.

It has been discovered that HSCs from ataxia telangiectasia mutated (ATM) gene-deficient mice are unable to self-renew in vivo (Ito et al. 2004). Reactive oxygen species (ROS) are constantly produced in living cells, mostly as a by-product of normal mitochondrial activity (Barzilai and Yamamoto 2004). ROS cause severe damage to DNA. ATM-deficient HSCs seem more sensitive to ROS than do wild-type HSCs. Surprisingly, this defect almost completely disappears on treatment with anti-oxidative agents such as *N*-acetyl-l-cysteine (Ito et al. 2004). Moreover, enforced expression of Bmi1 in ATM$^{-/-}$ HSCs restores self-renewal capacity as well. Expression of p16^{Ink4a} in ATM$^{-/-}$ HSCs is likely suppressed by overexpression of Bmi1 in HSCs. This study suggests that oxidative stress can cause senescence of HSCs via the p16-Rb pathway (Ito et al. 2004).

Differentiation in HSCs

In vitro colony formation by HSCs represents a paradigm for the study of epigenetic mechanisms. Certain transcriptional factors such as Pu.1 and Gata2 are expressed in HSCs. Probably these factors recruit coactivators or corepressors to the regulatory regions of genes, followed by assembly of chromatin remodeling machinery. A limited number of cytokines support these processes. A rather simple differentiation program initiated by these cytokines might govern all sequential epigenetic modifications and gene expressions. Although epigenetic modifications are considered stably inherited from parental cells to daughter cells, epigenetic asymmetry is also suggested for asymmetric cell division (Lansdorp 2007). An individual HSC has a single pattern of epigenetic modifications but gives rise, in its descendant mature blood cells, to cells with, one presumes, a great variety of patterns of epigenetic modifications. To learn what kind of program can establish such a diversity of epigenetic modifications from a single pattern of epigenetic modifications would be extremely interesting.

Differentiation Pathways

The bifurcation model for HSC differentiation has been widely accepted for quite a while now. This model simply and easily explains that HSCs differentiate exclusively into either common myeloid progenitors (CMPs) (Akashi et al. 2000) or common lymphoid progenitors (CLPs) (Kondo et al. 1997). Identification of several distinct progenitor populations has been useful for analysis of a number of knockout mice because it is easy to see which stages of differentiation are affected by deletion of genes. However, this model requires major revision (Arinobu et al. 2007). Although multipotent progenitor cells capable of giving rise to both CMPs and CLPs have been identified, differentiation potential of these progenitor cells was restricted to the neutrophil, macrophage, and B- and T-lymphoid lineages (Adolfsson et al. 2005). These progenitors thus were named lymphoid-primed multipotent progenitors. Potential to differentiate in the erythroid/megakaryocyte lineage was likely lost during differentiation stages between HSCs and lymphoid-primed multipotent progenitors. Erythroid/megakaryocytic differentiation seems lost prior to the emergence of CMPs. If this is the case, CMPs do not have full myeloid lineage neutrophil, macrophage, erythroid, and megakaryocyte differentiation potential. Cells with full neutrophil, macrophage, and erythroid/megakaryocyte potential were detected at the clonal level among HSCs and among populations closely related to HSCs, but not among other populations, including conventional CMPs (Ooehara et al., unpublished data, 2006). "True" CMPs, if they exist, might be in a population closely related to the HSC population (Arinobu et al. 2007).

Early thymic progenitors have the potential to give rise to myeloid lineage cells such as macrophages and dendritic cells but do not have the potential to differentiate along B-lymphoid lineage, suggesting that CLPs do not contribute to T-cell development (Bell and Bhandoola 2008; Wada et al. 2008). If this is the case, what are CPLs for? A new model is required, one based not only on population-hierarchy data but also on cell tracking analysis data that show direct relationships among populations. One hopes that the functional significance of each isolated population will be studied.

Molecular Control of Differentiation and Lineage Commitment

In our eyes, to understand cellular differentiation programs at the molecular level is one of the most difficult tasks in stem cell biology and even more generally in life science. A combination of transcription factors specifies a gene expression profile in a cell. Differentiation can be a process of selecting a certain set of transcription factors to be expressed. Expression levels of individual transcription factors likely bias lineage choices in differentiating cells. Transcriptional regulation is extremely complex because a transcription factor acts as a positive or negative regulator for gene expression, a transcription factor controls up- and downregulation of many genes, and expression and action of transcription factors are associated with epigenetic changes. What kind of program exists to control all of these events in a cell?

Transcription Factors

Although many gene knockout studies have shown that particular transcription factors play important roles in hematopoietic differentiation programs, no particular lineage-specific transcription factor alone controls lineage commitment or promotes differentiation along a certain lineage. Multiple transcription factors seemingly act together to direct differentiation in complex transcriptional networks.

Pu.1 is expressed from HSCs all the way to mature neutrophils and macrophages. This particularly interests us because all HSCs are almost certainly able to form myeloid colonies in culture within 2 weeks. Differentiation of neutrophils and macrophages is profoundly blocked in Pu.1 knockout mice (Scott et al. 1994; McKercher et al. 1996). Perhaps myeloid differentiation potential is the most important function if HSCs are to support the life of an organism. When Gata1 is expressed at a certain stage of HSC differentiation, the Pu.1:Gata1 expression ratio appears critical in specifying the choice of erythroblast/megakaryocyte lineage. Similarly, when C/EBPα and Pax5 are expressed at a certain stage of HSC differentiation, the C/EBPα:Pax5 expression ratio appears critical in specifying the choice of B-lymphoid lineage. Cross-regulation by key regulators is likely exerted via direct protein-protein interaction (the cross-antagonistic model) (Orkin 2000) in addition to transcriptional levels. Multiple transcription factors exhibit both positive and negative effects on one lineage program. It is more difficult to clarify the effects of multiple transcription factors if these transcription factors form large complexes. In addition, a stochastic versus deterministic debate always emerges on the issue of fate determination, making the matter more complicated.

Epigenetic Changes

Genomic DNA methylation and histone modifications are the major epigenetic modifications. DNA methylation is a postreplication modification that is predominantly found in cytosines of the dinucleotide sequence CpG. CpG islands are often found in promoters and first exons. Methylation of CpG-rich promoters prevents initiation of transcription. DNA methylation in centromeres and telomeres is essential for genomic stability. DNA methyltransferases Dnmt1, Dnmt3a, and Dnmt3b catalyze methylation of GpG. It has been suggested that the contribution of these methyltransferases to the maintenance of global DNA methylation varies and that its contribution is specific to cell type and developmental stage (Chen et al. 2003; Dodge et al. 2005). A primary role of Dnmt1, or of Dnmt3a and Dnmt3b, is the maintenance of the methylated genome or *de novo* methylation of the genome (Okano et al. 1999). Dnmt3a but not Dnmt3b has been shown to be required for spermatogenesis in mice (Kaneda et al. 2004). Dnmt3b$^{-/-}$ murine embryonic fibroblasts have been shown to undergo either premature senescence or spontaneous immortalization (Dodge et al. 2005).

We previously analyzed Dnmt3a$^{-/-}$, 3b$^{-/-}$, or 3b$^{-/-}$3b$^{-/-}$ adult HSCs from conditional knockout mice. Differentiation of HSCs was unaffected in these three types of mutant HSCs. However, self-renewal activity was severely impaired in

Dnmt3a$^{-/-}$b3$^{-/-}$ HSCs, although it remained intact in either Dnmt3a$^{-/-}$ or 3b$^{-/-}$ HSCs (Tadokoro et al. 2007). We used retroviral infection to introduce Cre recombinase into HSCs in this series of experiments. The Cre recombinase virus seems more or less toxic to self-renewal activity in HSCs. Presumably, a persistently high level of Cre recombinase expression adversely affects self-renewal. Therefore, we have attempted to reevaluate self-renewal activity in Dnmt3a$^{-/-}$3b$^{-/-}$ HSCs by using Mx1-Cre transgenic mice crossed with Dnmt3a/3b conditional knockout mice. Preliminary results show that long-term repopulating activity appears not so severely impaired in Dnmt3a$^{-/-}$3b$^{-/-}$ HSCs. Closer examination is needed to determine whether both Dnmt3a and 3b are really necessary for self-renewal in adult HSCs. We have recently analyzed HSCs in fetal liver of Dnmt3a or 3b straight knockout mice. Differentiation in Dnmt3a$^{-/-}$ or 3b$^{-/-}$ HSCs was within normal ranges, consistent with our previous observations. However, Dnmt3b, but not Dnmt3a, seems to play a crucial role in acquisition or maintenance of self-renewal potential in developing HSCs (Ibata et al., unpublished data, 2008). Dnmt3b may play a greater role in developing fetal HSCs than in established adult HSCs. Alternatively, Dnmt3a may play a role in establishment of fetal liver niches but not in bone marrow niches.

Nucleosomes consist of an octameric core of four types of histones: H2A, H2B, H3, and H4. Each octamer contains two copies of each histone. Chromatin is compacted by the incorporation of the linker histone H1 and its variants. Histone modification contributes to the dynamic structural characteristics of chromatin, which are inheritable from a parental cell to daughter cells. These characteristics affect transcription and, therefore, cellular identity (Sarma and Reinberg 2005). Histone tails (*N*-termini) are modified by a variety of mechanisms, such as acetylation, methylation, phosphorylation, and ubiquitination (Strahl and Allis 2000). The discovery of histone acetyltransferases (HATs) and histone deacetylases (HDACs) has facilitated studies of roles of histone acetylation in transcription regulation. Short preferred consensus motifs seem to exist for specificity of individual HATs and HDACs. Selected lysines in H3 and H4 are thereby acetylated. In general, histone acetylation is linked with transcriptional activation. Histone phosphorylation is a marker for activation of immediate early genes and a signal for mitotic chromatin condensation (Lachner et al. 2003). Histone lysine residues are mono-, di-, and trimethylated. Histone H3-K4 methylation is linked to transcriptional activation (Wysocka et al. 2005). H3-K9 methylation, in contrast, is present mainly in silenced chromatin domains. Histone modifications are associated with the function of PcG and trxG proteins. The Ezh-Eed complex (PRC2) can recruit HDACs, a step followed by local chromatin deacetylation. Because the PRC2 complex possesses histone lysine methyltransferase (HMTase) activity (Su et al. 2003), this complex can methylate H3-K27 (and probably also H3-K9) and induces stable recruitment of PRC1. PRC1 negatively regulates chromatin accessibility promoted by the ATP-dependent chromatin remodeling SWI/SNF complex. The trxG proteins MLL and Ash-1 also contain the SET domain. The SET domain proteins display HMTase activity (Dillon et al. 2005). The trxG complex performs H3-H4 methylation and triggers recruitment of ISW/SNF, resulting in an activated chromatin state (Lachner et al. 2003).

To date, little is known about the roles of histone modifications in HSCs. The differentiation of HSCs into specific progenitor cells and then into diverse blood cell types represents a particularly powerful system for the study of epigenetic mechanisms. We hope that a new paradigm for epigenetic analysis will be built in this research field.

HSC Homing and Intramedullary Mobilization

Although HSCs mainly reside in the bone marrow throughout adult life, HSCs in substantial numbers constantly leave their bone marrow niches and enter the circulation (Wright et al. 2001). In mice, the spleen functions as a hematopoietic organ. Some HSCs in the circulation go through the spleen and perhaps stay within the spleen for a while. A portion of circulating HSCs returns to the bone marrow. Mobilization is not always associated with cycling in HSCs.

After transplantation, HSCs, and perhaps progenitor cells, can extremely efficiently home to the bone marrow (Matsuzaki et al. 2004). Bone marrow homing is thus another important property of adult HSCs. Members of the integrin family, like $\alpha4\beta1$ or $\alpha6\beta1$ integrin, seem not to significantly participate in HSC homing (Brakebusch et al. 2002; Priestley et al. 2006), as previously thought (Potocnik et al. 2000; Qian et al. 2007). The homing receptor in HSCs has not yet been identified. Long-term reconstitution takes place only after successful homing and repopulation of bone marrow cells. To repopulate the whole bone marrow in the body, intra- and intermedullary mobilization is required for HSCs (Cao et al. 2004). CXCL12-abundant reticular cells are scattered widely throughout the bone marrow and are candidates for the cellular component of the HSC niche (Sugiyama et al. 2006). HSCs are thought present in close association with CXCL12-abundant reticular cells. Since CXCR4$^{-/-}$ HSCs exhibit poor reconstitution, CXCR4/CXCL12 interaction may be required to keep hibernating HSCs in the niche and to maintain the HSC pool. Alternatively, CXCR4/CXCL12 interaction may play a role in a process of intramedullary micromobilization by HSCs.

Perspectives

Stem Cell Therapy

In order to translate stem cell biology into stem cell therapy, we must know how to control the fate of stem cells. Otherwise, the possibility of unwanted outcomes is apparent, with harm due on the one hand to self-renewal capacity and on the other to pluripotency or multipotency. Malignant tumors originate from cells with uncontrolled self-renewal and deficiency of differentiation. Overproduction of only a particular lineage may result in a functional failure of the organ.

ESCs have become an extremely attractive source for adult stem cells since human ESCs and induced pluripotent stem cells were established (Thomson et al. 1998; Shamblott et al. 1998; Takahashi et al. 2007; Yu et al. 2007). In the field of hematology, efficient generation of HSCs from ESCs is the most important key for successful stem cell therapy. Ex vivo expansion of HSCs has been difficult. This problem should be overcome by use of ESCs. However, it remains difficult to generate HSCs from ESCs. ESCs supposedly give rise to HSCs through multiple steps (Matsumoto et al., 2009). In mice, enforced expression of HOXB4 supports generation of HSCs from ESCs (Kyba et al., 2002). Certain developmental stages of HSCs seem skipped by HOXB4. The resultant ESC-derived repopulating cells remain significantly inferior to adult HSCs from bone marrow, however, particularly in the context of repopulating activity or self-renewal potential. In humans, repopulating cells have not been successfully generated from ESCs by any means including enforced HOXB4 expression (Wang et al. 2005b). Therefore, we still have far to go to achieve practical use of human ESCs and induced pluripotent stem cells as a source of HSCs.

Cancer Stem Cells and Age-Related Changes in Stem Cells

The recognition of the hierarchical order among leukemic cells, like that in normal HSCs, evoked the concept of leukemic stem cells in the early 1990s (Bonnet and Dick, 1997; Lapidot et al., 1994). This concept has been gradually applied to a variety of malignant tumors, with invocation of "cancer stem cells." This has occurred against a shift in thought, with the hypothesis now entertained that cancer stem cells might have acquired properties of stem cells such as self-renewal capacity and hibernation capacity. In other words, "ordinary" cancer cells become cancer stem cells by stealing molecular mechanisms that operate in normal stem cells. Many candidate molecules have been proposed as essential for maintenance of both stem cells and cancer stem cells. These studies have suggested a potential new therapeutic strategy to target cancer stem cells. A new avenue toward eradication and complete cure of cancer will be seen if such essential molecules are specifically inhibited only in cancer stem cells.

Stem cell biology has provided the basis for the study of cancer stem cells. This may be reciprocated for stem cell biology by the study of cancer stem cells. The population origin of leukemic cells has recently been the focus of studies. Sten Eirik Jacobsen's group has identified $CD34^{+}CD38^{-}CD19^{-}$ cells as a bone marrow population in which leukemic stem cells arise, or blastic transformation takes place, in the case of chronic myeloid leukemia (CML) (Castor et al. 2005). This work is a good example of how stem cell biology made a remarkable contribution to the study of leukemogenesis. Although CML has been considered a stem cell disease, T-lymphoid lineage seems less often involved in CML than are myeloid and B-lymphoid lineages. This suggests that CML may not always originate in the most immature HSCs; if so, HSCs may remain intact even though leukemic cells occupy most of the bone marrow. It is possible that CD34⁻ HSCs remain spared from stem cell diseases, such as CML, myelodysplasia, and aplastic anemia. If normal hematopoietic cells are recovered from diseased bone marrow, they may contain candidate human HSCs. Most acquired aplastic anemia is considered to be autoimmune disease, which can in part be successfully treated with immunosuppressive therapy (Young et al. 2008). If HSCs indeed survive somewhere in almost empty bone marrow of this sort, to identify HSCs in aplastic bone marrow may be easier than to do so in normal bone marrow with a huge number of cells.

A multistep model has been accepted for leukemogenesis. Two major steps (class I and class II mutations) have been proposed in leukemia development (Gilliland 2002). Class I mutations are activating point mutations in signal molecules, which provide proliferation and survival signals. Class II mutations involve upregulation of hematopoietic transcription factors, which lead to impaired differentiation. Genes associated with translocations in human leukemia have often appeared to be important regulators of hematopoiesis. This model might be applied to manipulation of self-renewal signals in normal HSCs because self-renewal can be interpreted as a physiologic result of imbalance between proliferation and differentiation signals. For instance, if Jak2/Stat5 signaling in HSCs is transiently enhanced or if expression levels of transcriptional factors in HSCs are conditionally increased or decreased, these alterations might advantageously work to induce self-renewal in HSCs.

We are not sure whether HSCs age. Although aging can be defined as phenotypic change over a lifetime, aging in HSCs must be closely associated with their history of divisions rather than with the passage of time itself. HSCs may phenotypically change only when they approach maximum numbers of divisions. It has been suggested that $p16^{Ink4a}$ and $p21^{Cip1/Waf1}$ are upregulated, and that p38 MAPK is phosphorylated in aging HSCs (Janzen et al. 2006; Choudhury et al. 2007; Ito et al. 2006). We and others have observed a decline in self-renewal potential in serially transplanted mice, but not in normally aged mice (Sudo et al. 2000; Morrison et al. 1996). One may generally consider that stem cell function decreases as age increases, but this is not the case for HSCs in C57BL/6 mice. HSCs with highly repopulating activity exist in aged mice, and in numbers as large as those in young mice (Sudo et al. 2000; Morrison et al. 1996). These molecules thus may be related to the history of cell division and to DNA damage but not directly associated with aging. We have previously observed that differentiation potentials for myeloid and lymphoid lineages are imbalanced in aged HSCs. Interestingly, the imbalance appeared partially reversible (Sudo et al. 2000). We and others have proposed that self-renewal potential is extensive, but also limited, in HSCs. We need to know how many times and how often individual HSCs are allowed to divide in vivo.

Acknowledgments We are grateful to T. Shimizu and A. S. Knisely for their critical review of the manuscript. This work was supported by grants from the Ministry of Education, Culture, Sport, Science, and Technology, Japan.

References

Abramson S, Miller RG, Phillips RA (1977) The identification in adult bone marrow of pluripotent and restricted stem cells of the myeloid and lymphoid systems. J Exp Med 145:1567–1579

Adolfsson J, Mansson R, Buza-Vidas N, Hultquist A, Liuba K, Jensen CT, Bryder D, Yang L, Borge OJ, Thoren LA, Anderson K, Sitnicka E, Sasaki Y, Sigvardsson M, Jacobsen SE (2005) Identification of Flt3+ lympho-myeloid stem cells lacking erythro-megakaryocytic potential a revised road map for adult blood lineage commitment. Cell 121:295–306

Akala OO, Park IK, Qian D, Pihalja M, Becker MW, Clarke MF (2008) Long-term haematopoietic reconstitution by Trp53-/-p16Ink4a-/-p19Arf-/- multipotent progenitors. Nature 453:228–232

Akashi K, Traver D, Miyamoto T, Weissman IL (2000) A clonogenic common myeloid progenitor that gives rise to all myeloid lineages. Nature 404:193–197

Allsopp RC, Vaziri H, Patterson C, Goldstein S, Younglai EV, Futcher AB, Greider CW, Harley CB (1992) Telomere length predicts replicative capacity of human fibroblasts. Proc Natl Acad Sci U S A 89:10114–10118

Allsopp RC, Morin GB, Horner JW, DePinho R, Harley CB, Weissman IL (2003a) Effect of TERT over-expression on the long-term transplantation capacity of hematopoietic stem cells. Nat Med 9:369–371

Allsopp RC, Morin GB, DePinho R, Harley CB, Weissman IL (2003b) Telomerase is required to slow telomere shortening and extend replicative lifespan of HSCs during serial transplantation. Blood 102:517–520

Antonchuk J, Sauvageau G, Humphries RK (2002) HOXB4-induced expansion of adult hematopoietic stem cells ex vivo. Cell 109:39–45

Arai F, Hirao A, Ohmura M, Sato H, Matsuoka S, Takubo K, Ito K, Koh GY, Suda T (2004) Tie2/angiopoietin-1 signaling regulates hematopoietic stem cell quiescence in the bone marrow niche. Cell 118:149–161

Arai F, Hirao A, Suda T (2005) Regulation of hematopoiesis and its interaction with stem cell niches. Int J Hematol 82:371–376

Argiropoulos B, Humphries RK (2007) Hox genes in hematopoiesis and leukemogenesis. Oncogene 26:6766–6776

Arinobu Y, Mizuno S, Chong Y, Shigematsu H, Iino T, Iwasaki H, Graf T, Mayfield R, Chan S, Kastner P, Akashi K (2007) Reciprocal activation of GATA-1 and PU.1 marks initial specification of hematopoietic stem cells into myeloerythroid and myelolymphoid lineages. Cell Stem Cell 1:416–427

Barzilai A, Yamamoto K (2004) DNA damage responses to oxidative stress. DNA Repair (Amst) 3:1109–1115

Baxter EJ, Scott LM, Campbell PJ, East C, Fourouclas N, Swanton S, Vassiliou GS, Bench AJ, Boyd EM, Curtin N, Scott MA, Erber WN, Green AR (2005) Acquired mutation of the tyrosine kinase JAK2 in human myeloproliferative disorders. Lancet 365:1054–1061

Bell JJ, Bhandoola A (2008) The earliest thymic progenitors for T cells possess myeloid lineage potential. Nature 452:764–767

Bhatia M, Bonnet D, Murdoch B, Gan OI, Dick JE (1998) A newly discovered class of human hematopoietic cells with SCID-repopulating activity. Nat Med 4:1038–1045

Bonnet D, Dick JE (1997) Human acute myeloid leukemia is organized as a hierarchy that originates from a primitive hematopoietic cell. Nat Med 3:730–737

Bradley TR, Metcalf D (1966) The growth of mouse bone marrow cells in vitro. Aust J Exp Biol Med Sci 44:287–300

Brakebusch C, Fillatreau S, Potocnik AJ, Bungartz G, Wilhelm P, Svensson M, Kearney P, Korner H, Gray D, Fassler R (2002) Beta1 integrin is not essential for hematopoiesis but is necessary for the T cell-dependent IgM antibody response. Immunity 16:465–477

Brun AC, Bjornsson JM, Magnusson M, Larsson N, Leveen P, Ehinger M, Nilsson E, Karlsson S (2004) Hoxb4-deficient mice undergo normal hematopoietic development but exhibit a mild proliferation defect in hematopoietic stem cells. Blood 103:4126–4133

Cales C, Roman-Trufero M, Pavon L, Serrano I, Melgar T, Endoh M, Perez C, Koseki H, Vidal M (2008) Inactivation of the polycomb group protein Ring1B unveils an antiproliferative role in hematopoietic cell expansion and cooperation with tumorigenesis associated with Ink4a deletion. Mol Cell Biol 28:1018–1028

Calvi LM, Adams GB, Weibrecht KW, Weber JM, Olson DP, Knight MC, Martin RP, Schipani E, Divieti P, Bringhurst FR, Milner LA, Kronenberg HM, Scadden DT (2003) Osteoblastic cells regulate the haematopoietic stem cell niche. Nature 425:841–846

Cao YA, Wagers AJ, Beilhack A, Dusich J, Bachmann MH, Negrin RS, Weissman IL, Contag CH (2004) Shifting foci of hematopoiesis during reconstitution from single stem cells. Proc Natl Acad Sci U S A 101:221–226

Castor A, Nilsson L, Astrand-Grundstrom I, Buitenhuis M, Ramirez C, Anderson K, Strombeck B, Garwicz S, Bekassy AN, Schmiegelow K, Lausen B, Hokland P, Lehmann S, Juliusson G, Johansson B, Jacobsen SE (2005) Distinct patterns of hematopoietic stem cell involvement in acute lymphoblastic leukemia. Nat Med 11:630–637

Chen T, Ueda Y, Dodge JE, Wang Z, Li E (2003) Establishment and maintenance of genomic methylation patterns in mouse embryonic stem cells by Dnmt3a and Dnmt3b. Mol Cell Biol 23:5594–5605

Cheng T, Rodrigues N, Shen H, Yang Y, Dombkowski D, Sykes M, Scadden DT (2000a) Hematopoietic stem cell quiescence maintained by p21cip1/waf1. Science 287:1804–1808

Cheng T, Rodrigues N, Dombkowski D, Stier S, Scadden DT (2000b) Stem cell repopulation efficiency but not pool size is governed by p27(kip1). Nat Med 6:1235–1240

Choudhury AR, Ju Z, Djojosubroto MW, Schienke A, Lechel A, Schaetzlein S, Jiang H, Stepczynska A, Wang C, Buer J, Lee HW, von Zglinicki T, Ganser A, Schirmacher P, Nakauchi H, Rudolph KL (2007) Cdkn1a deletion improves stem cell function and lifespan of mice with dysfunctional telomeres without accelerating cancer formation. Nat Genet 39:99–105

Cobas M, Wilson A, Ernst B, Mancini SJ, MacDonald HR, Kemler R, Radtke F (2004) Beta-catenin is dispensable for hematopoiesis and lymphopoiesis. J Exp Med 199:221–229

Dillon SC, Zhang X, Trievel RC, Cheng X (2005) The SET-domain protein superfamily: protein lysine methyltransferases. Genome Biol 6:227

Dodge JE, Okano M, Dick F, Tsujimoto N, Chen T, Wang S, Ueda Y, Dyson N, Li E (2005) Inactivation of Dnmt3b in mouse embryonic fibroblasts results in DNA hypomethylation, chromosomal instability, and spontaneous immortalization. J Biol Chem 280:17986–17991

Dykstra B, Kent D, Bowie M, McCaffrey L, Hamilton M, Lyons K, Lee SJ, Brinkman R, Eaves C (2007) Long-term propagation of distinct hematopoietic differentiation programs in vivo. Cell Stem Cell 1:218–229

Ema H, Nakauchi H (2003) Self-renewal and lineage restriction of hematopoietic stem cells. Curr Opin Genet Dev 13:508–512

Ema H, Takano H, Sudo K, Nakauchi H (2000) In vitro self-renewal division of hematopoietic stem cells. J Exp Med 192:1281–1288

Ema H, Sudo K, Seita J, Matsubara A, Morita Y, Osawa M, Takatsu K, Takaki S, Nakauchi H (2005) Quantification of self-renewal capacity in single hematopoietic stem cells from normal and Lnk-deficient mice. Dev Cell 8:907–914

Ema H, Morita Y, Yamazaki S, Matsubara A, Seita J, Tadokoro Y, Kondo H, Takano H, Nakauchi H (2006) Adult mouse hematopoietic stem cells: purification and single-cell assays. Nat Protoc 1:2979–2987

Gilliland DG (2002) Molecular genetics of human leukemias: new insights into therapy. Semin Hematol 39:6–11

Goodell MA, Brose K, Paradis G, Conner AS, Mulligan RC (1996) Isolation and functional properties of murine hematopoietic stem cells that are replicating in vivo. J Exp Med 183:1797–1806

Goyama S, Yamamoto G, Shimabe M, Sato T, Ichikawa M, Ogawa S, Chiba S, Kurokawa M (2008) Evi-1 is a critical regulator for hematopoietic stem cells and transformed leukemic cells. Cell Stem Cell 3:207–220

Guenechea G, Gan OI, Dorrell C, Dick JE (2001) Distinct classes of human stem cells that differ in proliferative and self-renewal potential. Nat Immunol 2:75–82

Guo W, Lasky JL, Chang CJ, Mosessian S, Lewis X, Xiao Y, Yeh JE, Chen JY, Iruela-Arispe ML, Varella-Garcia M, Wu H (2008) Multi-genetic events collaboratively contribute to Pten-null leukaemia stem-cell formation. Nature 453:529–533

Hall MA, Curtis DJ, Metcalf D, Elefanty AG, Sourris K, Robb L, Gothert JR, Jane SM, Begley CG (2003) The critical regulator of embryonic hematopoiesis, SCL, is vital in the adult for megakaryopoiesis, erythropoiesis, and lineage choice in CFU-S12. Proc Natl Acad Sci U S A 100:992–997

Harrison DE, Jordan CT, Zhong RK, Astle CM (1993) Primitive hemopoietic stem cells: direct assay of most productive populations by competitive repopulation with simple binomial, correlation and covariance calculations. Exp Hematol 21:206–219

Haupt Y, Alexander WS, Barri G, Klinken SP, Adams JM (1991) Novel zinc finger gene implicated as myc collaborator by retrovirally accelerated lymphomagenesis in E mu-myc transgenic mice. Cell 65:753–763

Hock H, Hamblen MJ, Rooke HM, Schindler JW, Saleque S, Fujiwara Y, Orkin SH (2004) Gfi-1 restricts proliferation and preserves functional integrity of haematopoietic stem cells. Nature 431:1002–1007

Ichikawa M, Asai T, Saito T, Seo S, Yamazaki I, Yamagata T, Mitani K, Chiba S, Ogawa S, Kurokawa M, Hirai H (2004) AML-1 is required for megakaryocytic maturation and lymphocytic differentiation, but not for maintenance of hematopoietic stem cells in adult hematopoiesis. Nat Med 10:299–304

Ichikawa M, Goyama S, Asai T, Kawazu M, Nakagawa M, Takeshita M, Chiba S, Ogawa S, Kurokawa M (2008) AML1/Runx1 negatively regulates quiescent hematopoietic stem cells in adult hematopoiesis. J Immunol 180:4402–4408

Ito K, Hirao A, Arai F, Matsuoka S, Takubo K, Hamaguchi I, Nomiyama K, Hosokawa K, Sakurada K, Nakagata N, Ikeda Y, Mak TW, Suda T (2004) Regulation of oxidative stress by ATM is required for self-renewal of haematopoietic stem cells. Nature 431:997–1002

Ito K, Hirao A, Arai F, Takubo K, Matsuoka S, Miyamoto K, Ohmura M, Naka K, Hosokawa K, Ikeda Y, Suda T (2006) Reactive oxygen species act through p38 MAPK to limit the lifespan of hematopoietic stem cells. Nat Med 12:446–451

Iwama A, Oguro H, Negishi M, Kato Y, Morita Y, Tsukui H, Ema H, Kamijo T, Katoh-Fukui Y, Koseki H, van Lohuizen M, Nakauchi H (2004) Enhanced self-renewal of hematopoietic stem cells mediated by the polycomb gene product Bmi-1. Immunity 21:843–851

Jacobs JJ, Kieboom K, Marino S, DePinho RA, van Lohuizen M (1999) The oncogene and Polycomb-group gene bmi-1 regulates cell proliferation and senescence through the ink4a locus. Nature 397:164–168

James C, Ugo V, Le Couedic JP, Staerk J, Delhommeau F, Lacout C, Garcon L, Raslova H, Berger R, Bennaceur-Griscelli A, Villeval JL, Constantinescu SN, Casadevall N, Vainchenker W (2005) A unique clonal JAK2 mutation leading to constitutive signalling causes polycythaemia vera. Nature 434:1144–1148

Janzen V, Forkert R, Fleming HE, Saito Y, Waring MT, Dombkowski DM, Cheng T, DePinho RA, Sharpless NE, Scadden DT (2006) Stem-cell ageing modified by the cyclin-dependent kinase inhibitor p16INK4a. Nature 443:421–426

Jeannet G, Scheller M, Scarpellino L, Duboux S, Gardiol N, Back J, Kuttler F, Malanchi I, Birchmeier W, Leutz A, Huelsken J, Held W (2008) Long-term, multilineage hematopoiesis occurs in the combined absence of beta-catenin and gamma-catenin. Blood 111:142–149

Jordan CT, Lemischka IR (1990) Clonal and systemic analysis of long-term hematopoiesis in the mouse. Genes Dev 4:220–232

Kamminga LM, Bystrykh LV, de Boer A, Houwer S, Douma J, Weersing E, Dontje B, de Haan G (2006) The Polycomb group gene Ezh2 prevents hematopoietic stem cell exhaustion. Blood 107:2170–2179

Kaneda M, Okano M, Hata K, Sado T, Tsujimoto N, Li E, Sasaki H (2004) Essential role for de novo DNA methyltransferase Dnmt3a in paternal and maternal imprinting. Nature 429:900–903

Karlsson G, Blank U, Moody JL, Ehinger M, Singbrant S, Deng CX, Karlsson S (2007) Smad4 is critical for self-renewal of hematopoietic stem cells. J Exp Med 204:467–474

Kato S, Ando K, Nakamura Y, Muguruma Y, Sato T, Yabe H, Yabe M, Hattori K, Yasuda Y, Hotta T (2001) Absence of a CD34- hematopoietic precursor population in recipients of CD34+ stem cell transplantation. Bone Marrow Transplant 28:587–595

Kato Y, Iwama A, Tadokoro Y, Shimoda K, Minoguchi M, Akira S, Tanaka M, Miyajima A, Kitamura T, Nakauchi H (2005) Selective activation of STAT5 unveils its role in stem cell self-renewal in normal and leukemic hematopoiesis. J Exp Med 202:169–179

Katsumoto T, Aikawa Y, Iwama A, Ueda S, Ichikawa H, Ochiya T, Kitabayashi I (2006) MOZ is essential for maintenance of hematopoietic stem cells. Genes Dev 20:1321–1330

Kay HE (1965) How many cell-generations? Lancet 2:418–419

Kiel MJ, Yilmaz OH, Iwashita T, Yilmaz OH, Terhorst C, Morrison SJ (2005) SLAM family receptors distinguish hematopoietic stem and progenitor cells and reveal endothelial niches for stem cells. Cell 121:1109–1121

Kim I, He S, Yilmaz OH, Kiel MJ, Morrison SJ (2006) Enhanced purification of fetal liver hematopoietic stem cells using SLAM family receptors. Blood 108:737–744

Kirstetter P, Anderson K, Porse BT, Jacobsen SE, Nerlov C (2006) Activation of the canonical Wnt pathway leads to loss of hematopoietic stem cell repopulation and multilineage differentiation block. Nat Immunol 7:1048–1056

Koch U, Wilson A, Cobas M, Kemler R, Macdonald HR, Radtke F (2008) Simultaneous loss of beta- and gamma-catenin does not perturb hematopoiesis or lymphopoiesis. Blood 111:160–164

Kondo M, Weissman IL, Akashi K (1997) Identification of clonogenic common lymphoid progenitors in mouse bone marrow. Cell 91:661–672

Kozar K, Ciemerych MA, Rebel VI, Shigematsu H, Zagozdzon A, Sicinska E, Geng Y, Yu Q, Bhattacharya S, Bronson RT, Akashi K, Sicinski P (2004) Mouse development and cell proliferation in the absence of D-cyclins. Cell 118:477–491

Kralovics R, Passamonti F, Buser AS, Teo SS, Tiedt R, Passweg JR, Tichelli A, Cazzola M, Skoda RC (2005) A gain-of-function mutation of JAK2 in myeloproliferative disorders. N Engl J Med 352:1779–1790

Kyba M, Perlingeiro RC, Daley GQ (2002) HoxB4 confers definitive lymphoid-myeloid engraftment potential on embryonic stem cell and yolk sac hematopoietic progenitors. Cell 109:29–37

Lachner M, O'Sullivan RJ, Jenuwein T (2003) An epigenetic road map for histone lysine methylation. J Cell Sci 116:2117–2124

Lajtha LG (1979) Stem cell concepts. Differentiation 14:23–34

Lansdorp PM (2007) Immortal strands? Give me a break. Cell 129:1244–1247

Lapidot T, Sirard C, Vormoor J, Murdoch B, Hoang T, Caceres-Cortes J, Minden M, Paterson B, Caligiuri MA, Dick JE (1994) A cell initiating human acute myeloid leukaemia after transplantation into SCID mice. Nature 367:645–648

Larochelle A, Vormoor J, Hanenberg H, Wang JC, Bhatia M, Lapidot T, Moritz T, Murdoch B, Xiao XL, Kato I, Williams DA, Dick JE (1996) Identification of primitive human hematopoietic cells capable of repopulating NOD/SCID mouse bone marrow: implications for gene therapy. Nat Med 2:1329–1337

Lawrence HJ, Christensen J, Fong S, Hu YL, Weissman I, Sauvageau G, Humphries RK, Largman C (2005) Loss of expression of the Hoxa-9 homeobox gene impairs the proliferation and repopulating ability of hematopoietic stem cells. Blood 106:3988–3994

Lessard J, Sauvageau G (2003) Bmi-1 determines the proliferative capacity of normal and leukaemic stem cells. Nature 423:255–260

Magnusson M, Brun AC, Lawrence HJ, Karlsson S (2007a) Hoxa9/hoxb3/hoxb4 compound null mice display severe hematopoietic defects. Exp Hematol 35:1421–1428

Magnusson M, Brun AC, Miyake N, Larsson J, Ehinger M, Bjornsson JM, Wutz A, Sigvardsson M, Karlsson S (2007b) HOXA10 is a critical regulator for hematopoietic stem cells and erythroid/megakaryocyte development. Blood 109:3687–3696

Malumbres M, Sotillo R, Santamaria D, Galan J, Cerezo A, Ortega S, Dubus P, Barbacid M (2004) Mammalian cells cycle without the D-type cyclin-dependent kinases Cdk4 and Cdk6. Cell 118:493–504

Matsumoto K, Isagawa T, Nishimura T, Ogaeri T, Eto K, Miyazaki S, Miyazaki J, Aburatani H, Nakauchi H, Ema H (2009) Stepwise development of hematopoietic stem cells from embryonic stem cells. PLoS One 4, e4820

Matsuzaki Y, Kinjo K, Mulligan RC, Okano H (2004) Unexpectedly efficient homing capacity of purified murine hematopoietic stem cells. Immunity 20:87–93

McKercher SR, Torbett BE, Anderson KL, Henkel GW, Vestal DJ, Baribault H, Klemsz M, Feeney AJ, Wu GE, Paige CJ, Maki RA (1996) Targeted disruption of the PU.1 gene results in multiple hematopoietic abnormalities. Embo J 15:5647–5658

McKinney-Freeman SL, Lengerke C, Jang IH, Schmitt S, Wang Y, Philitas M, Shea J, Daley GQ (2008) Modulation of murine embryonic stem cell-derived CD41 + c-kit + hematopoietic progenitors by ectopic expression of Cdx genes. Blood 111:4944–4953

Metcalf D (2000) Summon up the blood. AlphaMed Press, Miamisburg, OH

Mikkola HK, Klintman J, Yang H, Hock H, Schlaeger TM, Fujiwara Y, Orkin SH (2003) Haematopoietic stem cells retain long-term repopulating activity and multipotency in the absence of stem-cell leukaemia SCL/tal-1 gene. Nature 421:547–551

Molofsky AV, Pardal R, Iwashita T, Park IK, Clarke MF, Morrison SJ (2003) Bmi-1 dependence distinguishes neural stem cell self-renewal from progenitor proliferation. Nature 425:962–967

Morita Y, Ema H, Yamazaki S, Nakauchi H (2006) Non-side-population hematopoietic stem cells in mouse bone marrow. Blood 108:2850–2856

Morrison SJ, Wandycz AM, Akashi K, Globerson A, Weissman IL (1996) The aging of hematopoietic stem cells. Nat Med 2:1011–1016

Nijnik A, Woodbine L, Marchetti C, Dawson S, Lambe T, Liu C, Rodrigues NP, Crockford TL, Cabuy E, Vindigni A, Enver T, Bell JI, Slijepcevic P, Goodnow CC, Jeggo PA, Cornall RJ (2007) DNA repair is limiting for haematopoietic stem cells during ageing. Nature 447:686–690

Ogawa M (1993) Differentiation and proliferation of hematopoietic stem cells. Blood 81:2844–2853

Oguro H, Iwama A, Morita Y, Kamijo T, van Lohuizen M, Nakauchi H (2006) Differential impact of Ink4a and Arf on hematopoietic stem cells and their bone marrow microenvironment in Bmi1-deficient mice. J Exp Med 203:2247–2253

Ohta H, Sawada A, Kim JY, Tokimasa S, Nishiguchi S, Humphries RK, Hara J, Takihara Y (2002) Polycomb group gene rae28 is required for sustaining activity of hematopoietic stem cells. J Exp Med 195:759–770

Ohta H, Sekulovic S, Bakovic S, Eaves CJ, Pineault N, Gasparetto M, Smith C, Sauvageau G, Humphries RK (2007) Near-maximal expansions of hematopoietic stem cells in culture using NUP98-HOX fusions. Exp Hematol 35:817–830

Okano M, Bell DW, Haber DA, Li E (1999) DNA methyltransferases Dnmt3a and Dnmt3b are essential for de novo methylation and mammalian development. Cell 99:247–257

Okuda T, van Deursen J, Hiebert SW, Grosveld G, Downing JR (1996) AML1, the target of multiple chromosomal translocations in human leukemia, is essential for normal fetal liver hematopoiesis. Cell 84:321–330

Orkin SH (2000) Diversification of haematopoietic stem cells to specific lineages. Nat Rev Genet 1:57–64

Orlando V (2003) Polycomb, epigenomes, and control of cell identity. Cell 112:599–606

Osawa M, Hanada K, Hamada H, Nakauchi H (1996) Long-term lymphohematopoietic reconstitution by a single CD34-low/negative hematopoietic stem cell. Science 273:242–245

Park IK, Qian D, Kiel M, Becker MW, Pihalja M, Weissman IL, Morrison SJ, Clarke MF (2003) Bmi-1 is required for maintenance of adult self-renewing haematopoietic stem cells. Nature 423:302–305

Pluznik DH, Sachs L (1965) The cloning of normal "mast" cells in tissue culture. J Cell Physiol 66:319–324

Porcher C, Swat W, Rockwell K, Fujiwara Y, Alt FW, Orkin SH (1996) The T cell leukemia oncoprotein SCL/tal-1 is essential for development of all hematopoietic lineages. Cell 86:47–57

Potocnik AJ, Brakebusch C, Fassler R (2000) Fetal and adult hematopoietic stem cells require beta1 integrin function for colonizing fetal liver, spleen, and bone marrow. Immunity 12:653–663

Priestley GV, Scott LM, Ulyanova T, Papayannopoulou T (2006) Lack of alpha4 integrin expression in stem cells restricts competitive function and self-renewal activity. Blood 107:2959–2967

Pronk CJ, Rossi DJ, Mansson R, Attema JL, Norddahl GL, Chan CK, Sigvardsson M, Weissman IL, Bryder D (2007) Elucidation of the phenotypic, functional, and molecular topography of a myeloerythroid progenitor cell hierarchy. Cell Stem Cell 1:428–442

Qian H, Georges-Labouesse E, Nystrom A, Domogatskaya A, Tryggvason K, Jacobsen SE, Ekblom M (2007) Distinct roles of integrins alpha6 and alpha4 in homing of fetal liver hematopoietic stem and progenitor cells. Blood 110:2399–2407

Ren S, Rollins BJ (2004) Cyclin C/cdk3 promotes Rb-dependent G0 exit. Cell 117:239–251

Reya T, Duncan AW, Ailles L, Domen J, Scherer DC, Willert K, Hintz L, Nusse R, Weissman IL (2003) A role for Wnt signalling in self-renewal of haematopoietic stem cells. Nature 423:409–414

Robb L, Elwood NJ, Elefanty AG, Kontgen F, Li R, Barnett LD, Begley CG (1996) The scl gene product is required for the generation of all hematopoietic lineages in the adult mouse. Embo J 15:4123–4129

Roegiers F, Jan YN (2004) Asymmetric cell division. Curr Opin Cell Biol 16:195–205

Rosendaal M, Hodgson GS, Bradley TR (1979) Organization of haemopoietic stem cells: the generation-age hypothesis. Cell Tissue Kinet 12:17–29

Rossi DJ, Bryder D, Seita J, Nussenzweig A, Hoeijmakers J, Weissman IL (2007) Deficiencies in DNA damage repair limit the function of haematopoietic stem cells with age. Nature 447:725–729

Sarma K, Reinberg D (2005) Histone variants meet their match. Nat Rev Mol Cell Biol 6:139–149

Sauvageau G, Thorsteinsdottir U, Eaves CJ, Lawrence HJ, Largman C, Lansdorp PM, Humphries RK (1995) Overexpression of HOXB4 in hematopoietic cells causes the selective expansion of more primitive populations in vitro and in vivo. Genes Dev 9:1753–1765

Scheller M, Huelsken J, Rosenbauer F, Taketo MM, Birchmeier W, Tenen DG, Leutz A (2006) Hematopoietic stem cell and multilineage defects generated by constitutive beta-catenin activation. Nat Immunol 7:1037–1047

Schofield R (1978) The relationship between the spleen colony-forming cell and the haemopoietic stem cell. Blood Cells 4:7–25

Schwaller J, Parganas E, Wang D, Cain D, Aster JC, Williams IR, Lee CK, Gerthner R, Kitamura T, Frantsve J, Anastasiadou E, Loh ML, Levy DE, Ihle JN, Gilliland DG (2000) Stat5 is essential for the myelo- and lymphoproliferative disease induced by TEL/JAK2. Mol Cell 6:693–704

Scott EW, Simon MC, Anastasi J, Singh H (1994) Requirement of transcription factor PU.1 in the development of multiple hematopoietic lineages. Science 265:1573–1577

Seita J, Ema H, Ooehara J, Yamazaki S, Tadokoro Y, Yamasaki A, Eto K, Takaki S, Takatsu K, Nakauchi H (2007) Lnk negatively regulates self-renewal of hematopoietic stem cells by modifying thrombopoietin-mediated signal transduction. Proc Natl Acad Sci U S A 104:2349–2354

Shamblott MJ, Axelman J, Wang S, Bugg EM, Littlefield JW, Donovan PJ, Blumenthal PD, Huggins GR, Gearhart JD (1998) Derivation of pluripotent stem cells from cultured human primordial germ cells. Proc Natl Acad Sci U S A 95:13726–13731

Sherr CJ, Roberts JM (1999) CDK inhibitors: positive and negative regulators of G1-phase progression. Genes Dev 13:1501–1512

Smith LG, Weissman IL, Heimfeld S (1991) Clonal analysis of hematopoietic stem-cell differentiation in vivo. Proc Natl Acad Sci U S A 88:2788–2792

Spangrude GJ, Heimfeld S, Weissman IL (1988) Purification and characterization of mouse hematopoietic stem cells. Science 241:58–62

Stepanova L, Sorrentino BP (2005) A limited role for p16Ink4a and p19Arf in the loss of hematopoietic stem cells during proliferative stress. Blood 106:827–832

Strahl BD, Allis CD (2000) The language of covalent histone modifications. Nature 403:41–45
Su IH, Basavaraj A, Krutchinsky AN, Hobert O, Ullrich A, Chait BT, Tarakhovsky A (2003) Ezh2 controls B cell development through histone H3 methylation and Igh rearrangement. Nat Immunol 4:124–131
Suda T, Suda J, Ogawa M (1984) Disparate differentiation in mouse hemopoietic colonies derived from paired progenitors. Proc Natl Acad Sci U S A 81:2520–2524
Sudo K, Ema H, Morita Y, Nakauchi H (2000) Age-associated characteristics of murine hematopoietic stem cells. J Exp Med 192:1273–1280
Sugiyama T, Kohara H, Noda M, Nagasawa T (2006) Maintenance of the hematopoietic stem cell pool by CXCL12-CXCR4 chemokine signaling in bone marrow stromal cell niches. Immunity 25:977–988
Tadokoro Y, Ema H, Okano M, Li E, Nakauchi H (2007) De novo DNA methyltransferase is essential for self-renewal, but not for differentiation, in hematopoietic stem cells. J Exp Med 204:715–722
Takahashi K, Tanabe K, Ohnuki M, Narita M, Ichisaka T, Tomoda K, Yamanaka S (2007) Induction of pluripotent stem cells from adult human fibroblasts by defined factors. Cell 131:861–872
Takano H, Ema H, Sudo K, Nakauchi H (2004) Asymmetric division and lineage commitment at the level of hematopoietic stem cells: inference from differentiation in daughter cell and granddaughter cell pairs. J Exp Med 199:295–302
Thomas T, Corcoran LM, Gugasyan R, Dixon MP, Brodnicki T, Nutt SL, Metcalf D, Voss AK (2006) Monocytic leukemia zinc finger protein is essential for the development of long-term reconstituting hematopoietic stem cells. Genes Dev 20:1175–1186
Thomson JA, Itskovitz-Eldor J, Shapiro SS, Waknitz MA, Swiergiel JJ, Marshall VS, Jones JM (1998) Embryonic stem cell lines derived from human blastocysts. Science 282:1145–1147
Tiedt R, Hao-Shen H, Sobas MA, Looser R, Dirnhofer S, Schwaller J, Skoda RC (2008) Ratio of mutant JAK2–V617F to wild-type Jak2 determines the MPD phenotypes in transgenic mice. Blood 111:3931–3940
Till JE, McCulloch EA (1961) A direct measurement of the radiation sensitivity of normal mouse bone marrow cells. Radiat Res 14:213–222
Till JE, McCulloch EA, Siminovitch L (1964) A Stochastic Model of Stem Cell Proliferation, Based on the Growth of Spleen Colony-Forming Cells. Proc Natl Acad Sci U S A 51:29–36
Tothova Z, Kollipara R, Huntly BJ, Lee BH, Castrillon DH, Cullen DE, McDowell EP, Lazo-Kallanian S, Williams IR, Sears C, Armstrong SA, Passegue E, DePinho RA, Gilliland DG (2007) FoxOs are critical mediators of hematopoietic stem cell resistance to physiologic oxidative stress. Cell 128:325–339
Trimarchi JM, Fairchild B, Wen J, Lees JA (2001) The E2F6 transcription factor is a component of the mammalian Bmi1-containing polycomb complex. Proc Natl Acad Sci U S A 98:1519–1524
Trowbridge JJ, Moon RT, Bhatia M (2006) Hematopoietic stem cell biology: too much of a Wnt thing. Nat Immunol 7:1021–1023
Tsai FY, Keller G, Kuo FC, Weiss M, Chen J, Rosenblatt M, Alt FW, Orkin SH (1994) An early haematopoietic defect in mice lacking the transcription factor GATA-2. Nature 371:221–226
Tulina N, Matunis E (2001) Control of stem cell self-renewal in *Drosophil*a spermatogenesis by JAK-STAT signaling. Science 294:2546–2549
van der Lugt NM, Domen J, Linders K, van Roon M, Robanus-Maandag E, te Riele H, van der Valk M, Deschamps J, Sofroniew M, van Lohuizen M, Berns A (1994) Posterior transformation, neurological abnormalities, and severe hematopoietic defects in mice with a targeted deletion of the bmi-1 proto-oncogene. Genes Dev 8:757–769
van Lohuizen M, Verbeek S, Scheijen B, Wientjens E, van der Gulden H, Berns A (1991) Identification of cooperating oncogenes in E mu-myc transgenic mice by provirus tagging. Cell 65:737–752
van Os R, Kamminga LM, Ausema A, Bystrykh LV, Draijer DP, van Pelt K, Dontje B, de Haan G (2007) A Limited role for p21Cip1/Waf1 in maintaining normal hematopoietic stem cell functioning. Stem Cells 25:836–843

Visser JW, Bauman JG, Mulder AH, Eliason JF, de Leeuw AM (1984) Isolation of murine pluripotent hemopoietic stem cells. J Exp Med 159:1576–1590
Wada H, Masuda K, Satoh R, Kakugawa K, Ikawa T, Katsura Y, Kawamoto H (2008) Adult T-cell progenitors retain myeloid potential. Nature 452:768–772
Wang JC, Warner JK, Erdmann N, Lansdorp PM, Harrington L, Dick JE (2005a) Dissociation of telomerase activity and telomere length maintenance in primitive human hematopoietic cells. Proc Natl Acad Sci U S A 102:14398–14403
Wang L, Menendez P, Shojaei F, Li L, Mazurier F, Dick JE, Cerdan C, Levac K, Bhatia M (2005b) Generation of hematopoietic repopulating cells from human embryonic stem cells independent of ectopic HOXB4 expression. J Exp Med 201:1603–1614
Willert K, Brown JD, Danenberg E, Duncan AW, Weissman IL, Reya T, Yates JR 3rd, Nusse R (2003) Wnt proteins are lipid-modified and can act as stem cell growth factors. Nature 423:448–452
Wodarz A, Huttner WB (2003) Asymmetric cell division during neurogenesis in Drosophila and vertebrates. Mech Dev 120:1297–1309
Wright DE, Wagers AJ, Gulati AP, Johnson FL, Weissman IL (2001) Physiological migration of hematopoietic stem and progenitor cells. Science 294:1933–1936
Wysocka J, Swigut T, Milne TA, Dou Y, Zhang X, Burlingame AL, Roeder RG, Brivanlou AH, Allis CD (2005) WDR5 associates with histone H3 methylated at K4 and is essential for H3 K4 methylation and vertebrate development. Cell 121:859–872
Xie T, Spradling AC (2000) A niche maintaining germ line stem cells in the *Drosophila* ovary. Science 290:328–330
Yamashita YM, Jones DL, Fuller MT (2003) Orientation of asymmetric stem cell division by the APC tumor suppressor and centrosome. Science 301:1547–1550
Yamazaki S, Iwama A, Takayanagi S, Morita Y, Eto K, Ema H, Nakauchi H (2006) Cytokine signals modulated via lipid rafts mimic niche signals and induce hibernation in hematopoietic stem cells. Embo J 25:3515–3523
Yamazaki S, Iwama A, Takayanagi S, Eto K, Ema H, Nakauchi H (2009) TGF-β as a candidate bone marrow niche signal to induce hematopoietic stem cell hibernation. Blood 113:1250–1256
Yang L, Bryder D, Adolfsson J, Nygren J, Mansson R, Sigvardsson M, Jacobsen SE (2005) Identification of Lin(-)Sca1(+)kit(+)CD34(+)Flt3- short-term hematopoietic stem cells capable of rapidly reconstituting and rescuing myeloablated transplant recipients. Blood 105:2717–2723
Yilmaz OH, Valdez R, Theisen BK, Guo W, Ferguson DO, Wu H, Morrison SJ (2006) Pten dependence distinguishes haematopoietic stem cells from leukaemia-initiating cells. Nature 441:475–482
Young NS, Scheinberg P, Calado RT (2008) Aplastic anemia. Curr Opin Hematol 15:162–168
Yu J, Vodyanik MA, Smuga-Otto K, Antosiewicz-Bourget J, Frane JL, Tian S, Nie J, Jonsdottir GA, Ruotti V, Stewart R, Slukvin II, Thomson JA (2007) Induced pluripotent stem cell lines derived from human somatic cells. Science 318:1917–1920
Yuan Y, Shen H, Franklin DS, Scadden DT, Cheng T (2004) In vivo self-renewing divisions of haematopoietic stem cells are increased in the absence of the early G1-phase inhibitor, p18INK4C. Nat Cell Biol 6:436–442
Zeng H, Yucel R, Kosan C, Klein-Hitpass L, Moroy T (2004) Transcription factor Gfi1 regulates self-renewal and engraftment of hematopoietic stem cells. Embo J 23:4116–4125
Zhang J, Niu C, Ye L, Huang H, He X, Tong WG, Ross J, Haug J, Johnson T, Feng JQ, Harris S, Wiedemann LM, Mishina Y, Li L (2003) Identification of the haematopoietic stem cell niche and control of the niche size. Nature 425:836–841
Zhao C, Blum J, Chen A, Kwon HY, Jung SH, Cook JM, Lagoo A, Reya T (2007) Loss of beta-catenin impairs the renewal of normal and CML stem cells in vivo. Cancer Cell 12:528–541
Zhou S, Morris JJ, Barnes Y, Lan L, Schuetz JD, Sorrentino BP (2002) Bcrp1 gene expression is required for normal numbers of side population stem cells in mice, and confers relative protection to mitoxantrone in hematopoietic cells in vivo. Proc Natl Acad Sci U S A 99:12339–12344

Hematopoietic Stem Cells and Their Niche

Hiroko Iwasaki and Toshio Suda

Abstract The stem cells' major capabilities (i.e., the pluripotency and the self-renewal) are the keys to sustain the lifelong functionality of the organ. Stem cells reside in the special microenvironment called niche. The niche and stem cells adhere to each other via adhesion molecules and exchange the molecular signals that maintain the stem cell features. It has been suggested that tumor tissue also contains such type of cells.

In this chapter, the hematopoietic system is exemplified to show the interaction between the stem cell and its niche, which is the most intensively studied and archetypical stem cell system. Different kinds of niche and the regulatory mechanisms are explained. Furthermore, the niche involvement in cancer stem regulation, tumor invasion, and metastasis, as well as the novel therapeutic approaches in association with the cancer stem cell niche are also discussed.

Introduction

Virtually every organ has stem cells. These stem cells possess two major roles: the pluripotency that gives rise to the mature cells to compose the specific organ or tissue, and the self-renewal capability to supply enough cells to maintain the organ's function. These stem cells tend to be found in specific areas in the organ, where special microenvironment maintains the stem cell functions. This special

H. Iwasaki (✉)
Department of Cell Differentiation, The Sakaguchi Laboratory of Developmental Biology, Keio University School of Medicine, 35 Shinanomachi, Shinjuku, Tokyo 160-8582 Japan
e-mail: hiroko-ko@umin.ac.jp

T. Suda
Director and Professor, Center for Integrated Medical Research, The Sakaguchi Laboratory of Developmental Biology, Keio University School of Medicine, 35 Shinanomachi, Shinjuku, Tokyo 160-8582 Japan
e-mail: sudato@sc.itc.keio.ac.jp

M. Kondo (ed.), *Hematopoietic Stem Cell Biology,* Stem Cell Biology and Regenerative Medicine, DOI 10.1007/978-1-60327-347-3_2,

microenvironment is called the niche. The stem cell and the niche cells adhere to each other via adhesion molecules and exchange the molecular signals that maintain the stem cell features.

The concept of the stem cell niche was proposed by R. Schofield in 1978 for the hematopoietic stem cell (HSC) in bone marrow (Schofield 1978). Schofield stated "the cellular environment which retains the stem cell I shall call a stem cell 'niche.' As long as the stem cell remains fixed its further maturation is prevented and it continues indefinitely to replicate as a stem cell, i.e., it exhibits immortality." Since this initial proposal, not only the HSCs but various other kinds of stem cells have also been identified, and their niches and the molecular mechanisms for stemness maintenance have been revealed.

The stem cells are not only present in healthy normal organs or tissues. Recent findings suggest that a group of cells exist in tumors that behave in a similar way to the normal stem cells. Such cells are called cancer stem cells and are considered to be deeply involved in the tumor proliferation, invasion, and metastasis. Their niche plays a regulatory role for the cancer stem cell maintenance.

As of today, the most advanced studies in the stem cell niche field can be found in the hematopoietic system. In this chapter, the interaction between the HSC and its niche will be discussed. Furthermore, the niche involvement in the cancer stem regulation, tumor invasion, and metastasis and the novel therapeutic approaches in association with the cancer stem cell niche will also be approached.

Stem Cell and Niche

The stem cell research, in terms of the maintenance of the pluripotency and self-renew capacity, was greatly advanced through the experiment using *Drosophila* and *Caenorhabditis elegance*. In 1997, the group of W. Deng and H. Lin demonstrated that the female *Drosophila* has "cap cells" at the tip of the ovary, to which the germ line stem cells adhere (Deng and Lin 1997). In mitosis, the daughter stem cells that divide to the direction in which they no longer adhere to the cap cells started the differentiation and grew into the cystoblast. In contrast, those that kept the adhesion to the cap cells after the cell division remained as the germ line stem cells. These results suggested that the cap cells provide the special environment for the germ line stem cell and play a role as the stem cell niche in the *Drosophila* ovary.

T. Xie and A. C. Spradling demonstrated in 2000 that the newly introduced cells to the *Drosophila* ovary performed as the germ line stem cell as long as they adhere to the cap cells, thus proving that the cap cells function as the true stem cell niche (Xie and Spradling 2000). They showed that the germ line stem cells require a signal mediated by dpp, a member of the TGF-β (transforming growth factor beta) super family, in order to maintain the adhesion to the cap cells (to maintain the stem cell) and control the frequency of the cell division. Additional signals, such as Hedgehog, Wingless, or Armadillo, were also suggested to be involved in the regulation of the germ line stem cells.

Interaction Between Hematopoietic Stem Cell and Niche

Osteoblastic Niche

In human hematopoiesis in bone marrow, the osteoblasts offer the microenvironment as the niche (i.e., osteoblastic niche) for the HSCs. HSC and the osteoblast bind each other via the adhesion molecules, such as N-cadherin (Fig. 1). The two independent research groups, led by L. Li and D. T. Scadden, respectively, revealed this new function of the osteoblast as the HSC niche in 2003. Both groups approached the issue in the investigation of the regulatory system for the number of the HSCs. Li's group reported that bone morphogenetic protein (BMP) signaling pathway through BMP receptor type IA (BMPRIA) expressed in osteoblasts controls the number of HSCs by regulating the niche size (Zhang et al. 2003). Scadden et al. showed that the osteoblastic cells stimulated by activated PTH (parathyroid hormone)/PTHrP(parathyroid hormone-related peptide) receptors (PPRs) and increased in number produced high levels of the Notch ligand jagged 1 and supported the increase in the number of HSCs (Calvi et al. 2003).

It is widely known that the stem cell in general is in a quiescent state (G0 phase in cell cycle) and that this quiescence prevents the stem cells from entering into the cell cycle and differentiation. In this regard, the evidence for the importance of Tie2 signaling in HSC–niche interaction has been demonstrated by Puri and Berstein by using chimeric mice generated between normal embryonic cells and cells lacking Tie family receptors (Tie1 and Tie2) (Puri and Berstein 2003). Although Tie receptors

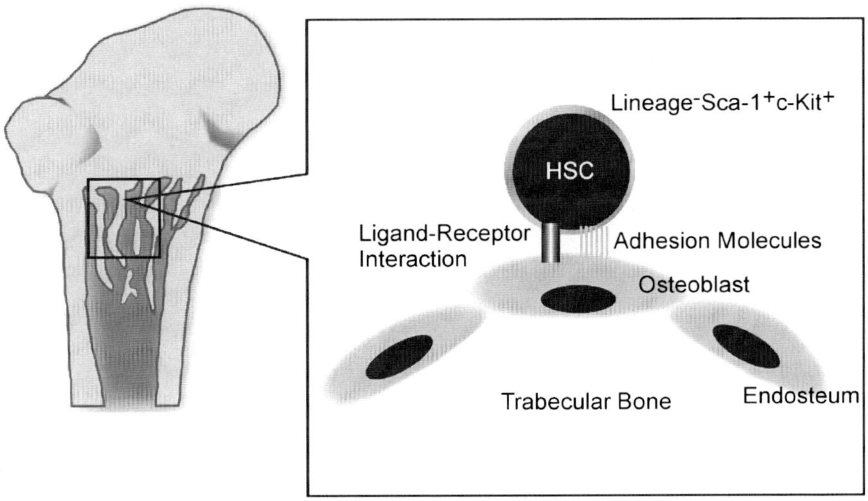

Fig. 1 Schematic model for hematopoietic stem cell (HSC) and osteoblastic niche. Osteoblastic niche lodges HSC via adhesion molecules, such as N-cadherin. Osteoblast and HSC exchange various signaling in order to maintain the stem cell functions

were not required for fetal hematopoiesis, including emergence of definitive HSCs, relocation to fetal liver, and differentiation, HSCs lacking these receptors were not maintained in the adult bone marrow microenvironment. Since Tie1-deficient cells, which express normal levels of Tie2, contribute to hematopoiesis, these findings indicate that Tie2 is required for postnatal bone marrow hematopoiesis. In addition, they analyzed chimeric mice generated between Tie1/Tie2 deficient embryonic cells and Rag2-deficient morula as the host, which does not produce mature lymphocytes, with a result that Tie-deficient cells could fully contribute to lymphopoiesis when no wild-type competing cells coexist. These findings indicated that Tie1 and Tie2 deficiency in HSCs leads not to the defect in differentiation but to survival in the microenvironment in adult bone marrow.

Arai et al. directly observed that angiopoietin-1 (Ang1) expressed in osteoblast interacts with Tie2, a type of receptor tyrosine kinase expressed in hematopoietic stem cell in bone marrow, and demonstrated that Tie2/Ang1 interaction activates β1-integrin and N-cadherin. This enhanced adhesion between the niche cell and the stem cell contributes to the maintenance of the quiescence of the stem cell and self-renew (Arai et al. 2004). Further studies are needed to elucidate the functional relationship between cell adhesion and cell cycle regulation.

Osteopontin (Opn), expressed by osteoblast, is also reported as a key component in HSC niche (Nilsson et al. 2005). Nilsson et al. demonstrated that the LSK (lineage$^-$Sca-1$^+$c-Kit$^+$) cells that were isolated from bone marrow of Opn knockout mouse after the continuous 4-week BrdU treatment were 100% BrdU positive, meaning that the HSCs divided at least once in 4 weeks in bone marrow of the Opn knockout mice. On the other hand, approximately 60% of the collected LSK cells were BrdU$^+$, as in the case of wild-type mice. They also showed that CD34$^+$ human hematopoietic progenitors were inhibited to bind to Opn when treated with the specific β_1 integrin-blocking antibody prior to the cell culture. Also, the LSK cells were observed to be rather randomly distributed within bone marrow after transplanted into Opn knockout mice, whereas they were likely to be located at the endosteal region, where osteoblasts reside, in the case of wild-type mice. These data suggested that Opn expressed by osteoblast contributes to the adhesion between the HSC and osteoblastic niche and negatively regulates HSC proliferation, contributing to the maintenance of the stem cell quiescence.

Thrombopoietin (TPO) and its ligand Mpl, the regulator of megakaryopoiesis, are also critical regulators in HSC maintenance in osteoblastic niche, as reported by two independent groups (Yoshihara et al. 2007; Qian et al. 2007). Yoshihara et al. found that the majority of Mpl$^+$ LSK cells were found in side population (SP), the most concentrated fraction of quiescent HSCs. Mpl$^+$ HSCs were observed in close contact with TPO-producing osteoblastic cells at the endosteal surface in trabecular bone area. On the other hand, Quin et al. demonstrated that TPO/Mpl signaling is required for postnatal HSC expansion but not in the prenatal phase, using adult *Thpo*$^{-/-}$ mice. In the knockout mice, the number of HSCs was 150-fold reduced, and *p57*^{Kip2}, a cell cycle regulator specifically expressed in the quiescent population of LT-HSC, was dramatically downregulated in HSCs. Meanwhile, Yoshihara et al. found that *p57*^{Kip2} was upregulated by stimulation with TPO and downregulated by

AMM2, an anti-Mpl neutralizing antibody. These data consistently indicate that the Mpl/TPO signaling regulates cell cycle of adult quiescent HSCs to maintain the pool at the physiologically reasonable level.

c-Myc is another important regulator of the HSC fate. Wilson et al. demonstrated, using conditional c-Myc deficient mice, that the number of HSCs in bone marrow was significantly higher, and that the HSC differentiation was prevented due to the upregulated adhesion molecules between HSC and the niche, such as N-cadherin and various integrins, resulting in the severe cytopenia. Conversely, overexpression of c-Myc led to loss of self-renewal activity and augmented differentiation of HSCs. The endogenous c-Myc mRNA level appeared to differ consistently with these results (i.e., the lower expression level in the self-renewing long-term HSCs and higher level in the differentiating short-term HSCs). c-Myc controls the balance between the HSC self-renewal and differentiation by adjusting the adhesion between HSCs and the niche (Wilson et al. 2004).

ECM Contribution to Osteoblastic Niche

Aside from the factors expressed in osteoblasts, various extracellular matrix (ECM) proteins in bone marrow are also confirmed to be involved in the regulation of the osteoblastic niche environment. Glycosaminoglycan hyaluronic acid (HA), for example, is one of the key components of the niche regulators. HA is the major component of the ECM and is found ubiquitously, including in bone marrow. Its receptor, CD44, is a multifunctional transmembrane protein expressed by a wide variety of cells, including the hematopoietic stem cells and progenitors. Matrosova et al. reported that the 5-fluorouracil (5-FU)-treated mice for bone marrow suppression exhibited quicker recovery in terms of the white blood cell and platelet counts when administered with HA during the recovery period, compared to phosphate buffered-saline as control (Matrosova et al. 2004). In order to examine the location of injected HA *in vivo*, the fluorescein isothiocyanate (FITC)-labeled HA was traced and found concentrated at bone marrow, binding to CD44. This suggests that HA plays a role as a regulator of supportive function of niche in bone marrow for hematopoietic stem cells.

Another report demonstrated that the xenografting of human $CD34^+$ hematopoietic stem/progenitor cells into the bone marrow of nonobese diabetic severe combined immunodeficient (NOD/SCID) mice was significantly impaired by masking the HSCs' cell surface with anti-CD44 antibody or HA (Avigdor et al. 2004), indicating that CD44-HA interaction is an important factor for homing and engraftment of the HSC to the niche.

Because the osteoblasts are located at the endosteal area, where higher calcium concentration is expected due to its active bone modeling/remodeling, Adams et al. hypothesized that the sensor for calcium must be expressed on HSCs and contributes to their homing and lodgment to bone marrow. They revealed that calcium-sensing receptor, CaR, was expressed on normal HSCs and that CaR-deficient mice had a normal number of primitive hematopoietic cells in the circulation and spleen but few of them were found in bone marrow. HSCs isolated from fetal liver of CaR-deficient mice could not localize at the endosteal niche in wild-type recipient mice because of the defective adhesion to the collagen I, one of the major components of the ECM secreted by osteoblast. Thus, it was suggested that the HSCs preferentially localize the calcium-rich endosteum and lodged to the niche via adhesion to the ECM surrounding the osteoblast (Adams et al. 2006).

Vascular Niche

As described above, the adhesion molecules and their signaling play a fundamental role in the HSC maintenance for the osteoblastic niche. There has been another kind of niche recently found along the endothelial cells of the sinusoidal vessels in bone marrow or spleen. This microenvironment is called the vascular niche. This new type of niche was revealed through a recent study to distinguish the stem and the progenitor cells in bone marrow using the SLAM family receptors. The most purified HSCs were fractioned as $CD150^{+}CD244^{-}CD48^{-}$ cells, and the majority of these cells were found associated with sinusoidal endothelium (Kiel et al. 2005).

Sugiyama et al. showed that the HSCs were found specifically adjacent to the cells expressing a high level of CXCL12 (also called stromal cell–derived factor-1 or SDF-1) and surrounding the sinusoidal endothelial cells. They named these cells CXCL12-abundant reticular (CAR) cells and demonstrated that the depletion of CXCR4 led to the reduction of the HSC pool and a poor survival rate after 5-FU-induced bone marrow suppression, thus suggesting that CXCL12-CXCR4 chemokine signaling plays an essential role in maintaining the quiescent HSC pool (Sugiyama et al. 2006). The connection between CAR cells and HSCs is also observed at the endosteum. It appears to be the universal component of the hematopoietic stem cell niche.

The homing and lodgment mechanism to the vascular niche, the integrin very late antigen 4 (VLA-4) and its ligand vascular cell adhesion molecule 1 (VCAM-1), was proposed to be the regulator (Papayannopoulou et al. 1995). The pretreatment of donor cells with either anti-VLA-4 or VCAM-1 antibody prior to the transplantation into the recipient mice resulted in impaired lodgment of HSCs to bone marrow and increased the number of hematopoietic progenitors in circulating blood and the spleen.

Taking advantage of the VLA-4-mediated interaction between HSCs and vascular niche in bone marrow, natalizumab, a human anti-VLA-4 antibody, has been suggested for treating patients with poor response to granulocyte colony-stimulating factor (G-CSF)-based protocols for the peripheral blood stem cell transplantation (Zohren et al. 2008).

Human anti-VLA-4 antibody natalizumab was originally developed and introduced for the treatment of autoimmune diseases, such as multiple sclerosis (MS) and Crohn's disease. The authors examined peripheral blood of patients with MS receiving natalizumab and found that the number of $CD34^{+}$ primitive hematopoietic cells, including

HSCs, circulating in peripheral blood was significantly higher in natalizumab-treated MS patients, compared to those in untreated MS patients or healthy volunteers. The increase in the number of $CD34^+$ cells was observed as soon as 1 hour after the infusion of natalizumab, and the colony-forming unit assay demonstrated that progenitors mobilized into peripheral blood formed significantly higher numbers of BFU-E and CFU-G/M/GM colonies compared to before infusion. The increase of $CD34^+$ cells in peripheral blood was the result of both mobilization effect and inhibited homing by natalizumab. The latter effect lasted longer than the half-life of natalizumab. Thus, anti-VLA-4 antibody natalizumab with mobilizing ability may have the potential to become an alternative in peripheral blood stem cell transplantation.

The functional difference between the osteoblastic and vascular niches is yet to be elucidated; however, it may be reasonable to approach this issue from a physiological aspect. One of the major differences of these two microenvironments is the oxygen level. In vascular niche, a higher oxygen level is expected than in osteoblastic niche. Under such a microenvironment, the cell cycle of the stem cell would resume (Parmar et al. 2007). Thus a model of the dynamics of the hematopoiesis can be hypothesized: the HSCs in G0 state located at the osteoblastic niche under the regional hypoxia would move to the vascular niche at a certain time, undergo differentiation, and supply the required pool of the mature cells into the peripheral bloodstream (Heissig et al. 2002). Once the supply fulfills the necessary mature cells, the HSCs at the vascular niche would move back to the osteoblastic niche, where they are maintained in the G0 state again. The logistics of the HSCs shuttled between these two types of niches may be the key for well-balanced hematopoiesis (Fig. 2).

The endogenous oxidative metabolism in HSCs was studied by Piccoli et al., using human $CD34^+$ HSCs forced to mobilize from bone marrow to peripheral blood in the G-CSF infusion regimen (Piccoli et al. 2005). The mitochondrial oxygen consumption level in HSCs was significantly reduced to approximately 10%, compared to regular types of cells, and it was found not because of inhibitory control but relatively poor content to mitochondrial cytochromes. The mitochondrial and extramitochondrial respiration was assessed by treatment with electron transport inhibitors, such as potassium cyanide, or enzymatic scavengers of reactive oxygen species (ROS), such as superoxide dismutase. The ratio of mitochondrial versus extramitochondrial respiratory activity was approximately 60:40, and by combining both treatments the endogenous respiratory activity was completely blocked. The major ROS producer in this cell is known to be cell membrane–bound NADPH oxidase, and the authors verified this by immunoprecipitation to show that NADPH oxidase exists on HSCs' membrane, forming the fully assembled functional complex coordinated with $p22^{phox}$, $p67^{phox}$, and $p47^{phox}$.

The amount of mitochondria was directly measured by confocal microscopy analysis using the fluorescent dye specifically accumulated in membrane potential-generating mitochondria, costained with CD34, an antigen expressed in HSCs or early hematopoietic progenitor cells. In general, the higher the expression is, the more primitive the cell is. The density and distribution of stained mitochondria have high variation among observed cells. Interestingly, the higher the CD34 staining signal density, the lower the mitochondria staining signal density.

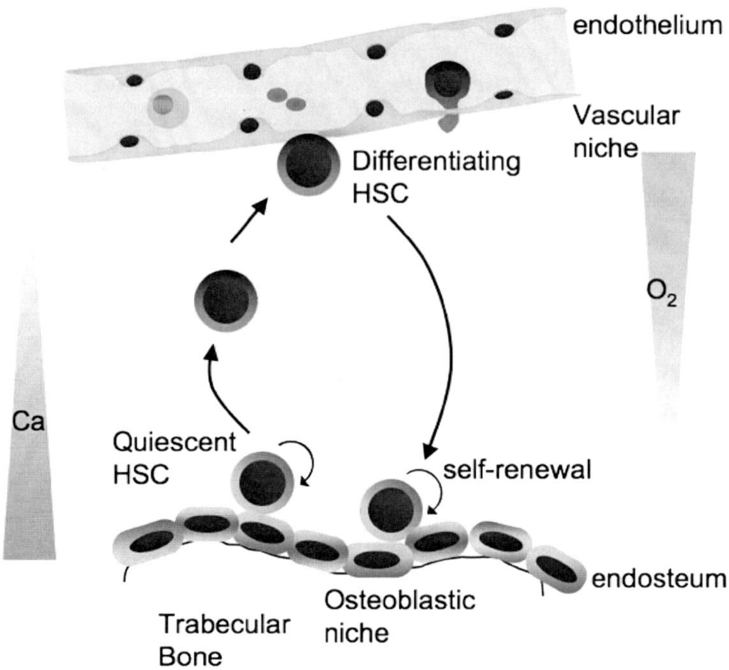

Fig. 2 Hematopoietic stem cell (HSC) logistic model between osteoblastic niche and vascular niche in bone marrow. HSC stays in quiescence in osteoblastic niche and activates the cell cycle upon leaving for vascular niche, where the differentiation is initiated. HSCs are guided by multiple factors, such as local calcium or oxygen concentration gradient between these two different kinds of niches

Collectively, a HSCs oxygen metabolism model was proposed, suggesting that primitive HSCs that have a low amount of mitochondria use membrane-bounded NADPH oxidase complex as oxygen sensor and/or ROS source, contributing to signaling inducing mitochondriogenesis, differentiation, or cell survival.

External Oxidative Stress and HSC

The evidence of the oxidative stress impact on the long-term HSCs' capability was demonstrated by Ito et al. (Ito et al. 2006). Treatment of the immature hematopoietic cells of LSK fraction with buthionine sulfoximine (BSO), even at low levels, affected the repopulation capacity of HSCs after transplantation, but not the number of colonies formed in culture. This result indicated that elevation of ROS induced by BSO did not have an impact on the colony-forming progenitors in LSK fraction but specifically lead to the defect of the HSC function. The elevation of ROS upregulated tumor suppressors $p16^{Ink4a}$ and $p19^{Arf}$, specifically in HSCs, and treatment with p38 MAPK inhibitor or an antioxidant *N*-acetyl-L-cyctein (NAC) blocked this ROS-induced increase of $p16^{Ink4a}$ and $p19^{Arf}$. This suggested that oxidative stress induces the HSC-specific phosphorylation of p38 MAPK, and this activation of p38 MAPK lead to the defect in the maintenance of HSC self-renewal capacity.

Ito et al. also investigated the ROS effect *in vivo* using the ataxia telangiectasia mutated *(Atm)* knockout mice, which exhibit the defect in oxidative stress regulation. In $Atm^{-/-}$ mice, ROS elevation and defective HSCs' self-renewal were observed (Ito et al. 2004) and, therefore, the activation of p38 MAPK was expected. After 2 days of incubation, $p16^{Ink4a}$ and $p19^{Arf}$ were upregulated in $Atm^{-/-}$ LSK cells as ROS level elevated. As predicted, p38 MAPK was activated only in LSK cells, and treatment with p38 MAPK inhibitor blocked this upregulation. Furthermore, prolonged treatment with NAC or p38 MAPK inhibitor extended the lifespan of the wild-type HSCs in serial transplantation. Collectively, the inactivation of p38 MAPK played a role as a protector for HSC against the loss of self-renewal capacity through ROS elevation.

Tothova et al. demonstrated that the Forkhead O family members (FoxOs) are also mediators of HSCs resistance against oxidative stress by generating *FoxO1*, *FoxO3*, and *FoxO4* knockout mice.(Tothova et al. 2007) The FoxOs play an important role in various physiological responses, such as induction of cell cycle arrest, stress resistance, or apoptosis. The *FoxO*-deficient mice exhibited a significant decrease in LSK size, myeloid lineage expansion, and lymphoid developmental abnormalities. The HSCs from *FoxO*-deficient bone marrow had defective self-renewal capacity accompanied by increased cell cycling and apoptotic HSCs as well as increased ROS level specifically observed in HSCs. On the other hand, NAC treatment rescued the LSK size, HSC cell cycling profile, and apoptosis, leading to the rescue of long-term HSC function in the cobblestone forming cell assay. Thus, FoxOs mediate quiescence and survival of HSCs by detoxification of ROS.

Seed or Soil? Niche Disruption and Disease

If a stem cell comes out of the niche, its differentiation program resumes and the mature cells are supplied to the organ. If the stem cell niche become corrupted for some reason, more stem cells would be forced to leave. This may lead to the ideal situation for the organ, at least temporarily, given plenty of healthy mature cells to sustain the function. However, what if the majority of the stem cell niche is permanently destroyed? If a disease has developed, is it because the niche is bad or because the stem cell itself is no longer healthy due to the niche malfunction?

Walkley et al. answered this question by introducing the retinoic acid receptor γ (*RARγ*) deficient mice (Walkley et al. 2007a). The mice exhibited myeloproliferative syndrome (MPS) with significantly increased granulocyte/macrophage progenitors and granulocytes in bone marrow, peripheral blood, and spleen. The phenotypes became more severe with age. A significant reduction of trabecular bone was observed, resulting in the osteoblasts depletion, and thus the niche was destroyed.

In order to determine whether this disease is caused by the bad seed (HSCs) or soil (niche), the transplantation of the wild-type bone marrow cells into lethally irradiated *RARγ* overexpressed and deficient mice was performed, revealing the result that the onset of MPS occurred exclusively in *RARγ* deficient mice. These data clearly proved that the MPS was not due to the altered HSC but the malfunctioning niche.

At the same time, they also investigated the effect of the *Rb* gene, a major cell cycle regulator, using the interferon-inducible *Mx-Cre* transgene and $pRb^{fl/fl}$ mice (Walkley et al. 2007b). When *Rb* was widely inactivated in the transgenic mice, including the bone marrow and HSCs, the mice exhibited profound myeloproliferation, loss of HSC from bone marrow, differentiation, and extramedullary hematopoiesis. The substantial loss of trabecular bone was also observed after the *Rb* inactivation. However, when either the transplanted HSC or recipient is the wild type, no such phenotype was observed.

The effect of *Rb* deletion was also examined using lysozyme-M-*Cre* mice, by which the gene deletion specific to the myeloid lineage (i.e., granulocytes, macrophages, and osteoclasts) is made possible. The myeloproliferation occurred when the *Rb*-inactivated hematopoietic cells, including the unaffected HSCs harvested from the lysozyme-M-*Cre* mice, were transplanted into the widely *Rb*-inactivated mice, but not when transplanted into the wild-type mice. Collectively, this *Rb* deletion model concluded that myeloproliferative disease is observed only when *Rb*-inactivated myeloid-derived cells interacted with *Rb*-inactivated microenvironment, regardless of the condition of the HSCs.

Both of these studies exhibited the loss of the osteoblastic niche and myeloproliferation, however, the causes were not identical. The status of the residual microenvironment, recovery system for the lost niche, or the relevance of the gene to hematopoiesis and microenvironment may influence the differences of the observation between these two cases. The bottom line is that the disruption of the niche or microenvironment can trigger the system disorder, albeit even with healthy stem cells. Knowing the pathological involvement of the niche can bring the novel therapeutic strategy with enhanced effectiveness, and further studies in this area are awaited.

Is There a Niche for Cancer Stem Cell?

Recently it has been reported by various studies that there is a functional microenvironment that supports cancer stem cells. This should also be considered as a niche, thus called cancer stem cell niche. A representative example is acute myeloid leukemia (AML) and its niche in bone marrow. Jin et al. showed that the anti-CD44 antibody-treated NOD/SCID mice transplanted with AML cells exhibited significantly lower rate of the disease onset (Jin et al. 2006). Also, Krause et al. showed impaired induction of chronic myeloid leukemia (CML)-like myeloproliferative disease among the recipient mice when transplanted with *BCR-ABL1*–transduced CML progenitors from CD44-mutant donors (Krause et al. 2006). These results indicate that, in both AML and CML cases, CD44 is essential for the homing and engraftment of the cancer stem cells to the niche. In another words, the

CD44-expressing leukemic stem cell adhere to the niche, binding to its ligand hyaluronic acid expressed by the cells on the surface of sinusoidal endothelium or endosteum in bone marrow, which is crucial for the niche maintenance of the stem cells. Interestingly, this molecular mechanism resembles that of healthy HSC and the vascular niche described earlier. Cancer and normal stem cells have much in common in the maintenance system in the niche.

Interestingly, cancer stem cell can evolve according to the microenvironment and initiate different type of cancer. Barabe et al. demonstrated that human hematopoietic cells infected with a retrovirus encoding mixed-lineage leukemia *(MLL)*-eleven-nineteen leukemia *(ENL)* fusion gene caused acute lymphoid leukemia (ALL) when transplanted into sublethally irradiated immunodeficient mice, but acute myeloid leukemia (AML) when cultured in a myeloid-promoting culture condition prior to transplantation for a prolonged period (Barabe et al. 2007). This indicates that the cancer stem cell with *MLL* fusion gene can initiate either ALL, AML, or both, depending on the microenvironment.

In contrast to the hematopoietic system, the stem cell research in the solid cancer field is relatively new. Among them, the stem cells study for brain tumors has made significant progress lately. Calabrese et al. showed brain tumor cells coexpressing Nestin and CD133, the fraction believed to contain the cancer stem cell, located closely to the capillaries in the brain tumor (Calabrese et al. 2007). When these cells were cocultured, the cancer stem cells selectively adhered to the endothelial cells. These suggested that the endothelial cells secreted factors necessary to maintain the cancer stem cells. Furthermore, the CD133-positive cells derived from human medulloblastoma developed brain tumor only when xenografted to the brain of the recipient nude mouse with endothelial cells. These data collectively suggest that the cancer stem cells of the brain tumor rely on the endothelial cells, which forms the vascular niche, for the maintenance of their self-renewal, differentiation, and proliferative capacity.

Roles of Niche Against Development, Maintenance, and Proliferation of Cancer

As described above, niche has the ability to control the pool, function, and even the fate of the stem cells. It is indeed an important aspect of niche to maintain the stem cells in a quiescent state; however, it also has to drive the adequate number of stem cells into the proliferation and differentiation path at the same time as far as the maintenance of the organ is concerned. This two-way signal of the niche against stem cell must be carefully regulated (Fig. 3). For example, in osteoblastic niche for hematopoietic stem cell in bone marrow, Tie2/Ang1 adhesion or BMP (bone morphogenetic protein), a member of TGF-β super family, are involved in maintaining the quiescence and proliferation suppression, while Wnt or Notch signaling promote the self-renewal, proliferation, and differentiation (Fig. 4) (Wilson and Trumpp 2006).

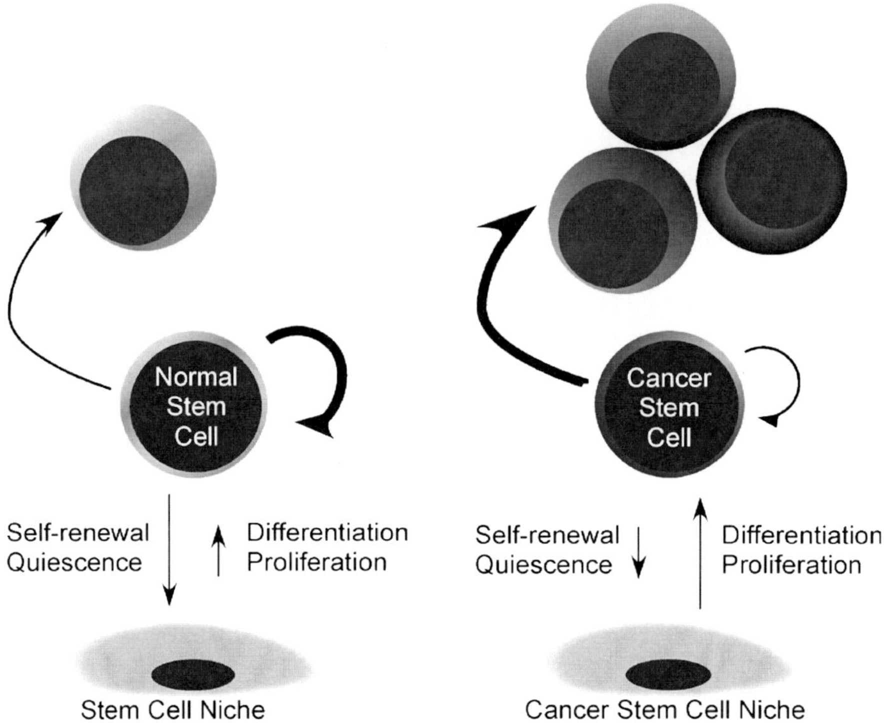

Fig. 3 Dynamics of two-way interaction between stem cell and niche: a comparison between a normal and a cancer stem cell. In normal stem cell and the niche, the signaling toward stem cell quiescence and self-renewal is dominant, while proliferative signaling is more dominant between a cancer stem cell and its niche

It is obvious that cancer tissue is highly proliferative. Does the stem cell regulatory system in the niche for cancer lean more toward the proliferative side than those for normal stem cells? Li and Neaves approached this problem via the dependence of stem cell on niche. They hypothesized that the behavior of cancer and normal stem cell are regulated by niche at different degrees by niche (Li & Neaves 2006). Cancer stem cells are derived through intrinsic mutation that leads to its highly proliferative activity. This highly proliferative state itself alters the signaling balance between niche and the stem cell. Namely, the niche function of quiescence maintenance becomes relatively ineffective, thus the function to support proliferation and differentiation become more dominant (Fig. 3). This model is supported by some clinical symptoms, one of which is the blast crisis of the CML. The Wnt/β-catenin signal, a causative pathway for self-renew, differentiation, and proliferation, is enhanced in the mutated nucleus of granulocyte-macrophage progenitors in the CML patient, due to the higher concentration of the β-catenin and irregularly activated TCF/LEF (T-cell factor/lymphocyte-enhancement factor) transcription activity (Jamieson et al. 2004). The blast crisis is thought to be triggered by this irregular TCF/LEF activity,

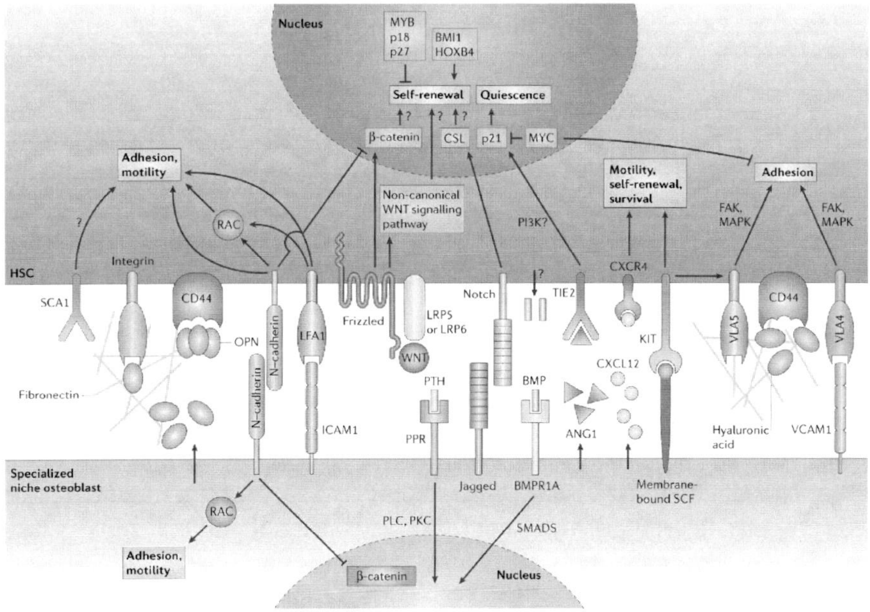

Fig. 4 A number of signaling pathways between HSC and niche are involved in stem cell regulation. Components in extracellular matrix are also important factors in niche formation. Normal and cancer stem cells share a majority of these regulatory signals but have different phenotypes

leading to the excess transcription of Wnt target gene products. Similarly, relevance of irregular Wnt/β-catenin activation has been reported in the cases of colon cancer or melanoma, in which the β-catenin itself has mutation. It is important to note that many signaling pathways involved in the normal stem cell and niche interaction are also found between the cancer stem cell and its niche, playing a role as the promoter of tumorigenesis and cancer proliferation. The identical set of proteins in a slightly different condition can deliver totally different results.

Mechanism of Cancer Metastasis Regulated by Niche

Niche not only cradles existing cancer stem cells, but also the future incoming cancer stem cell. Rather, it passively sends the inviting signal to the remote cancer stem cells.

MMPs (matrix metalloproteinases) are factors known not only for their contribution to the repair of inflammation or wounds but also for their involvement in cancer invasion and metastasis. A model of the molecular mechanism for remote metastasis and invasion in association with MMPs has been proposed for the lung (Hiratsuka et al. 2002). In this model, vascular endothelial growth factor (VGEF) secreted by primary cancer cells induces the MMP9 expression specifically in lung endothelial cells and macrophages via vascular endothelial growth factor receptor (VEGFR) tyrosine kinase, resulting in the formation of the cancer stem cell niche. This means that the cancer cells can produce their own favorable microenvironment, the future cancer stem cell niche, from a distance by secreting factors to influence the protein composition at that site.

This niche creation mechanism by cancer cells themselves has been verified for various other kinds of metastasis as well: the bone metastasis of prostate cancer has been shown to be supported by urokinase-type plasminogen activator (uPA) or prostate-specific antigen (PSA) secreted by the prostate cancer cells through altering the growth factors at the bone microenvironment, thus enhancing the proliferation of the osteoblasts that serve as the cancer stem cell niche (Logothetis and Lin 2005). Lung metastasis of breast cancer via secreted protein acidic and rich in cysteine (SPARC), osteonectin, or MMP2 is also found based on this mechanism (Minn et al. 2005).

Cells are not the only component of niche. The factors existing in the ECM can also be a part of the cancer mediator. In cancer stem cell in hypoxia, the oncogene *MET* is upregulated. MET binds to its ligand hepatocyte growth factor (HGF), which is found in the ECM, and enhances the transcription of plasminogen activator inhibitor type 1 (PAI-1) and cyclooxygenase 2 (COX2). This leads to enhanced blood coagulation and fibrin deposition. The fibrin deposition (fibrin nest) induces the vasculogenesis in the surrounding area, supporting the homing of cancer stem cell and its proliferation, thus serving as the cancer stem cell niche. The migration of the vascular vessel triggered by the fibrin nest also facilitates the metastasis to other sites (Boccaccio et al. 2005).

Kaplan et al. recently reported on a new type of cancer stem cell niche created by a third party. They demonstrated this mechanism using the lethally irradiated mice, transplanted with bone marrow-derived cell (BMDCs), followed by the lung cancer or melanoma cells. The transplanted BMDCs expressing VEGFR1 gathered at a site, such as lung, creating a group of cells. This microenvironment is highly receptive for cancer metastasis, and they call this microenvironment the "premetastatic niche" (Kaplan et al. 2005). The formation of premetastatic niche has been found for breast cancer, lung cancer, and gastrointestinal cancer in humans. The BMDCs creating the premetastatic niche are a type of hematopoietic progenitor expressing VLA-4 (integrin $\alpha_4\beta_1$) and/or Id3 (inhibitor of DNA binding 3). The primary cancer cells secrete factors (such as cytokines) that positively regulate the expression of fibronectin (a ligand of VLA-4) of fibroblasts located at the site of future premetastatic niche, inducing the homing of VEGFR1-expressing BMDCs. The mice transplanted with VEGFR1-depleted BMDCs or treated with anti-VEGFR1 antibody did not show the formation of premetastatic niche or metastasis, proving the theory of this new metastatic mechanism induced by cancer cells involving the third party.

Novel Cancer Therapy Targeting Cancer Stem Cell and Its Niche

Most of the current anticancer drugs act as inhibitors of DNA synthesis or cell division of the cancer cell, consequently suppressing the growth of tumor. Therefore, the cancer cells must be in the cell cycle, not in quiescence, in order to see the positive effect of chemotherapy. The cancer stem cells as well as normal stem cells are supposed to be in the quiescent state. Even if the tumor size is decreased after chemotherapy, it is simply because the regular cell-cycling cancer cells have been killed. One must expect the quiescent, tumorigenetic cancer stem cells to survive the chemotherapy.

A report by Graham et al. proved the cancer stem cells' insensitivity through the demonstration that most of the *BCR-ABL* expressing blood cells of patients with chronic myeloid leukemia arrested after the cell culture with addictive of STI571 (Gleevec®), except those in the quiescent state, namely the leukemic stem cells, survived and remained in quiescence (Graham et al. 2002). This suggests that degenerating the quiescent cancer stem cell is one of the most promising therapeutic methods in the next generation cancer treatments aimed at complete cure. Various attempts based on this concept have been investigated, including the alteration of cancer stem cell characteristics, manipulation of the niche environment to induce a resumption of the cell cycle, and differentiation of cancer stem cells. From the viewpoint of cancer stem cell control, complete inhibition of cell division of cancer stem cells, even if only minimally, may also have possibility.

HSC transplantation is routinely performed for patients with blood diseases, bone marrow diseases, or some types of cancer, such as leukemia. In transplantation, purified healthy HSCs derived from bone marrow, peripheral blood, or cord blood are injected intravenously into patient's bloodstream. These HSCs are then homed to niche in bone marrow and resume the healthy hematopoiesis.

Prior to transplantation, patients undergo conditioning with aggressive anticancer drugs and/or irradiation, which destroy their own hematopoietic system, including HSCs, and suppress the immune reactions. This way, the newly introduced healthy HSC can home to the empty niche in bone marrow and successfully reconstitute the recipient patient's hematopoietic system. This conditioning regimen, however, is highly toxic and life-threatening, and patients who do not respond to anticancer drug and/or irradiation or with low tolerance due to poor general status should be excluded from this treatment.

Recently, Czechowicz et al. reported that treatment with an antibody that blocks the HSC function led to the clearance of niche, suggesting the possibility of mild but effective conditioning regimen (Czechowicz et al. 2007). They demonstrated that administration of ACK2, an anti-c-Kit antibody, led to rapid and transient removal of almost all endogenous HSCs in mice, and subsequent transplantation of purified HSCs led to as high as 90% of donor chimerism. Without conditioning, typical donor chimerism in recipient mice was 3%, perhaps due to the limited number of empty niche available for engrafting donor HSCs. Moreover, the HSCs derived from ACK2-treated mice showed significantly (>90%) reduced engraftment capability, compared to control. The another anti-c-Kit antibody, 2B8, did not show the effect observed with ACK2. Thus, the mechanism of HSC clearance was considered as the complete inhibition of c-Kit signaling in HSC by ACK2.

Unlike the case of human anti-VLA-4 antibody natalizumab, as described earlier, ACK2 does not have the mobilization effect, which was confirmed by the lack of HSC detected in spleen, liver, and peripheral blood after infusion. Instead, ACK2 provides a short time window to allow donor HSC to graft by depleting the host HSC. This regimen may be an attractive alternative to the conventional conditioning regimen not only for those who should otherwise have been excluded in this medical procedure but also for tolerant patients.

Parthenolide (PTL) is a sesquiterpene lactone and major component in the herbal remedy feverfew. Recently, PTL was found to have a unique characteristic against tumors, including inhibition of DNA synthesis, cancer cell proliferation, and nuclear factor kappa B (NF-κB) activation, and increase intracellular reactive oxygen species (ROS). Guzman et al. successfully demonstrated that PTL induces stem cell–specific apoptosis among leukemic cells in primary acute myeloid leukemia (AML) and blast crisis chronic myeloid leukemia (bcCML) without affecting the normal hematopoietic cells (Guzman et al. 2005). This action was identified as a result of multiple effects of PTL, such as inhibition of upregulated NF-κB expression in tumor cells, proapoptotic activation of p53, and increased intracellular ROS.

Glioblastoma is considered one of the most malignant cancers, extremely proliferative, and resistant not only to chemotherapy but also to radiotherapy. The prognosis is very poor with the average survival duration less than a year. Piccirillo et al. showed a successful result for glioblastoma in attempting the novel therapy based on the above-mentioned concept. They discovered that the BMPs, especially BMP4, can induce the differentiation of the $CD133^+$ glioblastoma cells, which are considered to be the cancer stem cells, and ultimately decrease of cancer stem cell pool (Piccirillo et al. 2006). Moreover, xenograft of human glioblastoma cells to mice showed significant suppression of the tumor growth and improvement of the survival rate when treated along with BMP4, which would have been 100% lethal otherwise. In this case, BMP4 or the other BMPs did not kill the glioblastoma cells but let them extinct through the induced temporary differentiation and proliferation by altering the stem cell characteristics. This positive experimental result should stimulate the exploration of the novel therapeutic method targeting a cancer stem cell or its niche, showing an approach for a breakthrough in treatment of otherwise abandoned cancer patients.

For the niche-targeting therapy, Calabrese et al. suggested that alternation of vascular niche is effective to brain tumor treatment. Medulloblastoma cells express high levels of erythroblastic leukemia viral oncogenic homolog 2 (ERBB2), and also VEGF, consequently. ERBB2-overexpressing mice bearing brain tumor were treated with either ERBB2 inhibitor erlotinib (Tarceva®) or anti-VEGF antibody bevacizumab (Avastin®). The treatment significantly inhibited the tumor vasculature and suppressed tumor growth. The cancer stem cell expressing Nestin and CD133, the self-renewing cancer cells, were not found in the mice after either treatment. Similar treatment was applied to mice with glioma, and the results showed a significant suppression of vasculogenesis and tumor proliferation, as well as the decrease of the $Nestin^+CD133^+$ cancer stem cell pool. The small number of surviving $Nestin^+CD133^+$ cancer stem cells were all found in association with the tumor vessels. These data collectively suggest that the suppression of the vascular niche by inhibiting the vasculogenesis has a potential for novel medulloblastoma or glioma treatment.

Conclusion

Stem cell and its niche influence each other to maintain their capabilities, control the fate, and ultimately sustain the healthy organ or tissue. Besides the interactions between stem cell and niche, some intrinsic factors of stem cell have been also revealed as a regulator of stem cell capacity. A good example is *Bmi-1*, a member of polycomb group of genes involved in maintenance of transcriptional repression of target genes. HSCs derived from *Bmi-1*-deficient mice exhibited defective self-renewal phenotype, while maintaining the multilineage differentiation potential. The microenvironment in *Bmi-1*-deficient mice was found intact. The forced expression of *Bmi-1* led to the enhanced repopulation capacity *in vivo* and augmented expansion of the progenitors *ex vivo*. These data collectively suggest that the self-renewal capacity of HSCs is controlled by the intrinsic expression level of *Bmi-1* (Iwama et al. 2005).

Stem cell regulation and maintenance, whether intrinsic or extrinsic, are not simple single-factor phenomena, and the large part of detailed molecular mechanism is yet to be elucidated. Understanding the regulatory system could lead to a novel approach for the effective therapeutic methods that directly target stem cells (normal or cancer) or niche. As described in this chapter, the latest drugs targeting the cancer stem cell or its microenvironment employ this state-of-the-art strategy and are expected to minimize the complications and improve the patients' quality of life at the same time. The biology of stem cell and its niche is one of the most promising research fields of the next few decades, from both scientific and clinical viewpoints.

References

Adams GB, Chabner KT, Alley IR, Olson DP, Szczepiorkowski ZM, Poznansky MC, Kos CH, Pollak MR, Brown EM, Scadden DT (2006) Stem cell engraftment at the endosteal niche is specified by the calcium-sensing receptor. Nature 439:599–603

Arai F, Hirao A, Ohmura M, Sato H, Matsuoka S, Takubo K, Ito K, Koh GY, Suda T (2004) Tie2/angiopoietin-1 signaling regulates hematopoietic stem cell quiescence in the bone marrow niche. Cell 118:149–161

Avigdor A, Goichberg P, Shivtiel S, Dar A, Peled A, Samira S, Kollet O, Hershkoviz R, Alon R, Hardan I, Ben-Hur H, Naor D, Nagler A, Lapidot T (2004) CD44 and hyaluronic acid cooperate with SDF-1 in the trafficking of human $CD34^+$ stem/progenitor cells to bone marrow. Blood 15:2981–2989

Barabé F, Kennedy JA, Hope KJ, Dick JE (2007) Modeling the initiation and progression of human acute leukemia in mice. Science 316:600–604

Boccaccio C, Sabatino G, Medico E, Girolami F, Follenzi A, Reato G, Sottile A, Naldini L, Comoglio PM (2005) The *MET* oncogene drives a genetic programme linking cancer to haemostasis. Nature 434:396–400

Calabrese C, Poppleton H, Kocak M, Hogg TL, Fuller C, Hamner B, Oh EY, Gaber MW, Finklestein D, Allen M, Frank A, Bayazitov IT, Zakharenko SS, Gajjar A, Davidoff A, Gilbertson RJ (2007) A perivascular niche for brain tumor stem cells. Cancer Cell 11:69–82

Calvi LM, Adams GB, Weibrecht KW, Weber JM, Olson DP, Knight MC, Martin RP, Schipani E, Divieti P, Bringhurst FR, Milner LA, Kronenberg HM, Scadden DT (2003) Osteoblastic cells regulate the haematopoietic stem cell niche. Nature 425:841–846

Czechowicz A, Kraft D, Weissman IL, Bhattacharya D (2007) Efficient transplantation via antibody-based clearance of hematopoietic stem cell niches. Science 318:1296–1299

Deng W, Lin H (1997) Spectrosomes and fusomes anchor mitotic spindles during asymmetric germ cell divisions and facilitate the formation of a polarized microtubule array for oocyte specification in *Drosophila*. Dev Biol 189:79–94

Graham SM, Jørgensen HG, Allan E, Pearson C, Alcorn MJ, Richmond L, Holyoake TL (2002) Primitive, quiescent, Philadelphia-positive stem cells from patients with chronic myeloid leukemia are insensitive to STI571 in vitro. Blood 99:319–325

Guzman ML, Rossi RM, Karnischky L, Li X, Peterson DR, Howard DS, Jordan CT (2005) The sesquiterpene lactone parthenolide induces apoptosis of human acute myelogenous leukemia stem and progenitor cells. Blood 105:4163–4169

Heissig B, Hattori K, Dias S, Friedrich M, Ferris B, Hackett NR, Crystal RG, Besmer P, Lyden D, Moore MA, Werb Z, Rafii S (2002) Recruitment of stem and progenitor cells from the bone marrow niche requires MMP-9 mediated release of kit-ligand. Cell 109:625–637

Hiratsuka S, Nakamura K, Iwai S, Murakami M, Itoh T, Kijima H, Shipley JM, Senior RM, Shibuya M (2002) MMP9 induction by vascular endothelial growth factor receptor-1 is involved in lung-specific metastasis. Cancer Cell 2:289–300

Ito K, Hirao A, Arai F, Matsuoka S, Takubo K, Hamaguchi I, Nomiyama K, Hosokawa K, Sakurada K, Nakagata N, Ikeda Y, Mak TW, Suda T (2004) Regulation of oxidative stress by ATM is required for self-renewal of haematopoietic stem cells. Nature 431:997–1002

Ito K, Hirao A, Arai F, Takubo K, Matsuoka S, Miyamoto K, Ohmura M, Naka K, Hosokawa K, Ikeda Y, Suda T (2006) Reactive oxygen species act through p38 MAPK to limit the lifespan of hematopoietic stem cells. Nat Med 12:446–451

Iwama A, Oguro H, Negishi M, Kato Y, Nakauchia H (2005) Epigenetic regulation of hematopoietic stem cell self-renewal by polycomb group genes. Int J Hematol 81:294–300

Jamieson CH, Ailles LE, Dylla SJ, Muijtjens M, Jones C, Zehnder JL, Gotlib J, Li K, Manz MG, Keating A, Sawyers CL, Weissman IL (2004) Granulocyte-macrophage progenitors as candidate leukemic stem cells in blast-crisis CML. N Engl J Med 351:657–667

Jin L, Hope KJ, Zhai Q, Smadja-Joffe F, Dick JE (2006) Targeting of CD44 eradicates human acute myeloid leukemic stem cells. Nature Med 12:1167–1174

Kaplan RN, Riba RD, Zacharoulis S, Bramley AH, Vincent L, Costa C, MacDonald DD, Jin DK, Shido K, Kerns SA, Zhu Z, Hicklin D, Wu Y, Port JL, Altorki N, Port ER, Ruggero D, Shmelkov SV, Jensen KK, Rafii S, Lyden D (2005) VEGFR1-positive haematopoietic bone marrow progenitors initiate the pre-metastatic niche. Nature 438:820–827

Kiel MJ, Yilmaz OH, Iwashita T, Yilmaz OH, Terhorst C, Morrison SJ (2005) SLAM family receptors distinguish hematopoietic stem and progenitor cells and reveal endothelial niches for stem cells. Cell 121:1109–1121

Krause DS, Lazarides K, von Andrian UH, Van Etten RA (2006) Requirement for CD44 in homing and engraftment of BCR-ABL-expressing leukemic stem cells. Nat Med 12:1175–1180

Li L, Neaves WB (2006) Normal stem cells and cancer stem cells: the niche matters. Cancer Res 66:4553–4557

Logothetis CJ, Lin SH (2005) Osteoblasts in prostate cancer metastasis to bone. Nat Rev Cancer 5:21–28

Matrosova VY, Orlovskaya IA, Serobyan N, Khaldoyanidi SK (2004) Hyaluronic acid facilitates the recovery of hematopoiesis following 5-fluorouracil administration. Stem Cells 22:544–555

Minn AJ, Gupta GP, Siegel PM, Bos PD, Shu W, Giri DD, Viale A, Olshen AB, Gerald WL, Massagué J (2005) Genes that mediate breast cancer metastasis to lung. Nature 436:518–524

Nilsson SK, Johnston HM, Whitty GA, Williams B, Webb RJ, Denhardt DT, Bertoncello I, Bendall LJ, Simmons PJ, Haylock DN (2005) Osteopontin, a key component of the hematopoietic stem cell niche and regulator of primitive hematopoietic progenitor cells. Blood 15:1232–1239

Papayannopoulou T, Craddock C, Nakamoto B, Priestley GV, Wolf NS (1995) The VLA4/VCAM-1 adhesion pathway defines contrasting mechanism of lodgement of transplanted murine hemopoietic progenitors between bone marrow and spleen. Proc Natl Acad Sci USA 92:9647–9651

Parmar K, Mauch P, Vergilio JA, Sackstein R, Down JD (2007) Distribution of hematopoietic stem cells in the bone marrow according to regional hypoxia. Proc Natl Acad Sci USA 104:5431–5436

Piccirillo SG, Reynolds BA, Zanetti N, Lamorte G, Binda E, Broggi G, Brem H, Olivi A, Dimeco F, Vescovi AL (2006) Bone morphogenetic proteins inhibit the tumorigenic potential of human brain tumour-initiating cells. Nature 444:761–765

Piccoli C, Ria R, Scrima R, Cela O, D'Aprile A, Boffoli D, Falzetti F, Tabilio A, Capitanio N (2005) Characterization of mitochondrial and extra-mitochondrial oxyten consumption in human hematopoietic stem cells. J Biol Chem 280:26467–26476

Puri CM, Bernstein A (2003) Requirement for the TIE family of receptor tyrosine kinases in adult but not fetal hematopoiesis. Proc Natl Acad Sci USA 100:12753–12758

Qian H, Buza-Vidas N, Hyland CD, Jensen CT, Antonchuk J, Månsson R, Thoren LA, Ekblom M, Alexander WS, Jacobsen SE (2007) Critical role of thrombopoietin in maintaining adult quiescent hematopoietic stem cells. Cell Stem Cell 1:671–684

Schofield R (1978) The relationship between the spleen colony-forming cell and the haematopoietic stem cell. Blood Cells 4:7–25

Sugiyama T, Kohara H, Noda M, Nagasawa T (2006) Maintenance of the hematopoietic stem cell pool by CXCL12-CXCR4 chemokine signaling in bone marrow stromal cell niches. Immunity 25:977–988

Tothova Z, Kollipara R, Huntly BJ, Lee BH, Castrillon DH, Cullen DE, McDowell EP, Lazo-Kallanian S, Williams IR, Sears C, Armstrong SA, Passegué E, DePinho RA, Gilliland DG (2007) FoxOs are critical mediators of hematopoietic stem cell resistance to physiologic oxidative stress. Cell 128:325–339

Walkley CR, Olsen GH, Dworkin S, Fabb SA, Swann J, McArthur GA, Westmoreland SV, Chambon P, Scadden DT, Purton LE (2007a) A microenvironment-induced myeloproliferative syndrome caused by retinoic acid receptor γ deficiency. Cell 129:1097–1110

Walkley CR, Shea JM, Sims NA, Purton LE, Orkin SH (2007b) Rb regulates interactions between hematopoietic stem cells and their bone marrow microenvironment. Cell 129:1081–1095

Wilson A, Trumpp A (2006) Bone-marrow haematopoietic-stem-cell niches. Nat Rev Immunol 6:93–106

Wilson A, Murphy MJ, Oskarsson T, Kaloulis K, Bettess MD, Oser GM, Pasche AC, Knabenhans C, Macdonald HR, Trumpp A (2004) c-Myc controls the balance between hematopoietic stem cell self-renewal and differentiation. Genes Dev 18:2747–2763

Xie T, Spradling AC (2000) A niche maintaining germ line stem cells in the *Drosophila* ovary. Science 290:328–330

Yoshihara H, Arai F, Hosokawa K, Hagiwara T, Takubo K, Nakamura Y, Gomei Y, Iwasaki H, Matsuoka S, Miyamoto K, Miyazaki H, Takahashi T, Suda T (2007) Thrombopoietin/MPL signaling regulates hematopoietic stem cell quiescence and interaction with the osteoblastic niche. Cell Stem Cell 1:685–697

Zhang J, Niu C, Ye L, Huang H, He X, Tong WG, Ross J, Haug J, Johnson T, Feng JQ, Harris S, Wiedemann LM, Mishina Y, Li L (2003) Identification of the haematopoietic stem cell niche and control of the niche size. Nature 425:836–841

Zohren F, Toutzaris D, Klärner V, Hartung HP, Kieseier B, Haas R (2008) The monoclonal anti-VLA-4-antibody natalzumab mobilizes $CD34^+$ hematopoietic progenitor cells in humans. Blood 111:3893–3895

Hematopoietic Stem Cells and Somatic Stem Cells

Kah Yong Tan, Francis S. Kim, Amy J. Wagers, and Shane R. Mayack

Abstract Stem cells are unspecialized cells that can differentiate to generate more specialized cell types responsible for tissue-specific function. During development, the differentiation of pluripotent embryonic stem cells leads to the production of specialized somatic cells that are ultimately responsible for the structure and function of all adult tissues and organs. "Naturally" pluripotent cells exist only at the earliest stages of embryonic development. However, less differentiated stem and progenitor cells are also present in adult tissues. In contrast to pluripotent embryonic stem cells, adult stem cells generally only form a limited number of cell types corresponding to their tissues of origin. These multipotent or unipotent adult stem cells exist in many, though probably not all, adult tissues including bone marrow (BM), brain, spinal cord, dental pulp, blood vessels, skeletal muscle, epithelia of the skin and digestive system, cornea, and retina. "Multipotent" stem cells can generate several types of cells within a given tissue (e.g., hematopoietic/blood forming stem cells), whereas "unipotent" stem cells give rise to only one differentiated cell type (e.g., skeletal muscle stem cells). Though adult stem cells exist at a very low frequency in adult tissues, they serve a critical function in maintaining tissue homeostasis and generating replacement cells to repair tissues after injury. Here, we review the current literature on adult stem cells and specifically highlight the regulation of multipotent

K.Y. Tan
Section on Developmental and Stem Cell Biology, Joslin Diabetes Center, One Joslin Place, Boston, MA 02215

Department of Stem Cell and Regenerative Biology, Harvard University and Harvard Stem Cell Institute, 42, Church Street, Cambridge, MA 02138

Department of Molecular and Cellular Biology, Harvard University, 7 Divinity Avenue, Cambridge, MA 02138

F.S. Kim, A.J. Wagers and S.R. Mayack
Section on Developmental and Stem Cell Biology, Joslin Diabetes Center, One Joslin Place, Boston, MA 02215

Department of Stem Cell and Regenerative Biology, Harvard University and Harvard Stem Cell Institute, 42, Church Street, Cambridge, MA 02138

M. Kondo (ed.), *Hematopoietic Stem Cell Biology,* Stem Cell Biology and Regenerative Medicine, DOI 10.1007/978-1-60327-347-3_3,

hematopoietic stems cells in contrast with that of unipotent skeletal muscle stem cells, as well as discuss their respective roles in tissue homeostasis and repair.

Introduction

The word somatic derives from the Greek *sōma*, meaning "body." Hence, somatic cells are any cells forming the body of an organism and are distinct from germ line cells – the spermatozoa and ova – which fuse during fertilization to produce the zygote (Eilertsen et al. 2008; National Institutes of Health 2008).

Stem cells are unspecialized cells that can initiate a differentiation program to generate more specialized cell types that perform particular tasks within tissues (Garry et al. 2003; Giebel and Bruns 2008; Weissman 2000). The culmination of these events is the production of somatic cells that ultimately are responsible for the structure and function of adult tissues and organs. Stem cell differentiation is carefully regulated and only under proper conditions, which are often temporally and anatomically regulated, do stem cells begin to develop into specialized tissues and organs, ultimately forming and maintaining the soma (Sylvester and Longaker 2004; Naveiras and Daley 2006). Importantly, in addition to generating more specialized cells, stem cells also self-renew, dividing to give rise to more stem cells, and thereby maintaining an undifferentiated stem cell pool (National Institutes of Health 2008).

In general, stem cells are categorized based on their "potency" (i.e., the number of different types of cells to which they give rise) (Donovan and de Miguel 2003). Pluripotent stem cells, which include embryo-derived embryonic stem (ES) cells and induced pluripotent stem cells (iPS) generated by viral transduction of somatic cells, can form all cell types of the body (Byrne 2008; Kastenberg and Odorico 2008; Liu 2008; Vats et al. 2005; Ralston and Rossant 2005). It appears that "naturally" pluripotent cells exist only at the earliest stages of embryonic development, while stem cells in adult tissues generally form a more limited number of cell types, which typically correspond to their tissues of origin (Byrne 2008; Kastenberg and Odorico 2008; Liu 2008; Vats et al. 2005; Ralston and Rossant 2005). These multipotent or unipotent stem cells exist in many, though probably not all, tissues. Tissues in which adult stem cell populations have been identified include bone marrow (BM), brain, spinal cord, dental pulp, blood vessels, skeletal muscle, epithelia of the skin and digestive system, cornea, and retina. However in tissues such as liver and pancreas, the existence of a true stem cell population remains unclear (Vats et al. 2005; Tiedemann et al. 2001; Saini and Stewart 2006; Choumerianou et al. 2008). Multipotent stem cells can generate several types of cells within a given tissue (e.g., hematopoietic/blood forming stem cells), whereas unipotent stem cells give rise to only one differentiated cell type (e.g., skeletal muscle stem cells).

Somatic stem cells, like embryonic stem cells, share at least two characteristics (National Institutes of Health 2008). First, they can make identical copies of themselves for long periods of time. This ability is referred to as long-term self-renewal

(Weissman 2000; Weissman et al. 2001; Claudinot et al. 2005). Second, they can give rise to at least one mature cell type with characteristic morphology and specialized function. Typically, the initial differentiation of somatic stem cells generates an intermediate cell type or types prior to producing fully differentiated cells. This intermediate cell is called a progenitor cell. Progenitor cells in fetal or adult tissues are partly differentiated cells that divide and give rise to fully differentiated cells. These progenitor cells are usually regarded as strictly "committed" to differentiate along a particular cellular development pathway and importantly have lost the capacity to self-renew (Giebel and Bruns 2008; Weissman 2000; Vats et al. 2005; Weissman et al. 2001; Gros et al. 2005; Bonner-Weir et al. 2004; Masson et al. 2006) (Fig. 1).

Adult stem cells are typically rare (Joseph and Morrison 2005). For example, in the case of blood forming hematopoietic stem cells (HSCs), which reside predominantly in the BM in adult animals, only an estimated 1 in 100,000 cells is actually a blood-forming stem cell (Yilmaz et al. 2006; Kiel et al. 2005). But, despite their low frequency, stem cells serve a critical function in maintaining tissue homeostasis and generating replacement cells to repair tissues after injury (Broxmeyer et al. 2002; Bierkens et al. 1991; Fried et al. 1978; Cline and Golde 1978; Morrison et al. 1997).

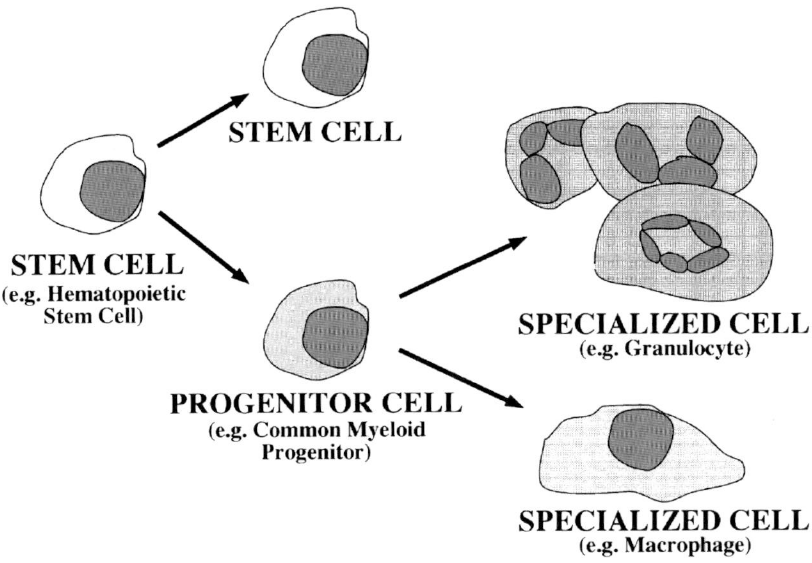

Fig. 1 Distinguishing features of progenitor cells and stem cells. A highly diagrammatic version of the hierarchy of stem cell self-renewal and differentiation. In a self-renewing division a stem cell makes identical copies of itself, maintaining the primitive stem cell pool and the capacity for long-term cell or tissue replenishment. Alternatively, stem cells divide and differentiate into the mature cells of a given lineage that are necessary for tissue function. Differentiation is generally accompanied by both changes in morphology and function that are distinguishable from stem cells as well as a loss of self-renewal potential

What Does It Means to Be an Adult Stem Cell? Criteria for Defining Adult Stem Cell Populations

Unlike ES cells, which are defined by their anatomical origin (the inner cell mass of the blastocyst) and potency (ability to generate all cell types in the body), somatic stem cells are found in many different adult tissues and it remains unclear if they share any defining anatomical constraints. The precise developmental origins of adult stem cells in mature tissues are also largely opaque. It has been proposed that stem cells are somehow set aside during fetal development and restrained from differentiating.

Adult stem cells then, instead of being defined by anatomical location, are defined by their concordant capacity for self-renewal and differentiation throughout the lifetime of an organism (Barker and Clevers 2007). This criterion, although fundamental to the nature of a stem cell, can sometimes be challenging to prove definitively in vivo. This is in part because it is difficult, particularly with respect to human stem cells, to design experiments that allow the fate of candidate adult stem cells to be defined and tracked in vivo over an individual's lifetime.

Clonogenicity, or the ability of a single cell to proliferate independently to form a colony, is one of the major properties commonly ascribed to stem cells. To be consistent with properties of a true stem cell, the progeny of these clonogenic progenitors should be able to completely reconstitute the mature cell types of the stem cell tissue of origin. Therefore, clonogenicity is typically demonstrated by showing that a single clone of a putative stem cell population can functionally repopulate the corresponding tissue by using highly purified or enriched populations of a putative tissue-specific stem cell or by colony-forming assays (Jackson et al. 2002).

Adult stem cells must also be able to give rise to fully differentiated cells that have mature phenotypes, are fully integrated into the tissue, and are capable of specialized functions that are appropriate for the tissue (Jackson et al. 2002). Stem cell biology often exploits two characteristics: appropriate cell morphology and demonstration that the resulting, differentiated cell types display surface markers that identify them as belonging to the tissue. More stringent studies further demonstrate that the differentiated cells that are derived from adult stem cells are truly functional and that stem cell daughters are fully integrated into the differentiated tissue in vivo and interact appropriately with neighboring cells. Strictly speaking, the gold standard often employed for demonstrating stem cell function is engraftment by a donor stem cell of an appropriate tissue to generate all of the daughter cell types of that tissue, including renewal of the tissue-specific stem cell pool.

In order to fully characterize the tissue regenerative and self-renewal capabilities of adult stem cells and, therefore, truly harness their potential, it is essential to demonstrate that candidate stem cell populations have bona fide stem cell characteristics and understand how these are regulated and maintained in various tissues.

In general, three experimental approaches are used to determine whether candidate adult stem cells give rise to specialized cells (Barker and Clevers 2007;

Barker et al. 2007). Adult stem cells can be labeled in vivo, and then they and their daughter cells can be tracked to determine cell fate potential and turnover time. Stem cell labeling in some tissues has been achieved using the label retaining cell (LRC) assay (Kiel et al. 2007). This technique exploits the fact that resident tissue stem cells are typically quiescent and will retain a DNA synthesis label (such as BrdU) for longer periods of time relative to their more differentiated daughters, which divide more often during a follow-up "chase" period and therefore more rapidly dilute the label. The LRC assay thus depends on the enrichment of stem cells within those cells, retaining the label over a long period of time. In some studies, long-term LRCs do indeed overlap with undifferentiated stem cells (Claudinot et al. 2005; Cotsarelis et al. 1990). However, this strategy generally does not label stem cells exclusively (Kiel et al. 2007) and may fail to label some stem cell populations (Barker et al. 2007).

Candidate adult stem cells can also be isolated, labeled in vitro, and then transplanted into recipient hosts to determine whether they can functionally reconstitute the cell types of a given tissue. Finally, candidate adult stem cells can be isolated, grown in vitro, and manipulated by adding growth factors or introducing genes that help determine what differentiated cell types they have the potential to yield. Such in vitro assays are less compelling than in vivo strategies for defining the intrinsic fate potential of putative stem cell populations, as differentiation potential can be profoundly influenced by culture conditions, and physiological assays of cell function are generally less available for cultured cells.

Another challenge in demonstrating that a particular cell population contains only stem cells is that it can be difficult to distinguish adult, tissue-specific stem cells from progenitor cells, which are also found throughout fetal or adult tissues. This is partially because progenitor cells are often more prevalent than true stem cells, which typically exist in a given tissue at a very low frequency. However, unlike stem cells, progenitor cells lack long-term self-renewal capacity and, particularly in comparison to multipotent stem cells, exhibit a more limited repertoire of differentiation potentials.

Adult Stem Cell Plasticity

There has been much debate centered on the question of whether stem cells in one adult tissue can generate the specialized cell types of another type of tissue, either a tissue derived from the same embryonic germ layer or from a different germ layer. For example, some previous studies have suggested that blood stem cells (derived from mesoderm) may be able to generate skeletal muscle (also derived from mesoderm) or neurons (derived from ectoderm) in addition to their well-established primary role in blood cell generation (Saini and Stewart 2006; Jackson et al. 2002; Huttmann et al. 2003; Horwitz 2003). The hypothesis that a stem cell from one tissue may "transdifferentiate" to form the differentiated progeny of another tissue has been referred to as stem cell plasticity. The implication that stem cells may transdifferentiate triggered a flurry of research, which in the end has provided very limited evidence for such "plasticity" events. In fact, numerous studies (Durr and Muller 2003; Lemoli et al. 2005; Phinney and Prockop 2007; Rovo and Gratwohl 2008) have demonstrated that stem cell transdifferentiation does not occur in vivo at physiologically meaningful levels; observations of apparent stem cell plasticity appear instead to reflect rare cell-cell fusion events or contamination of putative stem cell populations by itinerant cells circulating through a given tissue (Lemoli et al. 2005).

To date, despite many challenges, true adult stem cells have been identified in many different animals and in a variety of human tissues. As discussed, in order to be classified as an adult stem cell, there are many criteria, most importantly, that the cell should be capable of self-renewal and differentiation for the lifetime of the organism. One of the most well-studied and well-defined adult stem cell populations is the blood forming (or hematopoietic) stem cell, which provides homeostatic and regenerative replacement of all blood cell lineages.

Hematopoietic Stem Cells: The "Model" of Somatic Stem Cells

Embryonic Origins

In adults, hematopoiesis, which leads to the generation of all blood cell lineages, is maintained homeostatically by a rare population of multipotent HSCs found in the BM. In vivo, blood cells turn over on a daily basis. Thus, in order to continuously maintain blood cell homeostasis, there is a constant need for generation of HSCs in the BM where they differentiate to generate progenitor cells that ultimately produce mature types of blood cells (Weissman 2000; Weissman et al. 2001). How do HSCs, or other adult somatic stem cells, maintain the critical balance between self-renewal and differentiation necessary to maintain tissue homeostasis and support regeneration and repair, without depletion of the stem cell pool? Moreover, what is the embryonic origin of these primitive cells and are the same embryonic origins shared with those primitive stem cells that remain in adult tissues?

During embryogenesis, HSCs migrate through different anatomical sites that likely impart unique cues to the cells as they transition through different developmental stages (Cantor and Orkin 2001; Mikkola et al. 2005). In vertebrate development, distinct waves of hematopoietic activity (Lux et al. 2008; Palis et al. 1999) are believed to contribute to the generation and maintenance of precursors that seed both embryonic and adult hematopoiesis. Experiments probing the development of HSCs strongly suggest that blood cells are sequentially generated in several distinct sites (Cumano et al. 1996; Godin et al. 1995; Dieterlen-Lievre et al. 2002; Medvinsky and Dzierzak 1996; Yoder et al. 1997a, 1997b): the yolk sac (Lux et al. 2008; Yoder et al. 1997a, 1997b), the aorta-gonad-mesonephros (AGM) region (Cumano et al. 1996; Godin et al. 1995; Dieterlen-Lievre et al. 2002; Medvinsky and Dzierzak 1996), and the placenta (Mikkola et al. 2005; Medvinsky and Dzierzak 1996; Kumaravelu et al. 2002).

In mice, morphologically distinct primitive blood cells are first identifiable in the blood islands of the yolk sac at E7.5 (Yoder et al. 1997a, 1997b). The fetal liver is colonized by hematopoietic cells on E10.5 and thereafter becomes the principal site of fetal hematopoiesis (Houssaint 1981). The fetal thymus and spleen are also seeded by HSCs around the same time, E10.5 and E12.5, respectively (Palis et al. 1999; Cumano and Godin 2001; Godin et al. 1999; Dieterlen-Lievre et al. 1997; Robertson et al. 1999). Ultimately, the BM, which shortly after birth becomes the major site for blood production and source of long-lived adult HSCs, is colonized by fetal liver derived HSCs around E15 (Cumano and Godin 2001).

The term primitive hematopoiesis is given to the first yolk sac derived erythroid lineage and definitive hematopoiesis is applied to all lineages other than primitive erythroid (Palis et al. 1999; Robertson et al. 1999). However, while the yolk sac is generally accepted to be the source of primitive hematopoietic cells, the identification of the initial site of definitive hematopoiesis remains controversial, as does the origin of the HSCs, which ultimately are retained in adult tissues and give rise to all the adult blood lineages (Orkin and Zon 2008).

Recently, the emergence of fetal HSCs in the placenta, coincident with those seen in the AGM and yolk sac and preceding the seeding of HSCs in the fetal liver or circulating blood, has been demonstrated. While the onset of HSC activity in the placenta parallels the emergence of HSCs in the AGM (E10.5) (Cumano et al. 1996; Godin et al. 1995; Dieterlen-Lievre et al. 2002; Medvinsky and Dzierzak 1996), placental HSCs continue to expand until E13.5 and contain approximately 15-fold greater numbers of HSCs than the AGM. In fact, in studies performed in mice, the placental HSC pool exceeds in number the pool of HSCs in all other fetal hematopoietic organs other than the fetal liver (Mikkola et al. 2005; Kumaravelu et al. 2002). These data clearly indicate that the placenta has an important role in establishing hematopoiesis in mammals.

Although the final outcome of their differentiation—the generation of erythroid, lymphoid and myeloid lineages—is similar, fetal-derived hematopoietic precursor cells of various origins such as the yolk sac, AGM, or fetal liver differ from one another and from the adult HSCs of the BM. For example, the stem cells of the early mouse yolk sac express no histocompatibility antigens (major histocompatibility complex [MHC] class I or II), stem cell antigen (Sca-1), or CD45 (Ly5), but do express high levels of the adhesion molecule CD44, the heat stable antigen (HSA), and an epitope detected by antibody AA4.1 (Liu and Auerbach 1991; Waller et al. 1995a, 1995b; Liu et al. 1994). Stem cells of the fetal liver express MHC class I and II, but at a lower level than stem cells in the BM (Dieterlen-Lievre 1997). Like yolk sac and unlike BM, they exhibit high levels of HSA and the AA4.1 antigen, but unlike yolk sac stem cells they express both CD45 and Sca-1 (Waller et al. 1995a, 1995b; Ikuta and Weissman 1992).

Interestingly, not only are embryonic hematopoietic precursors derived from the yolk sac, AGM, or fetal liver different phenotypically (based on surface marker expression), they also display distinct differences in their differentiation potential within their anatomical location, suggesting that microenvironmental factors play an important role in controlling stem cell fate during development. Extrinsic control of stem cells via their microenvironment is an important regulatory mechanism that is carried over into other adult stem cell populations and acts at several levels of hematopoietic differentiation.

Adult Hematopoietic Stem Cells and Regulation of Hematopoiesis

During steady state hematopoiesis in adults, the HSC population is actually relatively quiescent, and this quiescent state has been correlated with maintenance of stem cell self-renewal (Bradford et al. 1997; Cheshier et al. 1999). For example, previous studies have demonstrated that among HSC, cells in the G_0 or G_1 phase of the cell cycle (2N DNA content) exhibit enhanced capacity for radioprotection and for long-term multilineage reconstitution when compared with cells in the more highly proliferative S or G_2/M phase of cell cycle (>2N DNA content) (Fleming et al. 1993; Eaves et al. 2001; Glimm and Eaves 1999; Glimm et al. 2000, 2001), directly leading to the notion that cellular quiescence may promote HSC engraftment function (Venezia et al. 2004; Cashman et al. 2002).

At a low but constant frequency, however, some HSCs are signaled to enter the cell cycle and either self-renew or give rise to a hierarchy of differentiating progenitor populations. These progenitors then undergo a massive proliferative expansion, which is required to replenish and maintain a fully replete blood system on a daily basis (Weissman 2000; Weissman et al. 2001). Though the precise lineage relationship among primitive hematopoietic populations has recently become a topic of some debate (and will be discussed more thoroughly elsewhere in this book), it is generally agreed that HSCs give rise to all of the mature blood lineages through complex production of a series of more differentiated intermediate cells.

An important aspect of adult HSCs is the ability to not only homeostatically maintain blood cell production, but to also respond efficiently to regenerate the blood in response to hematological stressors such as blood loss, infection, or exposure to cytotoxic agents. These agents generally induce expansion of the HSC population, primarily via an increase in the number and frequency of self-renewing divisions (Morrison et al. 1997; Wright et al. 2001). Thus, HSC proliferation must be highly adaptive to ensure both a durable production of differentiated progenitor populations and extensive, self-renewing proliferation during the increased demand for blood cell production that occurs during hematological stress.

In order to maintain a self-renewing stem cell pool and more mature progenitors, HSCs are faced not only with a decision to remain quiescent or to divide, but also at each cell division with the choice to self-renew or differentiate (Weissman 2000; Weissman et al. 2001; Cheung et al. 2006). Precise control of blood cell production requires a system of elegantly controlled triggers that signal the appropriate regulation of cell cycle progression and the ultimate fate of a given cell cycle event, which could be either a self-renewing proliferative event or one that leads to differentiation of an HSC into either an erythroid, myeloid, or lymphoid cell. Knowledge of the signaling networks that play a fundamental role in controlling the proliferation and fate of HSCs is essential to understand their function in maintaining long-term blood cell production. Though the detailed molecular events controlling proliferation and fate of HSCs are unclear, they are likely to be critically dependent on myriad complex cell-intrinsic and -extrinsic factors that collectively converge to regulate cell cycle. Therefore, cell cycle arguably stands as a critically important regulator of HSC function and will be the focus of our discussion. A number of additional important regulators of HSC function are discussed elsewhere in this book.

Cell Cycle Regulators in Controlling Hematopoietic Stem Cell Function

Because of the importance of controlling cell cycle events and proliferation in the regulation of HSC fate and function, many studies have investigated the contribution of specific components of the cell cycle machinery to various phases of HSC cell cycle and initiation of proliferation. Not surprisingly, both positive and negative regulators of the cell cycle machinery have been shown to be critical for various aspects of HSC and/or hematopoietic progenitor proliferation, including the Ckis p21, p27, p16, and p18 (Cheng et al. 2000a; Yu et al. 2006; Yuan et al. 2004; Passegue et al. 2005) and the D-type cyclins and their catalytic partners Cdk4-6 (Kozar et al. 2004; Malumbres et al. 2004). Several studies have analyzed the proliferation rates and distribution of cell cycle components in subsets of hematopoietic cells, linked or not linked to engraftment function (Fleming et al. 1993; Eaves et al. 2001; Glimm and Eaves 1999; Glimm et al. 2000, 2001; Venezia et al. 2004; Wagers 2008; Brack et al. 2007, 2008; Steinman 2002). These studies include multiple investigations employing gene expression arrays to define the global status of the cell cycle machinery in HSC and how expression of these factors change and potentially control stem cell function and fate (Passegue et al. 2005). In these investigations, the population of HSCs that displays the highest levels of engraftment capacity over the longest period of time (referred to as long-term HSC; LT-HSC) expressed the highest levels of most of the Ckis and the lowest levels of most of the cyclins. Differentiation of LT-HSCs into those cells that have a reduced, short-term engraftment capacity (ST-HSC), and then ultimately into even more differentiated multipotent progenitors, specifically correlated with increased expression of cyclin D1, whereas the expression of the other members of the D cyclin family did not change. Notably, multipotent progenitor differentiation was also accompanied by a decrease in the expression of at least four Ckis (p16, p18, p27, and p57) and with a specific increase in the expression of two mitotic cyclins: cyclin G2 and cyclin B1 (Passegue et al. 2005) (Fig. 2).

As mentioned, one hallmark of HSCs is their ability to rapidly proliferate in response to hematopoietic stressors, for example, myelosuppression, chemotherapy, or BM transplantation, in order to quickly generate progenitors as well as additional stem cells, which then rapidly return to a steady-state quiescence (Rosendaal et al. 1981;

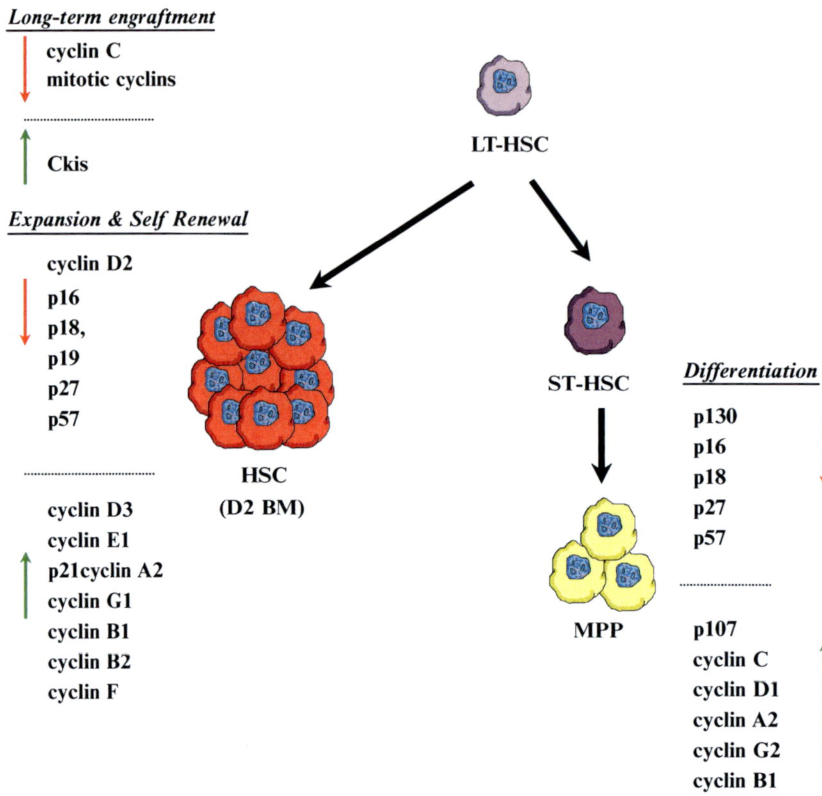

Fig. 2 Summary of changes in cell cycle machinery from long-term hematopoietic stem cells (LT-HSCs) to multipotent progenitor cells (MPPs). Multiple cell cycle factors control the proliferation status and phase of the cell cycle of HSCs and progenitors. Distinct changes in some cell cycle regulators (summarized in text and shown here as downregulated (↓) or upregulate (↑)) are specifically associated with differentiation, self-renewal, long-term reconstitution capacity and mobilization in HSCs and MPPs

Dixon and Rosendaal 1981). Proliferation in response to hematopoietic stress is thought to directly mimic the expansion of HSCs, which occurs naturally in vivo, and, therefore, understanding the molecular control of this process could provide substantial insight into the regulation of hematopoietic proliferation and stem cell fate. Thus, in parallel to the studies highlighted above, investigators have similarly studied the changes in cell cycle machinery that occur in response to activating triggers that induce HSC cell division, such as treatment with cytotoxic agents and cytokines such as cyclophosphamide and granulocyte colony-stimulating factor (Cy/G) (Morrison et al. 1997; Wright et al. 2001; Neben et al. 1993).

Treatment with a combination of Cy/G is a common therapeutic regimen used in human clinical transplantation to increase the numbers of HSCs available in the peripheral blood, a process called hematopoietic stem cell mobilization. In mice and humans, Cy/G treatment promotes HSC self-renewing proliferation in the BM

(Morrison et al. 1997; Wright et al. 2001; Neben et al. 1993). In fact, after administration of Cy followed by two daily doses of G (D2 Cy/G treatment), the BM HSC population dramatically expands, reaching 2–12 times the size of the HSC pool in untreated animals. These proliferating and self-renewing HSCs, when analyzed for changes in cell cycle, exhibited a substantial increase in the frequency of cells in S-G2/M phase and discrete alterations in regulatory and core components of the cell cycle machinery. Self-renewing proliferation in mobilized BM HSCs specifically correlated with an induction of cyclin D3 expression, while, conversely, cyclin D2 levels actually decreased and cyclin D1 levels remained constant. Mobilized BM HSCs also displayed reduced levels of all Ckis except for p21, which was increased (Passegue et al. 2005).

The functional importance that cell cycle regulators have on physiological hematopoiesis has been confirmed for multiple cell cycle factors by assessing the effects of genetic loss of function within the hematopoietic stem cell pool. Analyses of p21 and p18 knockout mice have demonstrated their critical functions in maintaining normal numbers of resting LT-HSCs capable of long-term engraftment and multi-lineage reconstitution (Cheng et al. 2000a, 2000b; Yuan et al. 2004), though the importance of these particular regulators may exhibit some strain-specificity in inbred mice (Cheng et al. 2000a, 2000b; Yuan et al. 2004). In contrast, p27-deficient mice exhibit normal HSC functions but show increased numbers of progenitor cells (Byrne 2008), whereas p16-deficient animals develop thymic hyperplasia (Sharpless et al. 2001) and p19- or p57-deficient mice show no overt hematopoietic abnormalities (Zhang et al. 1997; Zindy et al. 2000). These results suggest that distinct Cki proteins are critical for HSC function and control proliferation in discrete subsets of differentiating hematopoietic cells.

In addition to cell cycle machinery proteins, many other molecules, including multiple soluble/growth factors and cytokines, transcriptional regulators, and cell adhesion molecules, have been implicated in the intrinsic regulation of HSC function. The important roles for molecules such as Wnt/β-catenin (Eaves 2003; Rattis et al. 2004; Staal and Clevers 2005), Rb (Walkley et al. 2007), Bmi-1 (Park et al. 2003), HoxB4 (Giampaolo et al. 1995; Helgason et al. 1996; Lawrence et al. 1996; Sauvageau et al. 1995; Thorsteinsdottir et al. 1997), PU.1 (Dakic et al. 2005; Iwasaki et al. 2005), and SDF-1/CXCR4 (Voermans et al. 2001; Christensen et al. 2004) are discussed in detail in other chapters of this book.

Extrinsic Regulation of Hematopoietic Stem Cells

The regulation of stem cell self-renewal and differentiation is clearly crucial for normal hematopoietic function. A powerful concept of stem cell regulation that has received more attention recently and that organizes a number of important aspects of HSC regulation, including that of cell cycle control, is that of the stem cell niche as a determinant of stem cell activity (Jones and Wagers 2008; Martinez-Agosto et al. 2007; Calvi et al. 2003).

A niche is a distinct microenvironment that contributes to the regulation and maintenance of stem cell function. The concept of the niche was originally described by Schofield (1978), who posited that stem cells reside within fixed compartments, or niches, that are conducive to the maintenance of stem cell function. Therefore, a stem cell niche represents a defined anatomical compartment that provides signals to stem cells in the form of secreted and/or cell surface molecules to control stem cell proliferation, determine stem cell fate, and protect stem cells from exhaustion or death (Wilson and Trumpp 2006; Doetsch 2003; Kiel and Morrison 2006, 2008).

The specific cell types and factors involved in the extrinsic regulation of HSCs by their niche is being pieced together through complementary genetic and cell biological studies (Wilson and Trumpp 2006; Doetsch 2003; Kiel and Morrison 2006, 2008). In light of the importance and excitement centered around definitive studies of the HSC niche, both during homeostatic (Kiel et al. 2005; Wilson and Trumpp 2006) and regenerative hematopoiesis (Mayack and Wagers 2008; Adams and Scadden 2008), as well as parallels that might exist in the regulation of other somatic stem cell populations, an entire chapter of this book is dedicated to discussing the hematopoietic stem cell microenvironment.

Briefly, however, a number of recent studies have made progress toward the in situ localization of HSCs, and they indicate that HSCs can be found adjacent to both osteoblastic cells and endothelial cells in normal bone and that both these cell types contribute to the formation of the HSC niche (Kiel et al. 2005; Calvi et al. 2003; Wilson and Trumpp 2006; Mayack and Wagers 2008; Yin and Li 2006; Arai and Suda 2007; Porter and Calvi 2008). These initial studies thus implicate at least two cell types—bone derived osteoblasts and sinusoidal endothelial cells—as components of the HSC niche. This work has been followed by a series of studies that have further defined the cellular interactions between niche cells and HSCs and have identified a number of niche specific factors that are involved in regulating HSCs during both homeostatic and regenerative blood cell production. These niche-derived HSC regulatory factors include Tie2/Angiopoietin (Moore and Lemischka 2004; Arai et al. 2004), Notch/Jagged (Calvi et al. 2003; Gering and Patient 2008), PTH/PTHrP (Kronenberg 2007), Wnt/B-catenin (Rattis et al. 2004; Staal and Clevers 2005), and the SDF-1/CXCR4 signaling axis (Voermans et al. 2001; Christensen et al. 2004; Watt and Forde 2008; Dar et al. 2006).

Other Types of Somatic Stem Cells

HSCs, described above, were the first somatic stem cells to be isolated, and studies of these cells have provided many paradigms for the identification and analysis of stem cells in other adult tissues. Indeed, stem cells have now been identified and isolated from many other somatic tissues, and within these tissues too, stem cell–specific niches are essential in controlling stem cell activity. Yet, the demand for homeostatic and regenerative cell replacement clearly differs in distinct tissues, and these differences are reflected by variations in the fate potential and regulatory mechanisms of the stem cells that seed them.

Non-hematopoietic Stem Cells

In contrast to multipotent HSCs, some tissues are maintained by unipotent stem cells, which have a more restricted cell fate potential. Unipotent stem cells, such as those that maintain and repair the skeletal muscle, differentiate to produce only one mature specialized cell type (e.g., muscle fibers). Here we explore three different nonhematopoietic somatic stem cells (germ line stem cells [GSCs], interfollicular epidermal stem cells, and skeletal muscle stem cells), illustrating the similarities and differences that control their stem cell function.

Drosophila Male Germ Line Stem Cells

The *Drosophila* male GSC is one of the best understood examples of how cell fate is determined upon division of a unipotent stem cell and how this regulation is established by the stem cell niche (Gonczy and DiNardo 1996; Merida et al. 2008; Dansereau and Lasko 2008). In adult flies, around seven to nine GSCs lie at the apical tip of the testis, surrounding a group of somatic cells called a hub. This hub constitutes the male GSC stem cell niche.

Signals from the hub orient GSC division, ensuring that the spindle of a dividing GSC is always oriented perpendicular to the hub (Merida et al. 2008; Dansereau and Lasko 2008). This stereotypical anatomy sets up a situation in which a dividing GSC gives rise to one daughter cell that remains in contact with the hub and another cell that is displaced away from the hub. The daughter that maintains contact with the hub maintains stem cell identity, and while the displaced daughter initiates differentiation as a gonialblast, it undergoes four mitotic divisions to form clusters of interconnected spermatogonia cysts. These cysts then undergo terminal differentiation to form spermatocytes (Fig. 3).

How the Niche Determines Stem Cell Number

GSCs maintain cell polarity throughout the cell cycle, as their centrosome is consistently located in the region of the cell adjacent to the hub (Wallenfang and Matunis 2003; Takada et al. 2003). During mitosis, the centrosome duplicates, and pulse-chase experiments show that the mother centrosome is preferentially retained at the GSC-hub interface, whereas the daughter centrosome migrates away to the opposite side of the cell. This ensures that upon cell division one daughter cell

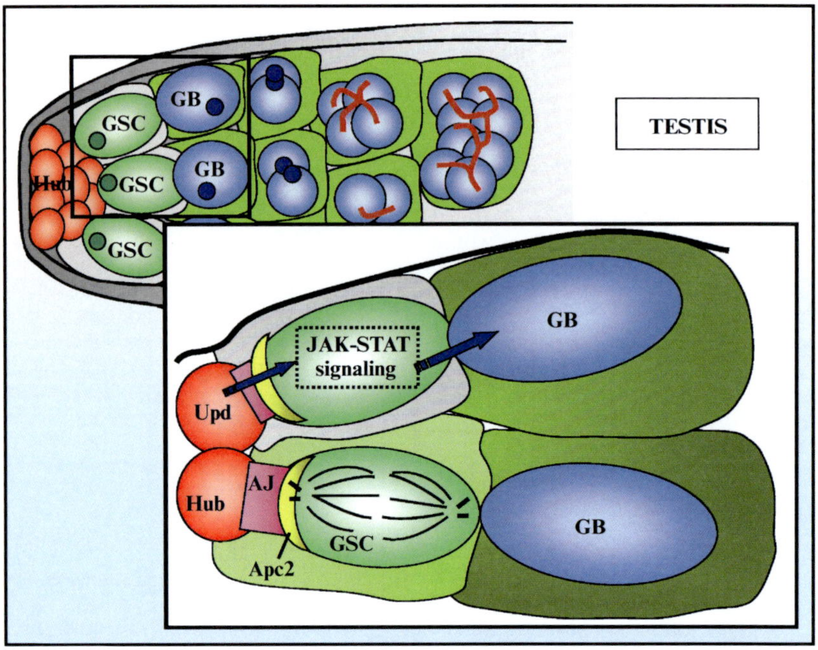

Fig. 3 Hub cells act as a stem cell niche for *Drosophila* male germline stem cells (GSCs). GSCs (*green*) surround a cluster of postmitotic somatic cells known as the apical hub (*red*), which act as a stem cell niche. GSCs divide asymmetrically, with the plane of division such that only one daughter cell will remain in contact with the hub and retain stem cell identity. The other daughter cell loses contact with the hub and forms a gonialblast (*blue*), which will differentiate further into spermatogonia

remains in contact with the hub and maintains stem cell identity, while the other daughter cell starts to differentiate in the absence of interactions with the hub.

Using powerful fly genetics, many of the molecular effectors of asymmetric cell division in the *Drosophila* testis have been identified. Centrosomin (cnn) is an integral centrosomal protein required to anchor astral microtubules to centrosomes (McCartney and Peifer 2003; Yamashita and Fuller 2005). In cnn mutants, spindle orientation is essentially randomized, and GSC divisions occur in both daughter cells, which remain in contact with the hub. In this situation, both daughter cells retain stem cell identity, resulting in a 20%–30% increase in the number of GSCs in cnn mutants compared to wild-type flies. Further increases in GSCs in cnn flies are probably prevented by the finite space available around the hub.

DE-cadherin and β-catenin co-localize at the hub-GSC interface, and this structure has been suggested to serve as a platform for binding the *Drosophila* homolog of adenomatous polyposis coli (Apc2) (McCartney and Peifer 2003). Apc2 also localizes to the hub-GSC interface. Since APC proteins interact with both β-catenin and with microtubules, Apc2 may effect spindle orientation in GSCs. Indeed, apc2 mutants show phenotypes similar to cnn mutants, including misoriented spindles and an increase in GSCs around the hub.

How the Niche Determines Stem Cell Identity

Hub cells express Upd, which acts over short distances to activate Janus kinase-signal transducer and activator of transcription (Jak-STAT) signaling in GSCs adjacent to the hub. Jak-STAT signaling is required to maintain GSC identity, as loss of function of either the Jak component hopscotch (hop) or the STAT homolog, Stat92E, causes a loss of GSCs due to inappropriate differentiation (Tulina and Matunis 2001; Kiger et al. 2001). Conversely, upregulating Jak-STAT signaling via overexpression of Upd causes an increase in GSC number in the testis, with GSCs found even more distant from the apical hub. This block in GSC differentiation results in decreased differentiating spermatogonia and thus impairs spermatogenesis. Strikingly, Jak-STAT signaling can cause dedifferentiation of transit-amplifying spermatogonia. At restrictive temperatures, GSCs containing a temperature-sensitive allele of *stat92E* start to differentiate, such that all cells contacting the hub form spermatogonia. However, restoration of Jak-STAT signaling by shifting the flies back to a permissive temperature causes spermatogonia adjacent to the hub to bud off from interconnected 8- or 16-cell cysts and reacquire GSC fate.

Interfollicular Epidermis

In contrast to *Drosophila* GSCs, the interfollicular epidermal stem cell niche has not been as well identified. Also, it has been difficult to isolate these stem cells prospectively. However, there is robust demonstration of functional stem cell activity when skin biopsies are taken to grow large sheets of epidermal cells for grafting, indicating the presence of interfollicular epidermal stem cells (Reynolds and Jahoda 1991; Janes et al. 2002; Niemann and Watt 2002; Watt 2002; O'Shaughnessy and Christiano 2001).

Adult skin epithelium consists of a diverse array of cell types, including stratified epidermis (also called interfollicular epidermis, IFE), hair follicles, and sebaceous glands. In contrast to the other tissues discussed in this chapter, several sources of stem cells potentially contribute to maintenance of the epidermal layer.

Multipotent Bulge Cells Do Not Normally Contribute to IFE

A region at the base of the sebaceous gland in the hair follicle, known as the bulge, contains stem cells that normally divide to form the various cell types that make up the hair follicle. These bulge cells retained their label after a bromodeoxyuridine (BrdU) pulse-chase, indicating that they cycle slowly, a property often associated with stem cells (Cotsarelis et al. 1990). Transplantation studies with clonal analysis have shown that these bulge stem cells are multipotent and can regenerate a complete new hair follicle containing self-renewing bulge stem cells as well as sebaceous glands and IFE. Under physiological conditions, the hair follicle cyclically degenerates,

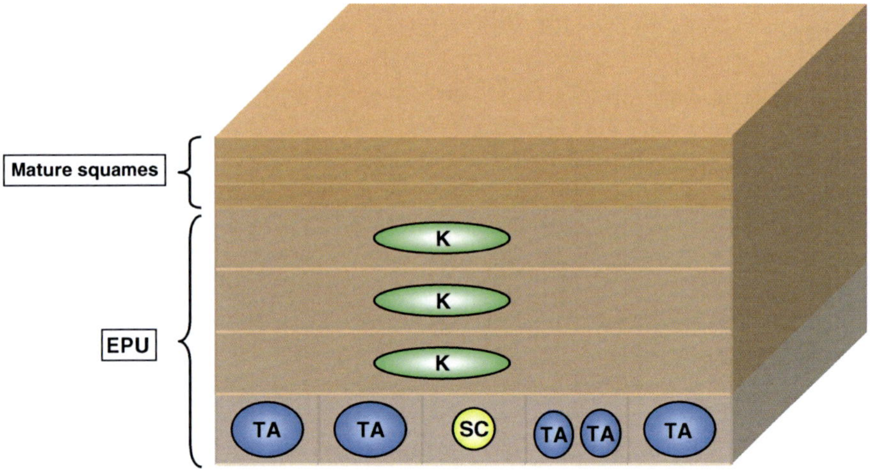

Fig. 4 Interfollicular epidermal stem cells maintain the epidermal layer. The interfollicular epidermis (IFE) consists of stacks of terminally differentiated keratinocytes on the outer surface, lying on a layer of basal cells. The basal cells consist of stem cells (*yellow*) and downstream transit-amplifying cells (*blue*). The transit-amplifying cells differentiate into keratinocytes, and then into squames, which slough off regularly

with the bulge stem cells emerging from a generally quiescent state to proliferate and regenerate the follicle. The bulge stem cell also regenerates the sebaceous gland after injury and has the potential to form IFE.

Under normal conditions, however, the bulge appears to contribute little to the IFE (Claudinot et al. 2005; Ito et al. 2005). This is supported by the fact that palmoplantar skin lacks hair follicles entirely but can regenerate and maintain epidermal homeostasis normally. Instead, most epidermis appears to be maintained by a basal layer of epidermal progenitor cells, or interfollicular epidermal stem cells. The IFE consists of stacks of terminally differentiated keratinocytes on the outer surface, lying on a layer of basal cells. The basal cells constitute the progenitor/stem cell population of IFE and continually divide throughout the lifetime of the organism to replenish the outer layer of keratinocytes, which are continuously sloughed off (Fig. 4). Lineage-tracing experiments show that these epithelial proliferative units (EPUs) can be maintained for prolonged periods of time, supporting the idea that basal cells are stem cells. Indeed, culture conditions for human IFE stem cells were defined in 1975 (Janes et al. 2002; Rowden et al. 1975; Rheinwald and Green 1975a, 1975b), leading to the routine use of autologous cultured skin grafts in the clinic to treat burn victims.

Unipotent Stem Cells in Clinical Therapies

Human epidermal cells can be isolated from a small skin biopsy and cultured on a feeder layer of 3T3 cells. Some of these cells form highly proliferative colonies that can be subcultured to generate sheets of epithelia. These epithelial sheets can be grafted onto wound sites, providing a permanent and self-renewing epidermis. Confirming the unipotency of the IFE stem cell, this new epidermis can participate in normal wound healing, but does not contain hair follicles or sebaceous glands.

Current models of epidermal homeostasis maintain that a self-renewing population of interfollicular epidermal stem cells differentiates to form transit-amplifying cells, which terminally differentiate after several rounds of cell division. However, in a recent study, Clayton et al. (2007) used inducible genetic labeling to analyze tail epidermis of adult mice, and the resulting size distributions of clusters of labeled cells with time fit a mathematical model where epidermis is maintained by a single type of progenitor cell. This cell can divide symmetrically or asymmetrically (as determined by staining labeled cells for cell cycle markers, as well as observed in asymmetric distribution of Numb in daughter cells). Such studies raise the possibility that normal tissue maintenance may rely on multiple modes of regeneration, calling into play cells of distinct developmental potency and proliferative potential.

Skeletal Muscle Stem Cells

Skeletal muscle regeneration in adults is mediated by myogenic precursor cells, contained within a population of muscle fiber-associated satellite cells. Satellite cells are mononucleated cells located between the basal lamina and the sarcolemma of multinucleated myofibers. Upon muscle injury, myogenic precursor cells (muscle stem cells) that reside in the satellite cell compartment proliferate and either form new myotubes or fuse with existing myotubes to repair the injured tissue. These cells also self-renew to maintain the muscle stem cell pool (Fig. 5).

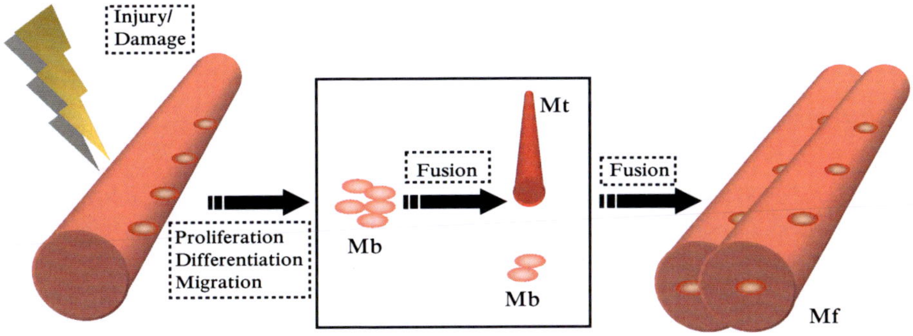

Fig. 5 Satellite cells are mononucleated cells located between the basal lamina and the sarcolemma of multinucleated myofibers. Upon muscle injury, muscle stem cells that reside in the satellite cell compartment proliferate and form myoblasts (Mb) that either form new myotubes (Mt) or fuse with existing myotubes, repairing the injured tissue. These cells also self-renew to maintain the muscle stem cell pool

Heterogeneity of the Muscle Satellite Cell Compartment

The satellite cell compartment contains a heterogeneous mix of cell types; in addition to putative muscle stem cells, it contains cells derived from the mesenchymal and hematopoietic lineages (including infiltrating inflammatory and immune cells), as well as more differentiated myoblasts that do not self-renew.

Single myogenic precursor cells isolated from the satellite cell compartment can proliferate and differentiate into multinucleated myotubes in vitro. Furthermore, transplantation of satellite cells into mouse muscle can regenerate both myofibers as well as the associated satellite cells, with long-term persistence and participation in repeated rounds of regeneration in response to additional muscle injury (Collins et al. 2005; Kuang et al. 2006; Cerletti et al. 2008). Moreover, irradiation experiments, which impair satellite cell proliferation in response to muscle damage (Morgan et al. 1990), strongly suggest that these cells are required for normal muscle repair. Thus, the satellite cell compartment contains stem cells capable of self-renewing and differentiating to form myotubes.

Satellite cells display a range of cell surface markers, different potentials for self-renewal, and differences in differentiation potential both along myogenic and nonmyogenic lineages. Immune cells can infiltrate beneath the basal lamina and populate the satellite cell compartment, both in the presence and absence of injury. Although some studies have suggested multipotency among muscle-forming satellite cells, cell isolation and transplantation studies indicate that nonmyogenic activities (i.e., fibrogenic, osteogenic, or adipogenic) of skeletal muscle satellite cells instead arise from distinct populations of cells that cohabit the satellite cell niche (Asakura et al. 2001; Seale et al. 2001). Other satellite cells have been shown to express markers of osteocytes or adipocytes, rather than those of muscle, and do not display appreciable myogenic capacity in vitro or in vivo. The balanced representation and complex interplay of these distinct cell lineages within the muscle may be critical for proper muscle homeostasis and regenerative activity.

Contribution of Other Cell Types to Muscle

A variety of nonmuscle lineage cell types have been reported to contribute to muscle formation. Transplanted HSCs have been reported to engraft muscle at low frequency (Gussoni et al. 1999), but this phenomenon appears to result from fusion of HSC-derived circulating myeloid cells with existing host myofibers, and there is no evidence for extensive repair or reconstitution of the satellite cell compartment from HSCs (Wagers et al. 2002). This poor muscle regenerative activity of blood lineage cells contrasts with satellite cells, where transplantation of as few as seven muscle fiber–associated satellite cells can generate over a hundred new myofibers, as well as vigorously self-renew to repopulate the host muscle with new satellite cells (Collins et al. 2005).

Another cell type currently being studied in the context of therapeutic muscle regeneration is the blood vessel–associated mesangioblast, a subset of blood vessel–associated cells originating from the embryonic dorsal aorta region. These cells contribute to multiple mesodermal tissues (including skeletal muscle) when transplanted into a chick embryo (Minasi et al. 2002). When mesangioblasts are injected into dystrophic mice or dogs, they can contribute to myotube formation and significantly improve muscle function (Sampaolesi et al. 2003, 2006). An advantage in this case is the amenability of these cells for systemic transplant.

Despite their capacity for myogenic contributions under certain experimental conditions, it is unlikely that these other cell types contribute to physiological maintenance and repair of muscle. Under pathological conditions, such as acute or systemic injury, such cells might be able to contribute to myotube formation and functional recovery. However, it is probable that long-term repair and maintenance of the injured muscle will require reconstitution of the satellite cell compartment, and there have been few examples of transplanted nonmyogenic cells reconstituting myogenic precursors in the satellite cell compartment. One report claimed that cultured outgrowths from adult human synovial membrane contribute to the satellite cell compartment when transplanted into the skeletal muscle of immunocompromised nude mouse (De Bari et al. 2003); however, the cells may represent a mixed population, and the identity of the myogenic subpopulation is yet unknown.

Stem Cell Markers

Most satellite cells express the paired box transcription factor Pax7, which is essential for maintenance of the satellite cell compartment. Pax7-null mice are born with skeletal muscle and a near-normal level of satellite cells, but Pax7-null satellite cells are progressively lost, likely via cell death, and as a result, Pax7-null mice exhibit defective skeletal muscle regeneration, progressive muscular dystrophy, and premature death (Kuang et al. 2006; Olguin and Olwin 2004; Seale et al. 2000; Zammit et al. 2006a, 2006b). Hence, Pax7 appears to be important although based on recent studies not absolutely required for the maintenance of satellite cells while the specification of satellite cells is Pax7 independent.

The myogenic population of satellite cells can express markers of quiescent satellite cells, such as Pax7, as well as markers such as MyoD and Myf-5, which are upregulated upon differentiation. Expression of Pax7 and MyoD reflects distinct stages of activation of myogenic precursor cells. Interestingly, upon injury, myogenic precursor cells can upregulate both Pax7 and MyoD expression. Some of these $Pax7^{+}$ $MyoD^{+}$ cells subsequently downregulate Pax7 expression to differentiate, while others downregulate MyoD expression, return to a quiescent state, and maintain satellite cell identity (Zammit et al. 2006a).

In mice, we recently identified a subset of satellite cells called skeletal muscle precursors (SMPs) that exhibit stem cell properties (ability for self-renewal and differentiation) (Cerletti et al. 2008). SMPs are discriminated via a combination of surface markers (CD45$^-$ Sca1$^-$ Mac1$^-$ CXCR4$^+$ β1-integrin$^+$, abbreviated CSM4B) and can be isolated from neonatal and mature muscle by flow cytometry. These cells express Pax7, are able to both self-renew and differentiate to form myotubes in vitro and in vivo, and exhibit extensive, functional engraftment activity of up to 94% when transplanted into diseased adult muscle. Engraftment of wild-type SMPs also restores functional capability of the repaired muscle. Furthermore, transplanted SMPs self-renew and reconstitute the muscle stem cell population and can be reisolated from the recipient animal for serial transplantation, hence demonstrating in vivo self-renewal.

Origin of Skeletal Muscle Stem Cells

The dorsal cells of the somite in the embryo form the dermomyotome, which gives rise to the dermis and myotome. As the somite matures, cells at the edges or lips of the dermomyotome start to express myogenic regulatory factors (Myf5 and MyoD), marking them as myogenic precursor cells. The Pax3 gene has been identified as an important regulator of this process; Pax3 expression induces expression of Myf5 in the somite (Bajard et al. 2006). After their specification, Pax3$^+$ myogenic precursors delaminate and migrate ventrally to form the myotome, the first differentiated skeletal muscle of the embryo. Some myogenic precursors at the limb-level somites also migrate from the dermomyotome into developing limb buds and later form limb muscle. A population of myogenic precursors in the dermomyotome that expresses both Pax3 and Pax7 gives rise to satellite cells in postnatal muscle (Gros et al. 2005; Relaix et al. 2005).

Pax3 expression is essential for the survival of myogenic precursor cells during embryogenesis (Gros et al. 2005; Relaix et al. 2005; Borycki et al. 1999), as Pax3-null mice die during embryogenesis and lack most skeletal muscle. Pax3 expression is downregulated in some muscles before birth, and satellite cells from different skeletal muscles in the adult exhibit varying levels of Pax3 expression.

Control of Self-Renewal, Proliferation, and Differentiation in Muscle Stem Cell

The Notch signaling pathway has been found to regulate developmental decisions in various tissues, both during embryogenesis and postnatally. In muscle, Notch signaling appears to regulate proliferation of muscle stem cells in response to injury. Quiescent satellite cells express the Notch receptor, and fibers adjacent to damaged muscle upregulate the Notch ligand, Delta. Furthermore, a temporally regulated switch from Notch to Wnt signaling occurs during muscle regeneration to signal differentiation of these precursors into myoblasts, as measured by expression of progressively more differentiated cell markers (Brack et al. 2008; Conboy and Rando 2002; Conboy et al. 2003).

Aging in Stem Cells

Somatic stem cells replenish many tissues throughout life. In general, they exhibit slow turnover and reside in specialized niches, protected from the environment, so that only a few are activated at a time. In these situations, the rest of the cells remain "on reserve" to maintain and repair tissues for the life of an organism. Aging of an organism is marked by the decline in overall tissue function and repair capacity and an increase in multiple, age-related conditions and diseases. Given the role for stem cells in maintaining tissue integrity and function throughout the lifetime of an organism, they are in theory a defense against aging, replacing cells lost through attrition. However, if the rejuvenating effect of stem cells were unaffected in aging, it would seem that senescing or otherwise age-related dysfunctional cells might be replaced indefinitely. In light of the prevalence of age-related disease, this appears not to be the case. Does aging also affect stem cell function? In fact, even in highly regenerative tissues such as the skin, the gut, and the hematopoietic system, an age-related decline in stem cell function is well established.

Regarding the hematopoietic system, data from mice support an age-related decline in stem cell function (Roobrouck et al. 2008; Warren and Rossi 2009; Chambers et al. 2007). When limited numbers of aged hematopoietic progenitors are transplanted into young recipients under competitive conditions, they show an overall reduction in long-term repopulating potential; in addition, lymphopoiesis is deficient, whereas myelopoiesis is enhanced, suggesting that older HSCs cope inadequately with the daily demands of blood production (Rossi et al. 2007a). Paradoxically, however, the total number of primitive progenitors has been reported to increase with age in mice. Recent studies investigating age-related molecular changes in HSC function suggest that increased expression of particular proto-oncogenes, as well as marked increases in inflammatory and stress responses and substantial alterations in the regulation of chromatin structure, may contribute to the major age-related alterations in HSC function and lymphopoiesis (Rossi et al. 2007b).

Similar to the aberrancies seen in the maintenance of the hematopoietic compartment, aged animals display a marked decrease in the regenerative capacity of their skeletal muscles, coupled with an increase in muscle fibrosis and adipose tissue deposition. These alterations in skeletal muscle function have been associated with a decline in Notch signaling (Conboy and Rando 2002, 2005; Conboy et al. 2003, 2005; Miller and Emerson 2003) and an increase in Wnt signaling (Wagers 2008; Brack et al. 2007, 2008). Treating aged mice in vivo with an activator of Notch signaling improved the muscle regeneration response following injury, whereas injection of an inhibitor of Notch signaling into young mouse muscle impaired regeneration. Furthermore, parabiosis experiments have shown that exposure of old muscle to soluble factors present in young mice rejuvenates the response to injury and improves muscle regeneration; while conversely, exposure to an old systemic environment can impede myogenesis in young animals (Wagers 2008; Brack et al. 2007, 2008). Together, these experiments suggest that the age of the environment is a major determinant of muscle stem cell function.

The precise effects of age on the regenerative capacity of adult stem cells, which should replenish tissues throughout life, as well as the role of microenvironmental factors in age-related stem cell dysfunction, are poorly understood. Age-related defects in stem cells can limit proper tissue turnover and hence may contribute to disease through a general loss of tissue regenerative capacities. However, while aging of the stem cell pool is certainly likely to contribute to an age-related decline in tissue function, multiple tissues use alternative means of maintaining tissue homeostasis and/or can use these alternative tissue repair mechanisms as strategies to counteract the loss of normal stem cell–mediated regeneration that can occur in aging and in injury and disease.

Tissue Regeneration by Facultative Repair Cells

Maintenance in Some Tissues is Primarily Mediated by Non–Stem Cells

As highlighted throughout this chapter, what usually comes to mind when one refers to an adult stem cell is a relatively undifferentiated cell that functions in maintaining, as well as rapidly replenishing, tissues through mechanisms of both long-term self-renewal and differentiation into more specialized, mature cell types. Such mechanisms of cell replacement are well documented for blood, skin, and intestine, where dedicated adult stem cells play a role in both tissue homeostasis as well as in regenerative tissue repair. However, there is mounting evidence that other more slowly renewing organs may rely primarily on different strategies to homeostatically maintain and renew themselves, in part through division of differentiated cells. For example, it has been shown in mice that insulin-producing β cells of the pancreas are replenished, at least most appreciably, by self-duplication of preexisting β cells, rather than through the differentiation of stem cells. β cell replication occurs both in normal adult life and after pancreatectomy (Dor et al. 2004). Similarly, regeneration of the liver after 2/3 partial hepatectomy (PHx) stimulates the division of already differentiated, mature hepatocytes (Higgins 1931; Rabes 1977; Rabes et al. 1976). However, if hepatocyte proliferation is inhibited, normal liver size and histology can be restored by the recruitment of what have been described as "facultative stem cells" (Alison et al. 1996).

Facultative stem cells have been defined as differentiated cells that are normally quiescent but respond to injury by proliferation to generate daughter cells that are either identical to themselves or can differentiate into one or more mature cell types (Rawlins and Hogan 2006). Thus, like more canonically defined stem cells, facultative stem cells appear to both self-renew and differentiate. However, the major difference between facultative stem cells and their dedicated adult stem cell counterparts is that the former may include already differentiated cells that are only infrequently recruited after injury to act like the latter.

Unlike stem cells of the blood, skin, and intestine, which have been more extensively characterized, there is some controversy surrounding the existence of facultative stem cells. For example, considering the pancreas, an organ containing a potential facultative stem cell, Bonner-Weir et al. proposed that after partial pancreatectomy, facultative stem cells that reside in pancreatic ducts are activated and subsequently generate new pancreatic lobes and islets (Bonner-Weir et al. 2004). Other studies (Dor et al. 2004) did not directly rule out the possible existence of a facultative pancreatic stem cell but argue that such a mechanism does not likely contribute significantly to β cell numbers or might be recruited only under unique, as of yet undefined conditions. The necessary experiments that more convincingly point toward a pancreatic facultative stem cell have yet to be conducted. Lineage-tracing analysis of ductal cells after pancreatectomy, which would utilize a heritable marker to discriminate the progeny of a particular cell of interest, would provide strong evidence as to whether a ductal facultative pancreatic stem cell exists or not.

The most extensively studied and highly debated facultative stem cell is that implicated in liver regeneration. The existence of hepatic stem cells was first proposed in the late 1950s by Leduc and Wilson (1958). In a study on liver regeneration in mice with severe nutritional injury, these authors observed that nonparenchymal cells of the distal cholangioles proliferate, and after nutritional injury they differentiate into hepatocytes and possibly interlobular bile ducts as well (Leduc and Wilson 1958). In summarizing their observations, Leduc and Wilson describe the cells of the cholangioles as possible "reserve cells of a primitive type capable of differentiating into either parenchymal or bile-duct cells." While at the time of their study the "terminology" of stem cell biology had not yet been established, they conclude that "prolonged and severe injury to the liver may make direct restoration by division of preexisting parenchymal cells impossible, and that, when this occurs, the new parenchyma is derived from the indifferent cholangiole cells" (Leduc and Wilson 1958). These concepts of "reserve" or "indifferent" cells encompass what is now characterized as a facultative stem cell population.

As alluded to earlier, the normal increase in proliferation of differentiated hepatocytes can be blocked after Partial hepatectomy (PHx), either experimentally through administration of certain carcinogenic chemicals including 2-acetylaminofluorene (2-AFF), ethionine, and 3-methyl-4-dimethyl aminobenzene (Farber 1956), or under pathophysiological conditions if tissue injury is too severe, as seen in humans with fulminant hepatitis (Gkretsi et al. 2007). In such instances, where proliferation of mature hepatocytes is inhibited, a potential facultative stem cell compartment located within the smallest branches of the intrahepatic biliary tree may be activated (Alison et al. 1996). It has been suggested that the actual cell that functions in these settings as a facultative stem cell is derived from the biliary epithelium and is named the "oval cell" for its oval-shaped nucleus, (Farber 1956). There is a marked proliferation of oval cells after PHx and chemical inhibition of hepatocytes, beginning in the periportal areas of the hepatic lobule. These proliferating cells take up [^{3}H]-thymidine and have been traced through retention of this label. Proliferating oval cells form irregular duct-like structures that are loosely associated with mature, preexisting bile ducts. Four to five days after their collective expansion, oval cells invade the adjacent lobular parenchyma and become basophilic hepatocytes, ultimately differentiating into mature hepatocytes that are responsible for restoring liver size and histology (Evarts et al. 1987, 1996; Sell 1994; Alison et al. 1993; Alison and Sarraf 1994; Sarraf et al. 1994).

Additional studies have expanded on this lineage from oval cell to hepatocyte after 2-AAF/PHx conditions. Through expression of the bile duct markers CK-7, CK-19, and OV-6, and the hepatocyte markers AFP and albumin in [^{3}H]-thymidine labeled oval cells over time, Evarts et al. suggested that oval cells are hepatocyte precursors (Evarts et al. 1987). Furthermore, the temporal expression of liver-enriched transcription factors in proliferating rat oval cells after activation by 2-AAF/PHx mirrors the expression pattern during liver development (Nagy et al. 1994). Other studies have shown that various stem cell genes, including c-kit, CD34, flt3 receptor, and LIF, are activated during oval cell proliferation (Fujio et al. 1994; Omori et al. 1996). These data provide circumstantial evidence that points toward oval cells as potential facultative hepatic stem cells though functional analysis is clearly essential in establishing their relevance to liver maintenance and repair.

In addition to proliferation and differentiation capabilities, an important criteria for establishing oval cells as hepatic stem cells is an analysis of their ability to restore liver mass after transplantation (Matsusaka et al. 2000; Braun and Sandgren 2000). In one study, Yasui et al. isolated oval cells from Long-Evans Cinnamon (LEC) rats, characterized them, and transplanted these cells into LEC and Nagase analbuminemic rats (NAR) where they observed that the transplanted cells transformed into hepatocytes (Yasui et al. 1997). LEC rats carry a defect in the Wilson disease gene (*ATP7B* gene) and eventually develop hepatitis and later both hepatocellular and cholangiocarcinomas; LEC rats have oval cells that histologically resemble those in rats administered 2-AFF and other carcinogens. Yasui et al. isolated oval cells in these rats by isopyknic centrifugation in a Percoll gradient and then observed that these isolated cells express markers of both hepatocytes and biliary duct epithelial cells but not albumin, which is expressed in hepatocytes. They then infected the oval cells with a LacZ-transducing retrovirus and transplanted them into the liver of LEC rats, subsequently observing that when they stained liver sections with X-gal 2 weeks later, hepatocyte-like cells but not biliary duct cells were labeled. To evaluate the feasibility and efficiency of the transplanted oval cells, Yasui et al. also transplanted oval cells into a double mutant NAR/LEC rat, in which they similarly observed that oval cells transformed into hepatocytes in vivo and stably expressed a liver-specific protein, albumin. Based on the number of cells transplanted, it was concluded that oval cells produce half as much albumin as the same number of normal hepatocytes.

As with other stem cells, the facultative hepatic oval cell is maintained by specific regulatory microenvironmental niche components. However, there is controversy over where the specific hepatic stem cell niche is. Careful studies in rats suggest that oval cells are predominantly derived from the canal of Hering, which are terminal ductules that are located between biliary caniculi and interlobular biliary ducts (Paku et al. 2001). However, others argue that any component of the biliary tree can give rise to oval cells (Alison et al. 1996; Sirica et al. 1990; Golding et al. 1995). Nagy et al. showed that chronic treatment with dexamethasone is able to inhibit proliferation of oval cells after PHx and administration of 2-AFF, but dexamethasone fails to inhibit the proliferation of larger interlobular bile ducts induced by ligation of the common bile duct (Nagy et al. 1998). Since it was shown that there is preferential uptake of BrdU by cells in the canal of Hering after administration of 2-AFF, it might be expected that dexamethasone could selectively inhibit proliferation of these cells. However, dexamethasone completely inhibited the 2-AFF-induced proliferation of biliary cells regardless of their location (Nagy et al. 1998). Thus, these data failed to support the notion that the terminal ductules are the sole source of oval cells, and as such, the current debate over the facultative hepatic stem cell niche continues.

Although the signaling pathways involved in oval cell emergence, expansion, and differentiation still remain unclear, there is growing evidence suggesting that Wnt/β-catenin signaling plays a critical role in various aspects of hepatic biology including liver development, growth, and regeneration, and diseases of the liver such as hepatocellular carcinoma (HCC) (Monga et al. 2001; Micsenyi et al. 2004).

Since Wnt signaling has been implicated as an important potential regulator of oval cell biology, β-catenin, the downstream effector of Wnt signaling, has also been studied in oval cell activation and proliferation particularly with regard to embryonic liver development and HCC. Using the standard 2-AFF/PHx model in rats and modulation of the Wnt/ β-catenin pathway components during extensive oval cell activation, studies have demonstrated that β-catenin plays an important functional role in oval cell activation. First, β-catenin expression colocalized with Thy-1, an oval cell marker, at 5 and 10 days post-PHx when oval cells would be expected to be activated and proliferating as a response to injury (Masson et al. 2006; Petersen et al. 1998a, 1998b). Although not all of the Thy-1 positive cells expressed β-catenin and vice versa, only morphologically identified Thy-1 positive oval cells showed cytoplasmic and nuclear expression of β-catenin along with the presence of its activated or dephosphorylated form at both 5 and 10 days post-PHx. In addition to the temporal expression of β-catenin during the oval cell response to PHx, activation of β-catenin coincided with ongoing oval cell proliferation at the same time points, collectively suggesting a possible role for β-catenin in the stimulation of the oval cell response (Apte et al. 2007). Although the exact mechanism by which Wnt/ β-catenin activates the oval cell response is unclear, after 2-AAF/PHx there is downregulation in the expression of Wif-1, thereby leading to Wnt-1-mediated activation of β-catenin (Monga et al. 2001; Apte et al. 2007). Interestingly, the expression of Wnt-1 was observed in the hepatocytes surrounding the oval cells but not in the oval cells themselves, suggesting that paracrine signaling of Wnt-1 extrinsically induces β-catenin activation in oval cells (Monga et al. 2001; Apte et al. 2007).

Since first appreciating that hepatocytes are capable of regenerating the liver after PHx, advances have been made in identifying the cells that are potentially important in liver repair and constitute a facultative hepatic stem cell population. However, in order to further advance the field of liver biology toward a clinical application, it will be important to establish conditions under which these cells can be prospectively isolated and expanded in culture and to devise acceptable strategies for their transplantation into patients. Elucidating the mechanisms that control both hepatocyte regeneration as well as oval cell activation when hepatocytes are damaged may collectively provide insight into possible future therapeutic applications.

Concluding Remarks

For many years, researchers and clinicians have been seeking to understand the body's ability to repair and replace the cells and tissues of some organs. After years of work pursuing the how and why of cell repair mechanisms, many scientists have now focused their attention on adult stem cells. Current evidence strongly suggests that stem cells are present in far more tissues and organs than once thought, and that these cells can provide a primary source not only of continuous homeostatic replenishment of adult cells, but also of induced cell replacement necessary for tissue repair and regeneration after injury and in response to disease.

Intense efforts are now under way to harness the potential of stem cells and to control their function in order to take advantage of their tissue regeneration capabilities. The ultimate goal of these efforts is to devise new and more effective treatments for a host of diseases and disabilities. For most tissues, that potential still lies ahead, and although many questions remain, clear progress has been made in somatic stem cell biology, and this work holds great and exciting therapeutic promise.

Critical Unanswered Questions in Adult Somatic Stem Cell Biology

- What are the developmental sources of adult stem cells in the body? Are they "leftover" embryonic stem cells, or do they arise independently during organogenesis?
- How many different kinds of adult stem cells exist, and in which tissues do they exist? Accumulating evidence indicates that although they occur in small numbers, adult stem cells are present in many differentiated tissues, though their formal existence in several tissues of clinical value remains elusive.
- Is it possible to manipulate adult stem cells to increase their ability to proliferate in vitro while maintaining stem cell qualities, so that adult stem cells can be used as a sufficient source of tissue for transplants and other therapeutic strategies?

- What are the intrinsic and extrinsic controls that keep stem cells from differentiating or that direct them along a particular differentiation pathway to form one specialized cell type rather than another?
- What are the factors responsible for stem cell responses to injury or damage that enable rapid activation and appropriate contribution to tissue repair and regeneration?
- Can the "stress" signals that command facultative stem cells to respond to tissue damage and gain specific regenerative qualities be harnessed for therapeutic value?

Glossary

Stem Cell An unspecialized cell capable of generating more specialized cell types with specialized functions. All stem cells have three general properties: they are capable of dividing and renewing themselves for long periods; they are unspecialized; and they can give rise to specialized cell types.

Embryonic Stem Cell An unspecialized cell that is derived from the inner cell mass of the blastocyst contained within the embryo. These cells are pluripotent and thus can develop into specialized cells that contribute to all of the mature tissues of an organism.

Somatic Stem Cell An unspecialized cell found among differentiated cells in a tissue or organ that can renew and differentiate to yield the major specialized cell types of the tissue or organ. Adult stem cells are multipotent or unipotent, and the primary role of adult stem cells is to maintain and repair the tissue in which they are found. The term "somatic stem cell" is often used interchangeably with adult stem cell.

Germ Line Cell Germ cells are responsible for the production of sex cells or gametes (ovum and spermatozoa) constituting a cell line through which genes are passed from generation to generation.

Somatic Cell Any cell (other than a germ cell) that contributes to the formation of the body, becoming differentiated into the various tissues and organs that comprise the mature organism.

Adult Stem Cell See Somatic Stem Cell.

Unipotent Stem Cell An unspecialized cell capable of developing into only one type of cell within a given tissue or organ.

Multipotent Stem Cell An unspecialized cell capable of developing into multiple specialized cells of a given tissue or organ. For example, multipotent blood stem cells give rise to the red cells, white cells, and platelets in the blood.

Pluripotent Stem Cell An unspecialized cell that is capable of developing into all of the differentiated cells of an organism.

Facultative Stem Cell An unspecialized cell that is capable of contributing to tissue repair and regeneration when normal tissue repair cells and mechanisms are blocked, often due to injury or disease.

Hematopoietic Stem Cell An unspecialized cell that can differentiate into all of the specialized cells of the blood including red blood cells, white blood cells, and platelets.

Progenitor Cell A cell in a transition state from an unspecialized cell to a fully differentiated mature cell with a specialized function; therefore, it is a more specialized cell than the stem cell that generated it but a less specialized cell than the cell that is the direct ancestor of it in a hierarchical cell lineage.

Stem Cell Niche The microenvironment in which stem cells are found, which interacts with stem cells to regulate stem cell function and fate. Several factors are important to regulate stem cell characteristics within the niche: cell-cell interactions, interactions between stem cells and adhesion molecules, extracellular matrix components, and soluble factors.

References

Adams GB, Scadden DT (2008) A niche opportunity for stem cell therapeutics. Gene Ther 15:96–99

Alison MR, Sarraf CE (1994) Liver cell death: patterns and mechanisms. Gut 35:577–581

Alison MR, Poulsom R, Jeffery R, Anilkumar TV, Jagoe R, Sarraf CE (1993) Expression of hepatocyte growth factor mRNA during oval cell activation in the rat liver. J Pathol 171:291–299

Alison MR, Golding MH, Sarraf CE (1996) Pluripotential liver stem cells: facultative stem cells located in the biliary tree. Cell Prolif 29:373–402

Apte U, Zeng G, Thompson MD, Muller P, Micsenyi A, Cieply B, Kaestner KH, Monga SP (2007) Beta-catenin is critical for early postnatal liver growth. Am J Physiol Gastrointest Liver Physiol 292:G1578–G1585

Arai F, Suda T (2007) Maintenance of quiescent hematopoietic stem cells in the osteoblastic niche. Ann N Y Acad Sci 1106:41–53

Arai F, Hirao A, Ohmura M, Sato H, Matsuoka S, Takubo K, Ito K, Koh GY, Suda T (2004) Tie2/angiopoietin-1 signaling regulates hematopoietic stem cell quiescence in the bone marrow niche. Cell 118:149–161

Asakura A, Komaki M, Rudnicki M (2001) Muscle satellite cells are multipotential stem cells that exhibit myogenic, osteogenic, and adipogenic differentiation. Differentiation 68:245–253

Bajard L, Relaix F, Lagha M, Rocancourt D, Daubas P, Buckingham ME (2006) A novel genetic hierarchy functions during hypaxial myogenesis: Pax3 directly activates Myf5 in muscle progenitor cells in the limb. Genes Dev 20:2450–2464

Barker N, Clevers H (2007) Tracking down the stem cells of the intestine: strategies to identify adult stem cells. Gastroenterology 133:1755–1760

Barker N, van Es JH, Kuipers J, Kujala P, van den Born M, Cozijnsen M, Haegebarth A, Korving J, Begthel H, Peters PJ, Clevers H (2007) Identification of stem cells in small intestine and colon by marker gene Lgr5. Nature 449:1003–1007

Bierkens JG, Hendry JH, Testa NG (1991) Recovery of the proliferative and functional integrity of mouse bone marrow in long-term cultures established after whole-body irradiation at different doses and dose rates. Exp Hematol 19:81–86

Bonner-Weir S, Toschi E, Inada A, Reitz P, Fonseca SY, Aye T, Sharma A (2004) The pancreatic ductal epithelium serves as a potential pool of progenitor cells. Pediatr Diabetes 5(Suppl 2):16–22
Borycki AG, Li J, Jin F, Emerson CP, Epstein JA (1999) Pax3 functions in cell survival and in pax7 regulation. Development 126:1665–1674
Brack AS, Conboy MJ, Roy S, Lee M, Kuo CJ, Keller C, Rando TA (2007) Increased Wnt signaling during aging alters muscle stem cell fate and increases fibrosis. Science 317:807–810
Brack AS, Conboy IM, Conboy MJ, Shen J, Rando TA (2008) A temporal switch from notch to Wnt signaling in muscle stem cells is necessary for normal adult myogenesis. Cell Stem Cell 2:50–59
Bradford GB, Williams B, Rossi R, Bertoncello I (1997) Quiescence, cycling, and turnover in the primitive hematopoietic stem cell compartment. Exp Hematol 25:445–453
Braun KM, Sandgren EP (2000) Cellular origin of regenerating parenchyma in a mouse model of severe hepatic injury. Am J Pathol 157:561–569
Broxmeyer HE, Bruns HA, Zhang S, Cooper S, Hangoc G, McKenzie AN, Dent AL, Schindler U, Naeger LK, Hoey T, Kaplan MH (2002) Th1 cells regulate hematopoietic progenitor cell homeostasis by production of oncostatin M. Immunity 16:815–825
Byrne JA (2008) Generation of isogenic pluripotent stem cells. Hum Mol Genet 17:R37–R41
Calvi LM, Adams GB, Weibrecht KW, Weber JM, Olson DP, Knight MC, Martin RP, Schipani E, Divieti P, Bringhurst FR, Milner LA, Kronenberg HM, Scadden DT (2003) Osteoblastic cells regulate the haematopoietic stem cell niche. Nature 425:841–846
Cantor AB, Orkin SH (2001) Hematopoietic development: a balancing act. Curr Opin Genet Dev 11:513–519
Cashman J, Dykstra B, Clark-Lewis I, Eaves A, Eaves C (2002) Changes in the proliferative activity of human hematopoietic stem cells in NOD/SCID mice and enhancement of their transplantability after in vivo treatment with cell cycle inhibitors. J Exp Med 196:1141–1149
Cerletti M, Jurga S, Witczak CA, Hirshman MF, Shadrach JL, Goodyear LJ, Wagers AJ (2008) Highly efficient, functional engraftment of skeletal muscle stem cells in dystrophic muscles. Cell 134:37–47
Chambers SM, Shaw CA, Gatza C, Fisk CJ, Donehower LA, Goodell MA (2007) Aging hematopoietic stem cells decline in function and exhibit epigenetic dysregulation. PLoS Biol 5:e201
Cheng T, Rodrigues N, Shen H, Yang Y, Dombkowski D, Sykes M, Scadden DT (2000a) Hematopoietic stem cell quiescence maintained by p21cip1/waf1. Science 287:1804–1808
Cheng T, Rodrigues N, Dombkowski D, Stier S, Scadden DT (2000b) Stem cell repopulation efficiency but not pool size is governed by p27(kip1). Nat Med 6:1235–1240
Cheshier SH, Morrison SJ, Liao X, Weissman IL (1999) In vivo proliferation and cell cycle kinetics of long-term self-renewing hematopoietic stem cells. Proc Natl Acad Sci USA 96:3120–3125
Cheung AM, Kwong YL, Liang R, Leung AY (2006) Stem cell model of hematopoiesis. Curr Stem Cell Res Ther 1:305–315
Choumerianou DM, Dimitriou H, Kalmanti M (2008) Stem cells: promises versus limitations. Tissue Eng Part B Rev 14:53–60
Christensen JL, Wright DE, Wagers AJ, Weissman IL (2004) Circulation and chemotaxis of fetal hematopoietic stem cells. PLoS Biol 2:E75
Claudinot S, Nicolas M, Oshima H, Rochat A, Barrandon Y (2005) Long-term renewal of hair follicles from clonogenic multipotent stem cells. Proc Natl Acad Sci USA 102:14677–14682
Clayton E, Doupe DP, Klein AM, Winton DJ, Simons BD, Jones PH (2007) A single type of progenitor cell maintains normal epidermis. Nature 446:185–189
Cline MJ, Golde DW (1978) Immune suppression of hematopoiesis. Am J Med 64:301–310
Collins CA, Olsen I, Zammit PS, Heslop L, Petrie A, Partridge TA, Morgan JE (2005) Stem cell function, self-renewal, and behavioral heterogeneity of cells from the adult muscle satellite cell niche. Cell 122:289–301
Conboy IM, Rando TA (2002) The regulation of Notch signaling controls satellite cell activation and cell fate determination in postnatal myogenesis. Dev Cell 3:397–409
Conboy IM, Rando TA (2005) Aging, stem cells and tissue regeneration: lessons from muscle. Cell Cycle 4:407–410

Conboy IM, Conboy MJ, Smythe GM, Rando TA (2003) Notch-mediated restoration of regenerative potential to aged muscle. Science 302:1575–1577
Conboy IM, Conboy MJ, Wagers AJ, Girma ER, Weissman IL, Rando TA (2005) Rejuvenation of aged progenitor cells by exposure to a young systemic environment. Nature 433:760–764
Cotsarelis G, Sun TT, Lavker RM (1990) Label-retaining cells reside in the bulge area of pilosebaceous unit: implications for follicular stem cells, hair cycle, and skin carcinogenesis. Cell 61:1329–1337
Cumano A, Godin I (2001) Pluripotent hematopoietic stem cell development during embryogenesis. Curr Opin Immunol 13:166–171
Cumano A, Dieterlen-Lievre F, Godin I (1996) Lymphoid potential, probed before circulation in mouse, is restricted to caudal intraembryonic splanchnopleura. Cell 86:907–916
Dakic A, Metcalf D, Di Rago L, Mifsud S, Wu L, Nutt SL (2005) PU.1 regulates the commitment of adult hematopoietic progenitors and restricts granulopoiesis. J Exp Med 201:1487–1502
Dansereau DA, Lasko P (2008) The development of germline stem cells in Drosophila. Methods Mol Biol 450:3–26
Dar A, Kollet O, Lapidot T (2006) Mutual, reciprocal SDF-1/CXCR4 interactions between hematopoietic and bone marrow stromal cells regulate human stem cell migration and development in NOD/SCID chimeric mice. Exp Hematol 34:967–975
De Bari C, Dell'Accio F, Vandenabeele F, Vermeesch JR, Raymackers JM, Luyten FP (2003) Skeletal muscle repair by adult human mesenchymal stem cells from synovial membrane. J Cell Biol 160:909–918
Dieterlen-Lievre F (1997) Intraembryonic hematopoietic stem cells. Hematol Oncol Clin North Am 11:1149–1171
Dieterlen-Lievre F, Godin I, Pardanaud L (1997) Where do hematopoietic stem cells come from? Int Arch Allergy Immunol 112:3–8
Dieterlen-Lievre F, Pardanaud L, Bollerot K, Jaffredo T (2002) Hemangioblasts and hemopoietic stem cells during ontogeny. C R Biol 325:1013–1020
Dixon R, Rosendaal M (1981) Contrasts between the response of the mouse haemopoietic system to 5-fluorouracil and irradiation. Blood Cells 7:575–587
Doetsch F (2003) A niche for adult neural stem cells. Curr Opin Genet Dev 13:543–550
Donovan PJ, de Miguel MP (2003) Turning germ cells into stem cells. Curr Opin Genet Dev 13:463–471
Dor Y, Brown J, Martinez OI, Melton DA (2004) Adult pancreatic beta-cells are formed by self-duplication rather than stem-cell differentiation. Nature 429:41–46
Durr M, Muller AM (2003) Plasticity of somatic stem cells: dream or reality? Med Klin (Munich) 98(2):3–6
Eaves CJ (2003) Manipulating hematopoietic stem cell amplification with Wnt. Nat Immunol 4:511–512
Eaves C, Glimm H, Eisterer W, Audet J, Maguer-Satta V, Piret J (2001) Characterization of human hematopoietic cells with short-lived in vivo repopulating activity. Ann N Y Acad Sci 938:63–70; discussion-1
Eilertsen KJ, Floyd Z, Gimble JM (2008) The epigenetics of adult (somatic) stem cells. Crit Rev Eukaryot Gene Expr 18:189–206
Evarts RP, Nagy P, Marsden E, Thorgeirsson SS (1987) A precursor-product relationship exists between oval cells and hepatocytes in rat liver. Carcinogenesis 8:1737–1740
Evarts RP, Hu Z, Omori N, Omori M, Marsden ER, Thorgeirsson SS (1996) Precursor-product relationship between oval cells and hepatocytes: comparison between tritiated thymidine and bromodeoxyuridine as tracers. Carcinogenesis 17:2143–2151
Farber E (1956) Similarities in the sequence of early histological changes induced in the liver of the rat by ethionine, 2-acetylamino-fluorene, and 3'-methyl-4-dimethylaminoazobenzene. Cancer Res 16:142–148
Fleming WH, Alpern EJ, Uchida N, Ikuta K, Spangrude GJ, Weissman IL (1993) Functional heterogeneity is associated with the cell cycle status of murine hematopoietic stem cells. J Cell Biol 122:897–902

Fried W, Barone SJ, Anagnostou A (1978) Effect of protein deprivation on hematopoietic stem cells and on peripheral blood counts. J Lab Clin Med 92:303–310
Fujio K, Evarts RP, Hu Z, Marsden ER, Thorgeirsson SS (1994) Expression of stem cell factor and its receptor, c-kit, during liver regeneration from putative stem cells in adult rat. Lab Invest 70:511–516
Garry DJ, Masino AM, Meeson AP, Martin CM (2003) Stem cell biology and therapeutic applications. Curr Opin Nephrol Hypertens 12:447–454
Gering M, Patient R (2008) Notch in the niche. Cell Stem Cell 2:293–294
Giampaolo A, Pelosi E, Valtieri M, Montesoro E, Sterpetti P, Samoggia P, Camagna A, Mastroberardino G, Gabbianelli M, Testa U et al (1995) HOXB gene expression and function in differentiating purified hematopoietic progenitors. Stem Cells 13(Suppl 1):90–105
Giebel B, Bruns I (2008) Self-renewal versus differentiation in hematopoietic stem and progenitor cells: a focus on asymmetric cell divisions. Curr Stem Cell Res Ther 3:9–16
Gkretsi V, Mars WM, Bowen WC, Barua L, Yang Y, Guo L, St-Arnaud R, Dedhar S, Wu C, Michalopoulos GK (2007) Loss of integrin linked kinase from mouse hepatocytes in vitro and in vivo results in apoptosis and hepatitis. Hepatology 45:1025–1034
Glimm H, Eaves CJ (1999) Direct evidence for multiple self-renewal divisions of human in vivo repopulating hematopoietic cells in short-term culture. Blood 94:2161–2168
Glimm H, Oh IH, Eaves CJ (2000) Human hematopoietic stem cells stimulated to proliferate in vitro lose engraftment potential during their S/G(2)/M transit and do not reenter G(0). Blood 96:4185–4193
Glimm H, Eisterer W, Lee K, Cashman J, Holyoake TL, Nicolini F, Shultz LD, von Kalle C, Eaves CJ (2001) Previously undetected human hematopoietic cell populations with short-term repopulating activity selectively engraft NOD/SCID-beta2 microglobulin-null mice. J Clin Invest 107:199–206
Godin I, Dieterlen-Lievre F, Cumano A (1995) Emergence of multipotent hemopoietic cells in the yolk sac and paraaortic splanchnopleura in mouse embryos, beginning at 8.5 days postcoitus. Proc Natl Acad Sci USA 92:773–777
Godin I, Garcia-Porrero JA, Dieterlen-Lievre F, Cumano A (1999) Stem cell emergence and hemopoietic activity are incompatible in mouse intraembryonic sites. J Exp Med 190:43–52
Golding M, Sarraf CE, Lalani EN, Anilkumar TV, Edwards RJ, Nagy P, Thorgeirsson SS, Alison MR (1995) Oval cell differentiation into hepatocytes in the acetylaminofluorene-treated regenerating rat liver. Hepatology 22:1243–1253
Gonczy P, DiNardo S (1996) The germ line regulates somatic cyst cell proliferation and fate during Drosophila spermatogenesis. Development 122:2437–2447
Gros J, Manceau M, Thome V, Marcelle C (2005) A common somitic origin for embryonic muscle progenitors and satellite cells. Nature 435:954–958
Gussoni E, Soneoka Y, Strickland CD, Buzney EA, Khan MK, Flint AF, Kunkel LM, Mulligan RC (1999) Dystrophin expression in the mdx mouse restored by stem cell transplantation. Nature 401:390–394
Helgason CD, Sauvageau G, Lawrence HJ, Largman C, Humphries RK (1996) Overexpression of HOXB4 enhances the hematopoietic potential of embryonic stem cells differentiated in vitro. Blood 87:2740–2749
Higgins CC (1931) Solitary Cysts of the Kidney. Ann Surg 93:868–879
Horwitz EM (2003) Stem cell plasticity: the growing potential of cellular therapy. Arch Med Res 34:600–606
Houssaint E (1981) Differentiation of the mouse hepatic primordium. II. Extrinsic origin of the haemopoietic cell line. Cell Differ 10:243–252
http://stemcells.nih.gov/info/basics (2008) Stem cell basics. In Stem cell information. National Institutes of Health, Bethesda
Huttmann A, Li CL, Duhrsen U (2003) Bone marrow-derived stem cells and "plasticity". Ann Hematol 82:599–604
Ikuta K, Weissman IL (1992) Evidence that hematopoietic stem cells express mouse c-kit but do not depend on steel factor for their generation. Proc Natl Acad Sci USA 89:1502–1506
Ito M, Liu Y, Yang Z, Nguyen J, Liang F, Morris RJ, Cotsarelis G (2005) Stem cells in the hair follicle bulge contribute to wound repair but not to homeostasis of the epidermis. Nat Med 11:1351–1354

Iwasaki H, Somoza C, Shigematsu H, Duprez EA, Iwasaki-Arai J, Mizuno S, Arinobu Y, Geary K, Zhang P, Dayaram T, Fenyus ML, Elf S, Chan S, Kastner P, Huettner CS, Murray R, Tenen DG, Akashi K (2005) Distinctive and indispensable roles of PU.1 in maintenance of hematopoietic stem cells and their differentiation. Blood 106:1590–1600

Jackson KA, Majka SM, Wulf GG, Goodell MA (2002) Stem cells: A minireview. J Cell Biochem Suppl 38:1–6

Janes SM, Lowell S, Hutter C (2002) Epidermal stem cells. J Pathol 197:479–491

Jones DL, Wagers AJ (2008) No place like home: anatomy and function of the stem cell niche. Nat Rev Mol Cell Biol 9:11–21

Joseph NM, Morrison SJ (2005) Toward an understanding of the physiological function of Mammalian stem cells. Dev Cell 9:173–183

Kastenberg ZJ, Odorico JS (2008) Alternative sources of pluripotency: science, ethics, and stem cells. Transplant Rev (Orlando) 22:215–222

Kiel MJ, Morrison SJ (2006) Maintaining hematopoietic stem cells in the vascular niche. Immunity 25:862–864

Kiel MJ, Morrison SJ (2008) Uncertainty in the niches that maintain haematopoietic stem cells. Nat Rev Immunol 8:290–301

Kiel MJ, Yilmaz OH, Iwashita T, Yilmaz OH, Terhorst C, Morrison SJ (2005) SLAM family receptors distinguish hematopoietic stem and progenitor cells and reveal endothelial niches for stem cells. Cell 121:1109–1121

Kiel MJ, He S, Ashkenazi R, Gentry SN, Teta M, Kushner JA, Jackson TL, Morrison SJ (2007) Haematopoietic stem cells do not asymmetrically segregate chromosomes or retain BrdU. Nature 449:238–242

Kiger AA, Jones DL, Schulz C, Rogers MB, Fuller MT (2001) Stem cell self-renewal specified by JAK STAT activation in response to a support cell cue. Science 294:2542–2545

Kozar K, Ciemerych MA, Rebel VI, Shigematsu H, Zagozdzon A, Sicinska E, Geng Y, Yu Q, Bhattacharya S, Bronson RT, Akashi K, Sicinski P (2004) Mouse development and cell proliferation in the absence of D-cyclins. Cell 118:477–491

Kronenberg HM (2007) PTH regulates the hematopoietic stem cell niche in bone. Adv Exp Med Biol 602:57–60

Kuang S, Chargc SB, Seale P, Huh M, Rudnicki MA (2006) Distinct roles for Pax7 and Pax3 in adult regenerative myogenesis. J Cell Biol 172:103–113

Kumaravelu P, Hook L, Morrison AM, Ure J, Zhao S, Zuyev S, Ansell J, Medvinsky A (2002) Quantitative developmental anatomy of definitive haematopoietic stem cells/long-term repopulating units (HSC/RUs): role of the aorta-gonad-mesonephros (AGM) region and the yolk sac in colonisation of the mouse embryonic liver. Development 129:4891–4899

Lawrence HJ, Sauvageau G, Humphries RK, Largman C (1996) The role of HOX homeobox genes in normal and leukemic hematopoiesis. Stem Cells 14:281–291

Leduc EH, Wilson JW (1958) Injury to liver cells in carbon tetrachloride poisoning; histochemical changes induced by carbon tetrachloride in mouse liver protected by sulfaguanidine. AMA Arch Pathol 65:147–157

Lemoli RM, Bertolini F, Cancedda R, De Luca M, Del Santo A, Ferrari G, Ferrari S, Martino G, Mavilio F, Tura S (2005) Stem cell plasticity: time for a reappraisal? Haematologica 90:360–381

Liu SV (2008) iPS cells: A more critical review. Stem Cells Dev 17:391–397

Liu CP, Auerbach R (1991) In vitro development of murine T cells from prethymic and preliver embryonic yolk sac hematopoietic stem cells. Development 113:1315–1323

Liu A, Zhu P, Li X, Zhang L, Chi F, Wu Z, Song Y, Zhang Y (1994) Ultrastructural study of the hemopoietic microenvironment in human fetal spleen. Chin Med Sci J 9:157–161

Lux CT, Yoshimoto M, McGrath K, Conway SJ, Palis J, Yoder MC (2008) All primitive and definitive hematopoietic progenitor cells emerging before E10 in the mouse embryo are products of the yolk sac. Blood 111:3435–3438

Malumbres M, Sotillo R, Santamaria D, Galan J, Cerezo A, Ortega S, Dubus P, Barbacid M (2004) Mammalian cells cycle without the D-type cyclin-dependent kinases Cdk4 and Cdk6. Cell 118:493–504

Martinez-Agosto JA, Mikkola HK, Hartenstein V, Banerjee U (2007) The hematopoietic stem cell and its niche: a comparative view. Genes Dev 21:3044–3060

Masson NM, Currie IS, Terrace JD, Garden OJ, Parks RW, Ross JA (2006) Hepatic progenitor cells in human fetal liver express the oval cell marker Thy-1. Am J Physiol Gastrointest Liver Physiol 291:G45–G54

Matsusaka S, Toyosaka A, Nakasho K, Tsujimura T, Sugihara A, Takanashi T, Uematsu K, Terada N, Okamoto E (2000) The role of oval cells in rat hepatocyte transplantation. Transplantation 70:441–446

Mayack SR, Wagers AJ (2008) Osteolineage niche cells initiate hematopoietic stem cell mobilization. Blood 112:519–531

McCartney BM, Peifer M (2003) Stem cells in the news: CNN and APC make headlines. Dev Cell 5:532–534

Medvinsky A, Dzierzak E (1996) Definitive hematopoiesis is autonomously initiated by the AGM region. Cell 86:897–906

Merida I, Avila-Flores A, Merino E (2008) Diacylglycerol kinases: at the hub of cell signalling. Biochem J 409:1–18

Micsenyi A, Tan X, Sneddon T, Luo JH, Michalopoulos GK, Monga SP (2004) Beta-catenin is temporally regulated during normal liver development. Gastroenterology 126:1134–1146

Mikkola HK, Gekas C, Orkin SH, Dieterlen-Lievre F (2005) Placenta as a site for hematopoietic stem cell development. Exp Hematol 33:1048–1054

Miller JB, Emerson CP, Jr (2003) Does the road to muscle rejuvenation go through Notch? Sci Aging Knowledge Environ 2003:pe34

Minasi MG, Riminucci M, De Angelis L, Borello U, Berarducci B, Innocenzi A, Caprioli A, Sirabella D, Baiocchi M, De Maria R, Boratto R, Jaffredo T, Broccoli V, Bianco P, Cossu G (2002) The meso-angioblast: a multipotent, self-renewing cell that originates from the dorsal aorta and differentiates into most mesodermal tissues. Development 129:2773–2783

Monga SP, Pediaditakis P, Mule K, Stolz DB, Michalopoulos GK (2001) Changes in WNT/beta-catenin pathway during regulated growth in rat liver regeneration. Hepatology 33:1098–1109

Moore KA, Lemischka IR (2004) "Tie-ing" down the hematopoietic niche. Cell 118:139–140

Morgan JE, Hoffman EP, Partridge TA (1990) Normal myogenic cells from newborn mice restore normal histology to degenerating muscles of the mdx mouse. J Cell Biol 111:2437–2449

Morrison SJ, Wright DE, Weissman IL (1997) Cyclophosphamide/granulocyte colony-stimulating factor induces hematopoietic stem cells to proliferate prior to mobilization. Proc Natl Acad Sci USA 94:1908–1913

Nagy P, Bisgaard HC, Thorgeirsson SS (1994) Expression of hepatic transcription factors during liver development and oval cell differentiation. J Cell Biol 126:223–233

Nagy P, Kiss A, Schnur J, Thorgeirsson SS (1998) Dexamethasone inhibits the proliferation of hepatocytes and oval cells but not bile duct cells in rat liver. Hepatology 28:423–429

Naveiras O, Daley GQ (2006) Stem cells and their niche: a matter of fate. Cell Mol Life Sci 63:760–766

Neben S, Marcus K, Mauch P (1993) Mobilization of hematopoietic stem and progenitor cell subpopulations from the marrow to the blood of mice following cyclophosphamide and/or granulocyte colony-stimulating factor. Blood 81:1960–1967

Niemann C, Watt FM (2002) Designer skin: lineage commitment in postnatal epidermis. Trends Cell Biol 12:185–192

O'Shaughnessy RF, Christiano AM (2001) Stem cells in the epidermis. Skin Pharmacol Appl Skin Physiol 14:350–357

Olguin HC, Olwin BB (2004) Pax-7 up-regulation inhibits myogenesis and cell cycle progression in satellite cells: a potential mechanism for self-renewal. Dev Biol 275:375–388

Omori N, Evarts RP, Omori M, Hu Z, Marsden ER, Thorgeirsson SS (1996) Expression of leukemia inhibitory factor and its receptor during liver regeneration in the adult rat. Lab Invest 75:15–24

Orkin SH, Zon LI (2008) Hematopoiesis: an evolving paradigm for stem cell biology. Cell 132:631–644

Paku S, Schnur J, Nagy P, Thorgeirsson SS (2001) Origin and structural evolution of the early proliferating oval cells in rat liver. Am J Pathol 158:1313–1323
Palis J, Robertson S, Kennedy M, Wall C, Keller G (1999) Development of erythroid and myeloid progenitors in the yolk sac and embryo proper of the mouse. Development 126:5073–5084
Park IK, Qian D, Kiel M, Becker MW, Pihalja M, Weissman IL, Morrison SJ, Clarke MF (2003) Bmi-1 is required for maintenance of adult self-renewing haematopoietic stem cells. Nature 423:302–305
Passegue E, Wagers AJ, Giuriato S, Anderson WC, Weissman IL (2005) Global analysis of proliferation and cell cycle gene expression in the regulation of hematopoietic stem and progenitor cell fates. J Exp Med 202:1599–1611
Petersen BE, Zajac VF, Michalopoulos GK (1998a) Hepatic oval cell activation in response to injury following chemically induced periportal or pericentral damage in rats. Hepatology 27:1030–1038
Petersen BE, Goff JP, Greenberger JS, Michalopoulos GK (1998b) Hepatic oval cells express the hematopoietic stem cell marker Thy-1 in the rat. Hepatology 27:433–445
Phinney DG, Prockop DJ (2007) Concise review: mesenchymal stem/multipotent stromal cells: the state of transdifferentiation and modes of tissue repair–current views. Stem Cells 25:2896–2902
Porter RL, Calvi LM (2008) Communications between bone cells and hematopoietic stem cells. Arch Biochem Biophys 473:193–200
Rabes HM (1977) Kinetics of hepatocellular proliferation as a function of the microvascular structure and functional state of the liver. Ciba Found Symp 55:31–53
Rabes HM, Wirsching R, Tuczek HV, Iseler G (1976) Analysis of cell cycle compartments of hepatocytes after partial hepatecomy. Cell Tissue Kinet 9:517–532
Ralston A, Rossant J (2005) Genetic regulation of stem cell origins in the mouse embryo. Clin Genet 68:106–112
Rattis FM, Voermans C, Reya T (2004) Wnt signaling in the stem cell niche. Curr Opin Hematol 11:88–94
Rawlins EL, Hogan BL (2006) Epithelial stem cells of the lung: privileged few or opportunities for many? Development 133:2455–2465
Relaix F, Rocancourt D, Mansouri A, Buckingham M (2005) A Pax3/Pax7-dependent population of skeletal muscle progenitor cells. Nature 435:948–953
Reynolds AJ, Jahoda CA (1991) Hair follicle stem cells? A distinct germinative epidermal cell population is activated in vitro by the presence of hair dermal papilla cells. J Cell Sci 99(Pt 2):373–385
Rheinwald JG, Green H (1975a) Serial cultivation of strains of human epidermal keratinocytes: the formation of keratinizing colonies from single cells. Cell 6:331–343
Rheinwald JG, Green H (1975b) Formation of a keratinizing epithelium in culture by a cloned cell line derived from a teratoma. Cell 6:317–330
Robertson S, Kennedy M, Keller G (1999) Hematopoietic commitment during embryogenesis. Ann N Y Acad Sci 872:9–15; discussion-6
Roobrouck VD, Ulloa-Montoya F, Verfaillie CM (2008) Self-renewal and differentiation capacity of young and aged stem cells. Exp Cell Res 314:1937–1944
Rosendaal M, Dixon R, Panayi M (1981) Haemopoietic stem cells: possibility of toxic effects of 5-fluorouracil on spleen colony formation. Blood Cells 7:561–574
Rossi DJ, Bryder D, Weissman IL (2007a) Hematopoietic stem cell aging: mechanism and consequence. Exp Gerontol 42:385–390
Rossi DJ, Seita J, Czechowicz A, Bhattacharya D, Bryder D, Weissman IL (2007b) Hematopoietic stem cell quiescence attenuates DNA damage response and permits DNA damage accumulation during aging. Cell Cycle 6:2371–2376
Rovo A, Gratwohl A (2008) Plasticity after allogeneic hematopoietic stem cell transplantation. Biol Chem 389:825–836
Rowden G, Lewis MG, Sheikh KM, Summerlin WT (1975) Long-term organ culture of human skin: an ultrastructural and immunochemical study. J Pathol 117:139–149

Saini A, Stewart CE (2006) Adult stem cells: the therapeutic potential of skeletal muscle. Curr Stem Cell Res Ther 1:157–171
Sampaolesi M, Torrente Y, Innocenzi A, Tonlorenzi R, D'Antona G, Pellegrino MA, Barresi R, Bresolin N, De Angelis MG, Campbell KP, Bottinelli R, Cossu G (2003) Cell therapy of alpha-sarcoglycan null dystrophic mice through intra-arterial delivery of mesoangioblasts. Science 301:487–492
Sampaolesi M, Blot S, D'Antona G, Granger N, Tonlorenzi R, Innocenzi A, Mognol P, Thibaud JL, Galvez BG, Barthelemy I, Perani L, Mantero S, Guttinger M, Pansarasa O, Rinaldi C, Cusella De Angelis MG, Torrente Y, Bordignon C, Bottinelli R, Cossu G (2006) Mesoangioblast stem cells ameliorate muscle function in dystrophic dogs. Nature 444:574–579
Sarraf C, Lalani EN, Golding M, Anilkumar TV, Poulsom R, Alison M (1994) Cell behavior in the acetylaminofluorene-treated regenerating rat liver. Light and electron microscopic observations. Am J Pathol 145:1114–1126
Sauvageau G, Thorsteinsdottir U, Eaves CJ, Lawrence HJ, Largman C, Lansdorp PM, Humphries RK (1995) Overexpression of HOXB4 in hematopoietic cells causes the selective expansion of more primitive populations in vitro and in vivo. Genes Dev 9:1753–1765
Schofield R (1978) The relationship between the spleen colony-forming cell and the haemopoietic stem cell. Blood Cells 4:7–25
Seale P, Sabourin LA, Girgis-Gabardo A, Mansouri A, Gruss P, Rudnicki MA (2000) Pax7 is required for the specification of myogenic satellite cells. Cell 102:777–786
Seale P, Asakura A, Rudnicki MA (2001) The potential of muscle stem cells. Dev Cell 1:333–342
Sell S (1994) Liver stem cells. Mod Pathol 7:105–112
Sharpless NE, Bardeesy N, Lee KH, Carrasco D, Castrillon DH, Aguirre AJ, Wu EA, Horner JW, DePinho RA (2001) Loss of p16Ink4a with retention of p19Arf predisposes mice to tumorigenesis. Nature 413:86–91
Sirica AE, Mathis GA, Sano N, Elmore LW (1990) Isolation, culture, and transplantation of intrahepatic biliary epithelial cells and oval cells. Pathobiology 58:44–64
Staal FJ, Clevers HC (2005) WNT signalling and haematopoiesis: a WNT-WNT situation. Nat Rev Immunol 5:21–30
Steinman RA (2002) Cell cycle regulators and hematopoiesis. Oncogene 21:3403–3413
Sylvester KG, Longaker MT (2004) Stem cells: review and update. Arch Surg 139:93–99
Takada S, Kelkar A, Theurkauf WE (2003) Drosophila checkpoint kinase 2 couples centrosome function and spindle assembly to genomic integrity. Cell 113:87–99
Thorsteinsdottir U, Sauvageau G, Hough MR, Dragowska W, Lansdorp PM, Lawrence HJ, Largman C, Humphries RK (1997) Overexpression of HOXA10 in murine hematopoietic cells perturbs both myeloid and lymphoid differentiation and leads to acute myeloid leukemia. Mol Cell Biol 17:495–505
Tiedemann H, Asashima M, Grunz H, Knochel W (2001) Pluripotent cells (stem cells) and their determination and differentiation in early vertebrate embryogenesis. Dev Growth Differ 43:469–502
Tulina N, Matunis E (2001) Control of stem cell self-renewal in Drosophila spermatogenesis by JAK-STAT signaling. Science 294:2546–2549
Vats A, Bielby RC, Tolley NS, Nerem R, Polak JM (2005) Stem cells. Lancet 366:592–602
Venezia TA, Merchant AA, Ramos CA, Whitehouse NL, Young AS, Shaw CA, Goodell MA (2004) Molecular signatures of proliferation and quiescence in hematopoietic stem cells. PLoS Biol 2:e301
Voermans C, van Hennik PB, van der Schoot CE (2001) Homing of human hematopoietic stem and progenitor cells: new insights, new challenges? J Hematother Stem Cell Res 10:725–738
Wagers AJ (2008) Wnt not, waste not. Cell Stem Cell 2:6–7
Wagers AJ, Sherwood RI, Christensen JL, Weissman IL (2002) Little evidence for developmental plasticity of adult hematopoietic stem cells. Science 297:2256–2259
Walkley CR, Shea JM, Sims NA, Purton LE, Orkin SH (2007) Rb regulates interactions between hematopoietic stem cells and their bone marrow microenvironment. Cell 129:1081–1095
Wallenfang MR, Matunis E (2003) Developmental biology. Orienting stem cells. Science 301:1490–1491

Waller EK, Olweus J, Lund-Johansen F, Huang S, Nguyen M, Guo GR, Terstappen L (1995a) The "common stem cell" hypothesis reevaluated: human fetal bone marrow contains separate populations of hematopoietic and stromal progenitors. Blood 85:2422–2435
Waller EK, Huang S, Terstappen L (1995b) Changes in the growth properties of CD34+, CD38– bone marrow progenitors during human fetal development. Blood 86:710–718
Warren LA, Rossi DJ (2009) Stem cells and aging in the hematopoietic system. Mech Ageing Dev 130:46–53
Watt FM (2002) The stem cell compartment in human interfollicular epidermis. J Dermatol Sci 28:173–180
Watt SM, Forde SP (2008) The central role of the chemokine receptor, CXCR4, in haemopoietic stem cell transplantation: will CXCR4 antagonists contribute to the treatment of blood disorders? Vox Sang 94:18–32
Weissman IL (2000) Stem cells: units of development, units of regeneration, and units in evolution. Cell 100:157–168
Weissman IL, Anderson DJ, Gage F (2001) Stem and progenitor cells: origins, phenotypes, lineage commitments, and transdifferentiations. Annu Rev Cell Dev Biol 17:387–403
Wilson A, Trumpp A (2006) Bone-marrow haematopoietic-stem-cell niches. Nat Rev Immunol 6:93–106
Wright DE, Cheshier SH, Wagers AJ, Randall TD, Christensen JL, Weissman IL (2001) Cyclophosphamide/granulocyte colony-stimulating factor causes selective mobilization of bone marrow hematopoietic stem cells into the blood after M phase of the cell cycle. Blood 97:2278–2285
Yamashita YM, Fuller MT (2005) Asymmetric stem cell division and function of the niche in the Drosophila male germ line. Int J Hematol 82:377–380
Yasui O, Miura N, Terada K, Kawarada Y, Koyama K, Sugiyama T (1997) Isolation of oval cells from Long-Evans Cinnamon rats and their transformation into hepatocytes in vivo in the rat liver. Hepatology 25:329–334
Yilmaz OH, Kiel MJ, Morrison SJ (2006) SLAM family markers are conserved among hematopoietic stem cells from old and reconstituted mice and markedly increase their purity. Blood 107:924–930
Yin T, Li L (2006) The stem cell niches in bone. J Clin Invest 116:1195–1201
Yoder MC, Hiatt K, Dutt P, Mukherjee P, Bodine DM, Orlic D (1997a) Characterization of definitive lymphohematopoietic stem cells in the day 9 murine yolk sac. Immunity 7:335–344
Yoder MC, Hiatt K, Mukherjee P (1997b) In vivo repopulating hematopoietic stem cells are present in the murine yolk sac at day 9.0 postcoitus. Proc Natl Acad Sci USA 94:6776–6780
Yu H, Yuan Y, Shen H, Cheng T (2006) Hematopoietic stem cell exhaustion impacted by p18 INK4C and p21 Cip1/Waf1 in opposite manners. Blood 107:1200–1206
Yuan Y, Shen H, Franklin DS, Scadden DT, Cheng T (2004) In vivo self-renewing divisions of haematopoietic stem cells are increased in the absence of the early G1-phase inhibitor, p18INK4C. Nat Cell Biol 6:436–442
Zammit PS, Partridge TA, Yablonka-Reuveni Z (2006a) The skeletal muscle satellite cell: the stem cell that came in from the cold. J Histochem Cytochem 54:1177–1191
Zammit PS, Relaix F, Nagata Y, Ruiz AP, Collins CA, Partridge TA, Beauchamp JR (2006b) Pax7 and myogenic progression in skeletal muscle satellite cells. J Cell Sci 119:1824–1832
Zhang P, Liegeois NJ, Wong C, Finegold M, Hou H, Thompson JC, Silverman A, Harper JW, DePinho RA, Elledge SJ (1997) Altered cell differentiation and proliferation in mice lacking p57KIP2 indicates a role in Beckwith-Wiedemann syndrome. Nature 387:151–158
Zindy F, van Deursen J, Grosveld G, Sherr CJ, Roussel MF (2000) INK4d-deficient mice are fertile despite testicular atrophy. Mol Cell Biol 20:372–378

Developmental Biology of Mammalian T-Cell Progenitors: From Early Lymphoid Progenitors to Thymus-Colonizing Cells

Sophie Ezine, Laetitia Gautreau, Aude Parcelier, and Bruno Canque

Abstract This review gives a general overview of the recent advances in the field of fetal hematopoiesis and T-cell development. Although it is now well recognized that early lymphoid progenitors first emerge in the fetal liver (FL) where active B lymphopoiesis take places, the identity of early T-cell precursors, as well as the mechanisms of thymus colonization, has long been controversial. Here we discuss the experimental evidence supporting the concept of prothymocyte in both human and mouse species, as well as its major implications regarding lineage relationships among the immune system and general hematopoietic organization.

Introduction

Blood is a complex liquid tissue that develops early during embryonic development and undergoes permanent renewal throughout life. It is composed of three major cell lineages: erythrocytes, which ensure oxygen transport and tissue delivery, platelets, which prevent bleeding as well as inappropriate activation of blood coagulation, and leukocytes, which are in charge of the immune defense. At first sight, it may seem paradoxical that such highly divergent functions (i.e., oxygen delivery, blood coagulation, and immune defense) are condensed in a unique system. But all blood cells share the need to circulate to fulfill their function at the organism level. They originate from a minute population of rare hematopoietic stem cell (HSCs) characterized by three major functional properties: multipotency, the capacity to generate all blood lineages and stably reconstitute multilineage

S. Ezine and L. Gautreau
Inserm U591-Institut Necker, Université Paris Descartes, 156 rue de Vaugirard, 75730, Paris Cedex 15, France
e-mail: sophie.ezine@inserm.fr; Laetitia.Gautreau@univ-

A. Parcelier and B. Canque
UMR CNRS 7151-Institut Universitaire d'Hématologie (IUH), Centre Hayem, 1 avenue Claude Vellefaux, 75475, Paris cedex 10, France
e-mail: aude_parcelier@yahoo.fr; bruno.canque@gmail.com

M. Kondo (ed.), *Hematopoietic Stem Cell Biology,* Stem Cell Biology and Regenerative Medicine, DOI 10.1007/978-1-60327-347-3_4,

hematopoiesis upon adoptive transfer; quiescence, the capacity to remain in a dormant state during long periods of time; and self-renewal, the capacity to undergo symmetric mitotic divisions preserving multipotency. The immune system itself displays an unusually high degree of complexity, comprising two major effector arms dedicated to either innate or adaptive immune responses, as well as a constellation of mature cell effectors. Macrophages, dendritic cells, granulocytes, and mast cells are endowed with phagocytic activity, antigen processing, and presentation, as well as with pro-inflammatory and/or regulatory functions. Natural killer (NK) cells ensure non–major histocompatibility complex–restricted, cell-mediated cytotoxicity and cytokine-producing effector functions. In contrast, the capacity to support antigen-specific immune response is restricted to prototypic T lymphocytes (TLs) and B lymphocytes (BLs), which are among the most highly evolved cell types of the organism. For example, they share with neurons the expression of a broad repertoire of clonotypic receptors generated by combinatorial diversity, as well as a unique capacity to retain stable traces of previous immunologic events through the production of long-lived memory B and T cells. Despite more than 50 years of research in the field, the organization of the immune system (i.e., the developmental and phyletic relationships between diverse leukocyte populations) is still a matter of controversy (Katsura 2002; Laiosa et al. 2006). Until recently, it was widely admitted that HSCs first differentiate into two unique founder populations dedicated to either granulomonocytic and erythromegacaryocytic lineages or to the lymphoid lineage. The first population was subsequently referred to as common myeloid progenitors (CMPs) and the second, from which only TLs, BLs, and NK cells differentiate, as common lymphoid precursors (CLPs). This view was supported also by most in vitro differentiation assays developed throughout the 1970s that only allowed the generation of erythroid and granulomonocytic colonies that did not contain lymphoid cells (Dexter and Testa 1976). The field of in vitro lymphoid differentiation began to change in the early 1980s, when it was first shown by Whitlock and Witte that mouse BLs can be efficiently generated by culturing mouse bone marrow (BM) progenitors onto stromal feeder layers (Collins and Dorshkind 1987; Whitlock et al. 1987; Whitlock and Witte 1982). At the same time, Kingston et al. provided evidence that alymphoid embryonic thymus lobes support complete T-cell development, providing thus an ex vivo model to study T-cell development (Kingston et al. 1985). Both the fetal thymic organ cultures (FTOCs) and B-cell differentiation assays were then successfully subsequently adapted to humans (Berardi et al. 1997; Galy et al. 1995; Plum et al. 1994). This period coincides with the characterization and large-scale production of major hematopoietic growth factors and cytokines. The latest advance in the field of in vitro TLs differentiation is the demonstration by Schmitt Nakano et al. that, after stable transduction with Notch-ligand Delta-Like 1, OP9 stromal cells, established from macrophage colony-stimulating factor (M-CSF) deficient mice (Nakano et al. 1994), support the differentiation hematopoietic progenitors into mature functional TLs (La Motte-Mohs et al. 2004; Schmitt and Zuniga-Pflucker 2002). The rapid generalization of in vitro lymphoid differentiation assays, which permits investigators to characterize extremely rare HSC subsets and probe their differentiation potential at the single cell level, has profoundly modified the vision of hematopoietic development and lineage relationships, since there is now evidence that the classical structure of the hematopoietic system, based on the early dichotomy between so-called myeloid and lymphoid lineages, needs to be reconsidered.

A cellular basis for the CMP versus CLP paradigm was first provided in the late 1990s with the identification in postnatal BM of discrete populations of hematopoietic progenitors with apparently exclusive lymphoid- or myeloid-restricted potentials (Galy et al. 1995; Kondo et al. 1997; Manz et al. 2002). But this conception of hematopoietic organization suffers what might be called an "immunological sin" because of the split of the immune system into two distinct lymphoid and myeloid lineages, making granulomonocytes more closely related to erythromegakaryocytes than to TLs and BLs with which they continuously interact to initiate and coordinate the immune response. In more recent years, the picture proved to be more complex than anticipated. Subsequent works on early mouse lymphoid progenitors led to the finding that mouse $Lin^-Sca\text{-}1^{lo}Kit^+IL\text{-}7R^+$ CLPs and their human $CD34^+CD10^+Lin^-$ counterparts actually display a predominant BL differentiation potential, suggesting that they correspond to early B precursors rather than to prototypic CLPs (Allman et al. 2003; Haddad et al. 2004). Studies on early fetal lymphoid progenitors failed to identify fetal liver (FL) or BM populations fitting the criterion for CLPs. Instead, they could show that early lymphoid progenitors (ELPs) always retain some capacity to differentiate into macrophages (Kawamoto et al. 1997; Lacaud et al. 1998; Mebius et al. 2001; Yokota et al. 2003). In line with these observations, it is now well recognized that early thymic progenitors (ETPs) still preserve the capacity to generate macrophages in vivo under homeostatic conditions (Bell and Bhandoola 2008; Wada et al. 2008). The recent subcategorization of postnatal mouse BM HSCs, based on Flt3 expression, supports an alternative model of phyletic and developmental relationships among the hematopoietic system. Adolfsson et al. showed that, based on Flt3 expression, the $Lin^-Sca\text{-}1^+Kit^+$ (LSK) compartment segregates into two distinct populations endowed with strikingly different potentials (Adolfsson et al. 2001, 2005). Conversely, the $Flt3^-$ LSK fraction was found to be enriched in multipotent long-term repopulating stem cells, and $Flt3^+$ LSK progenitors yielded high numbers of lymphoid and granulomonocytic cells but essentially lacked erythromegakaryocytic potential. These differences were further documented at the molecular level, since expression of GATA1, GATA2 transcription factors, as well as erythropoietin (EpoR) and thrombopoietin (mpl) receptors was restricted to $Flt3^-$ LSKs, whereas their $Flt3^+$ counterparts harbor RAG1, RAG2, terminal deoxynucleotidyl transferase (Tdt), as well as immunoglobulin-H (IgH) sterile transcripts (Mansson et al. 2007). Yoshida et al. obtained overall similar results in a transgenic mouse model expression the green fluorescent protein (GFP) under control of an Ikaros promoter-enhancer (Yoshida et al. 2006). Luc et al. subsequently proposed a revised map of hematopoietic development based on the initial separation between lymphoid-primed multipotent progenitors (LMPPs) as dedicated to the generation of immune cells, and myeloid progenitors supporting erythroid and megakaryocytic development (Luc et al. 2007). This alternative model of hematopoietic organization is further supported by the phenotypes observed in either conventional or conditional knockout mice with targeted invalidation of key transcription factors. Indeed, conversely to PU.1$^{-/-}$ mice that selectively lack lymphoid and granulomonocytic cells, GATA1$^{-/-}$ mice display a severe block in erythroid and megakaryocytic development (Pevny et al. 1991; Scott et al. 1994, 1997; Shivdasani et al. 1997). Hematopoietic development should thus no longer be considered as a series of branching points that determine successive binary choices, but as a dynamic equilibrium between a few master regulators controlling the opening or closing of diverse lineage-affiliated loci. Most importantly, most current evidence suggests that developmental hematopoiesis actually obeys Darwinian law, since it proceeds gradually through selection and progressive stabilization of lineage-specific gene expression programs. Inasmuch as the developmental biology of HSCs has been discussed elsewhere (Cumano and Godin 2007; Dzierzak and Speck 2008; Orkin and Zon 2008; Zon 2008), this review focuses on three major aspects of developmental lymphopoiesis: (1) the emergence of lymphoid differentiation potential during early embryonic and fetal development; (2) the dynamics of lymphoid progenitors in FL and BM; and (3) the developmental regulation of thymus-colonizing cells.

Emergence of Lymphoid Differentiation Potential During Early Embryonic Development

In vertebrate embryos, the first wave of extra-embryonic hematopoiesis begins immediately after gastrulation in the yolk sac (YS) and is referred to as primitive hematopoiesis. It is characterized by a limited lineage diversification and results in the production of primitive erythrocytes and platelets-forming megakaryocytes (Tober et al. 2007). Hence, primitive YS hematopoietic precursor cells (HPCs) are considered bilineage precursors since they completely lack lymphoid potential. Although this issue has long remained controversial, most current evidence indicates that primitive erythrocytes and endothelial cells share a common mesodermal blood-forming precursor, the so-called hemangioblast (Sabin 1920). In mice, this population emerges in the posterior region of primitive streak by E6.7–8 and rapidly migrates toward the YS to form hemangioblastic chords that evolve into blood islands; prototypic blood islands are made of nucleated erythroblasts surrounded by a crown of endothelial cells (Choi et al. 1998; Haar and Ackerman 1971; Huber et al. 2004; Moore and Metcalf 1970). Whether primitive erythroblasts and endothelial cells derive from progenitors restricted to each differentiation pathway or differentiate from intermediate hemogenic endothelial cells is still unclear. Besides hemangioblasts, there is nonetheless substantial evidence that unilineage angioblastic precursors lacking hematopoietic potential are also present in the early YS, which indicates that at least a fraction of endothelial cells actually differentiate from already lineage-restricted precursors (Furuta et al. 2006). Primitive YS progenitors differ from prototypic HSCs in two major aspects: first, they are unable to engraft into an irradiated host, which suggests either a lack of self-renewal or an inability to seed the BM niches due to absence of specific homing receptors or defective interaction with the microenvironment; second, as mentioned above, they are totally devoid of lymphoid differentiation potential (Dzierzak and Speck 2008). Early YS nonetheless contains a population of immune cells referred to as primitive macrophages whose origin remains unclear; this population is often considered an entirely separate hematopoietic lineage (Herbomel et al. 1999).

The second wave of hematopoiesis, sometimes referred to as "pro-definitive" or "mesodefinitive" hematopoiesis (Dzierzak and Speck 2008), takes place by day 8.25 of development. At that time, clonogenic HPCs display high proliferative capacity and broader differentiation potential. Most importantly, they have the ability to generate all immune cell types: lymphocytes, granulocytes, macrophages, as well as mast cells (Palis et al. 1999). This population does not differentiate from hemangioblasts; instead it derives from transiently hemogenic endothelial cells. Whether YS progenitors participate in definitive hematopoiesis remains a controversial issue. Because their emergence by E8.5 coincides with the onset of intraembryonic hematopoiesis in the aortogonadomesonephros (AGM) region, as well as with heartbeat initiation and establishment of blood circulation that rapidly distributes HSCs throughout the conceptus, this question cannot be easily addressed (Cumano et al. 2001; de Bruijn et al. 2000, 2002). For the past 10 years, it was widely admitted that only intraembryonic, AGM-derived HSCs had the capacity to support multilineage definitive hematopoiesis. But two recent reports based on elegant genetic mouse models suggest that this might not be the case since definitive HSCs could also emerge in the YS. Using SLC8A1/NCX1$^{-/-}$ mice that die by E11 due to the lack of cardiac beat, Lux et al. provided convincing evidence that definitive HPCs undergo approximately tenfold expansion between 8.25 and 9.5 days postcoitum and proposed that all definitive HSCs emerging before E10 are of YS origin (Lux et al. 2008). But because neither lymphoid potential nor the engraftment capacity of the second wave of YS progenitors was assessed, this study does not allow us to decipher the presence of prototypic definitive HSCs at this level. Samokhvalov et al. also provided some evidence that the YS might actually represent a reliable source of definitive HSCs (Samokhvalov et al. 2007). Based on in vivo genetic pulse labeling of Runx1-expressing YS cells by E7.5 (i.e., 1 day before the onset of intraembryonic AGM hematopoiesis), they showed that their progeny can still be detected in the blood of 9- to 12-month-old mice. Although these findings strongly suggest that second wave YS hematopoiesis is supported by prototypic definitive HSCs, one cannot formally exclude a leakage of the pulse-labeling system based on a tamoxifen-induced Cre/loxP recombination. Comparative analysis of the engraftment capacity of YS- and AGM-derived HSCs also yielded interesting results. Whereas both populations stably repopulate irradiated newborn mice (Yoder et al. 1997), it appears that only AGM-derived cells can accomplish this in adult mice (Cumano et al. 2001; Muller et al. 1994). Thus, even if the YS still generates definitive HSCs, they probably differ from those derived from the AGM. It is believed that these two populations correspond to successive hematopoietic developmental stages (Dzierzak and Speck 2008). Most interestingly, it is suggested that AGM-specific microenvironmental cues, such as Notch signaling and HoxB4 expression, may play a major role in the differentiation and repopulation potentials of definitive HSCs (Kumano et al. 2003; Kyba et al. 2002). The ventral aspect of the aorta is only transiently hemogenic with intra-aortic hematopoietic clusters being detected between days 10.5 and 12.5 of development (Cumano and Godin 2007). The transient nature of this process, due to the rapid exhaustion of vascular endothelial cell hemogenic activity, remains poorly understood. Pouget et al. nonetheless provided strong data that, during early avian embryonic development, hemogenic endothelial cells from the splanchnic mesoderm are progressively replaced by a population of endothelial cells of somatic origin, lacking hemogenic activity (Pouget et al. 2006).

The finding that embryonic hematopoiesis primarily consists essentially in the production of definitive HSCs and does not resume the classical opposition between extraembryonic primitive YS and intraembryonic definitive AGM hematopoiesis has profoundly changed the conception of the developmental biology of the hematopoietic system. As early as 1999, de Bruijn et al. showed that, by E10–12, vitelline and umbilical arteries produce significant numbers of definitive HSCs (de Bruijn et al. 2000), and it is now recognized that midgestation mouse placenta also harbors large numbers of HSCs, a substantial fraction of which actually emerge directly from the wall of large chorioallantoic vessels (Gekas et al. 2005; Ottersbach and Dzierzak 2005; Rhodes et al. 2008). Altogether, these findings argue for a multisite generation of definitive HSCs from transiently hemogenic endothelium of the large vessels. Definitive mouse HSCs possess lymphoid differentiation, pot is considered a marker of multipotency, but whether they display substantial preliver lymphopoietic activity remains controversial. The detection of lymphoid-specified T- and B-cell precursors in the AGM suggests that this might be the case (Ohmura et al. 1999). But because T-cell differentiation potential was assessed in FTOCs, and because of the absence of molecular characterization, one cannot exclude that T cells actually differentiated stochastically from multipotent HSCs, rather than from already lineage-restricted precursors. This point should probably be reinvestigated using more powerful two-dimensional OP9-DL1 cultures and molecular tools (Schmitt and Zuniga-Pflucker 2002). Conversely, most current evidence indicates that disclosure of a preliver lymphoid potential in AGM-derived progenitors is by no means indicative of active ongoing lymphopoietic activity. Lineage tracing studies using RAG1/GFP knockin mice add some support to this view, since only basal transcription levels of the RAG1 locus are detected in AGM progenitors (Yokota et al. 2006).

Only limited data on embryonic human hematopoiesis are available, but most current evidence indicates that the mouse schema cannot be mechanically transposed to primates. Extraembryonic YS hematopoiesis starts in blood islands that form by day 18.5 of development (i.e., approximately 4 days before heart beat initiation), and it remains active until development weeks 4–5 (Bloom and Bartelmez 1940; Migliaccio et al. 1986). Once again, the key point is that decline in YS hematopoietic activity coincides with the onset of FL erythropoiesis. Pioneering studies have shown that the human YS contains substantial numbers of CFU-GEMM (colony-forming unit granulocyte, erythrocyte, macrophage, megakaryocyte), BFU-E (burst-forming unit erythrocyte), CFU-E (colony-forming unit erythrocyte), as well as CFU-GM (colony-forming unit granulocyte macrophages) (Huyhn et al. 1995; Migliaccio et al. 1986). Early YS progenitors nonetheless constantly lack T- and B-cell potential, which indicates that, as for their mouse counterparts, they do not comprise prototypic definitive HSCs (Tavian et al. 2001). The restricted or constrained developmental potentials of YS mesodermal hematopoietic intermediates is also exemplified by studies that show their capacity to generate CD34$^+$ HPCs in explant cultures and is markedly lower than that of para-aortic splanchnopleuric cells. In humans, intraembryonic aortic hematopoiesis begins by day 19 of development, 3 days prior to establishment of blood circulation and 1 week before apparition of intra-aortic hematopoietic clusters (Tavian et al. 1999, 2001). Explant cultures of subaortic mesoderm performed by Tavian et al. revealed the presence at this level of yet uncharacterized prehemogenic mesodermal precursors. Although this remains an open question, these observations suggest that definitive HSCs originate from transiently hemogenic endothelial cells that differentiate from more primitive pluripotent mesodermal precursors. Most interestingly, during the same period (i.e., between 30 and 36 days of development), intraluminal hematopoietic cell clusters are detected at the surface of the ventral endothelium of the vitelline, which further argues for the multisite origin of definitive HSCs (Tavian et al. 1999). The lack of detection of significant lymphoid potential in cultures of YS explants harvested from fetuses between days 19 and 40 of gestational age was somewhat unexpected and is still intriguing (Tavian et al. 2001). This suggests that even after the onset of cardiac contraction, when hematopoietic cells redistribute throughout the embryo, the human YS might not be permissive for maintenance or amplification of definitive HSCs. Thus, converse to mice, evidence for a second wave of human YS hematopoiesis is still lacking. In this context, the observation that NK cells are easily obtained from YS explants remains intriguing. Interestingly, several years ago, Dahl also detected a significant natural cytotoxic activity in cultures of early mouse YS cells E10, but this process could not be ascribed to a particular cell phenotype (Dahl 1980, 1983) and, due to lack of robust in vitro differentiation assay, whether YS progenitors actually display NK differentiation potential remains elusive. The human data nonetheless suggest that NK cells might develop independently of the other T- and B-lymphoid lineages, raising the provocative hypothesis that, during early human embryonic development, innate and adaptive immune systems should emerge successively.

From Multipotent Stem Cells to Early Lymphoid Precursors

The liver represents the predominant hematogenic site throughout gestation and the first capable of supporting a fully diversified hematopoiesis. The first wave of colonization by hematopoietic cells occurs immediately after heartbeat initiation; it is supported by YS-derived progenitors that transiently produce erythrocytes, megakaryocytes, and granulomacrophages (Cumano and Godin 2007; Lux et al. 2008).

Later, as YS hematopoiesis vanishes, the FL is seeded by definitive adult-repopulating stem cells that progressively dissociate from the intravascular clusters to enter the bloodstream and ensure subsequent waves of colonization. Immediately after settlement in the liver parenchyma, they undergo a rapid expansion and start to differentiate while acquiring surface expression of diverse lineage-specific markers (Kumaravelu et al. 2002; Morrison et al. 1995; Strasser et al. 1989). Little is known regarding the early FL environment, but there is some evidence that its capacity to support hematopoiesis might actually relate to the transient presence of an original population of mesenchymal cells with mixed epithelial and mesodermal features that also disappears by the time liver hematopoietic capacity declines (Chagraoui et al. 2003). Almost a decade ago, Lacaud et al. used AA4.1 (which recognizes the CD93/C1QR1 marker) and anti-FcγRII/III (CD16/CD32) antibodies to search for early lymphoid progenitors among E12–13 mouse FL mononuclear cells. This led to the identification of an original AA4.1$^+$FcγRII/III$^+$IL-7R$^-$ population endowed with T-, B-, and macrophage restricted potential (Lacaud et al. 1998). At the same time, Sca1^{lo}kit^{lo}IL-7R$^+$B220$^{lo/-}$ CLPs were also detected at the early FL (Kondo et al. 1997; Mebius et al. 2001). Fetal CLPs apparently display a broader developmental potential compared to their postnatal homologs; after transfer into newborn mice they have the capacity to generate T-, B-, and NK lymphocytes, as well as CD8α$^-$ and CD8α$^+$ dendritic cell subsets, and CD45$^+$CD4$^+$CD3$^-$ lymphoid tissue–inducer cells (Mebius et al. 2001). Fetal CLPs do not overlap with AA4.1$^+$FcγRII/III$^+$ precursors since they lack FcγRII/III expression but, most importantly, they rapidly differentiate into CD19$^+$ B cells when cultured onto S17 stromal cells. Although the ontogenic relationship between the two populations remains unclear, most current evidence suggests that the AA4.1$^+$FcγRII/III$^+$ population relates to previously Lin$^-$Sca-1$^+$c-kit$^+$Flt3$^+$ LMPPs also found in the FL (Mansson et al. 2007), whereas fetal CLPs are closer to pre–pro-B cells (Rumfelt et al. 2006).

In recent years, several groups have developed knockin or transgenic mouse models where expression of reporter genes is controlled by diverse lymphoid-specific, promoter-enhancer sequences (Benz and Bleul 2005; Gounari et al. 2002; Igarashi et al. 2002). These models allow a precise in vivo tracking, as well as a fine cellular and molecular characterization of lymphoid precursor populations. Using their RAG1/GFP knockin model Igarashi et al. have shown that E13 FLs mononuclear cells segregate into two phenotypically and functionally distinct populations. The RAG1/GFP^{hi} fraction, which represents a vast majority of the RAG1/GFP-expressing cells, is also positive for CD45R/B220 marker and it rapidly differentiates into CD19$^+$ BLs in vitro, an indication that most correspond to early B-cell precursors. Analysis of the RAG1/GFP^{lo} fraction revealed a more complex pattern and a significant degree of heterogeneity. Approximately half of the cells were CD127/IL-7R$^+$, which suggests that at least some of them correspond to fetal CLPs (Mebius et al. 2001; Yokota et al. 2003). As expected, from their more immature phenotype (c-kit$^{+/hi}$Sca1$^{+/-}$), cells of the RAG1/GFP^{lo} fraction retained a significant degree of multipotency, since they could still generate TLs in reaggregated FTOCs or OP9-DL1 cocultures. Intriguingly, most immature Sca-1$^+$c-kit^{hi} RAG1/GFP^{lo} cells still displayed a predominant B-lymphopoietic activity when transferred into sublethally irradiated alymphoid $RAG1^{-/-}$ mice. One may thus consider that activation of the RAG1 locus actually marks the very first stages of FL B-cell development (Igarashi et al. 2002, Yokota et al. 2003; 2006), which is in accordance with the burst of B-cell development that takes place in the early FL (Raff et al. 1976; Strasser et al. 1989). Of note, FL pre-B cells expressing cytoplasmic immunoglobulins are detected as early as E13, and $sIgM^{+}$ mature B cells appear 4 days later (Dorshkind and Montecino-Rodriguez 2007). Fetal B-cell development is also characterized by the production of two major B-1 and B-2 BL subsets. B-1/$CD5^{+}$ BLs were identified 25 years ago by Hayakawa et al.; it mainly resides in spleen, intestine, and the peritoneal and pleural cavities and displays an original phenotype ($CD11b^{+}sIgM^{hi}sIgD^{low}$) (Hayakawa et al. 1983). This population is characterized by low to undetectable Tdt expression levels, resulting in a limited junctional diversity, as well as by a restricted use of V_H repertoire (Hardy and Hayakawa 1991). Also, compared to their B-2 B homologs, B-1 BLs recognize T-cell independent carbohydrate antigens and spontaneously produce IgMs; they are thus usually considered effectors of the innate immune system. The key point is that this population develops predominantly during fetal life and persists during the postnatal period due to self-renewal of mature effectors (Hardy and Hayakawa 2001). In line with these observations, it is well established that the B-lineage developmental potential of FL mononuclear cells is markedly skewed toward the B-1 B subset (Hardy and Hayakawa 1991; Kantor et al. 1992). The developmental relationships between B-1 and B-2 BLs has long remained controversial as two mutually exclusive "selection" and "lineage" differentiation models have been proposed (Dorshkind and Montecino-Rodriguez 2007). The selection model posits that both populations derived from a common precursor and that ultimate lineage choice was driven by antigen ligation at the $sIgM^{+}$ stage, strengthening its strong dependency on antigen recognition repertoire. This view also received significant experimental support, and recent reports indicate that B-1 and B-2 BLs most probably correspond to separate lineages. The dual origin of B-1 and B-2 lymphocytes has been provided recently by Montecino-Rodriguez et al. Based on CD19 and B220/CD45R expression, they could identify two populations of B-cell precursors and show that, by adoptive transfer experiments, bipotent B/macrophage $CD19^{+}B220^{-/low}$ precursors differentiate exclusively into B-1 B cells, whereas their $CD19^{-}B220^{+}$ homologs comprise a majority of B-2 B-cell precursors (Montecino-Rodriguez et al. 2006). Most interestingly, there is also evidence that development of B-1 and B-2 subsets depends on different gene regulatory networks. For example, targeted invalidation of transcription factor PU.1 induces a complete arrest in B-2 B-cell development without affecting the B-1 subset (Ye et al. 2005). Similarly, Witt et al. have recently shown that an activated form of Notch2 promotes the development B-1 B cells at the expense of their B2 homologs, arguing thus for binary choice between the two subsets (Witt et al. 2003). The intriguing observation that $CD19^{+}CD45R^{-/low}$ B-1 B-cell precursors retain the ability to generate macrophages is in accordance with an older report from Cumano et al. that the FL contains bipotent B/macrophage precursors ($AA4.1^{+}Sca\text{-}1^{+}B220^{+}Mac\text{-}1^{-}$) (Cumano et al. 1992), as well as with the well-documented B/macrophage promiscuity. This phenomenon was first reported

20 years ago by showing that, upon LPS stimulation, several H-Ras transformed FL cell lines could differentiate into either BLs or macrophages (Davidson et al. 1988; Holmes et al. 1986). A molecular basis for that process has been recently provided by Xie et al. who showed that enforced expression of either C/EBPα or C/EBPβ is sufficient to promote the transdifferentiation of mature BLs into prototypic macrophages (Xie et al. 2004). Little is known regarding NK cell development in the early mouse FL. Although it is well established that NK-restricted or mixed T/NK precursors are present at this level, only few if any mature NK1.1$^+$ lymphocytes are detected among FL mononuclear cells that otherwise lack natural cytotoxic activity (Boggs et al. 1998). Thus, most current evidence suggests that the mouse FL might not provide a favorable environment for NK cell development.

Extrathymic Emergence of T-Cell Precursors

Mouse thymus anlagen develop by embryonic day 11 from a bilateral primordium arising from the third pharyngeal pouch endoderm. At E12.5, the thymus primordium separates from the pharynx and begins it migration toward the anterior chest cavity, sometimes living behind functional cervical remnants (Terszowski et al. 2006). Although the thymus has long been considered as being composed of two functionally distinct epithelial cell populations of dual ectodermal and endodermal origin, recent studies based on elegant cellular or genetic approaches have demonstrated that thymus epithelial cells (TECs) actually share a common single endodermal progenitor (Bleul et al. 2006; Gordon et al. 2004; Rossi et al. 2006). Inasmuch as the origin and functional specialization of TECs has been recently reviewed by Blackburn and Manley (2004), this point will not be thoroughly discussed. Thymus colonization remains a central question in developmental immunology and it has been the subject of long-standing interest and prolonged controversy (Hammar 1905; Kolliker 1879). Despite many efforts (Gregoire 1932), it was not before the mid-1960s that the extrinsic origin of thymocytes was formally demonstrated in the pioneering works of Moore and Owen (1967). A few years later, using their now classical quail-chick chimera model, Le Douarin and Jotereau further confirmed that thymus colonization is actually ensured by blood-born precursors (Le Douarin and Jotereau 1975). Since then, the identity of fetal thymus-colonizing cells, the so-called prothymocytes, has been debated and there are conflicting results as to whether they correspond to early lymphoid precursors, such as ELPs or CLPs, or to existing T-lineage–specified precursors. To approach this question, two major aspects of fetal thymus colonization must be taken into account. First, the rapid influx of hematopoietic cells into the developing thymus makes the hypothesis of "random" colonization highly unlikely, and there is now ample experimental evidence that thymic recruitment depends on specific ligand-receptor interactions. Second, definitive commitment toward the T lineage occurs only after thymus entry, through established direct contact with TECs, implying that thymus seeding is ensured by still multipotent precursors. Indeed, most current evidence indicates that immature thymocytes display a triple T, NK, and DC potential (Haddad et al. 2006; Peault et al. 1994). In line with these observations, it is now well established that with expression of as well as diverse Notch target genes, whose expression is acquired only after thymus entry upon contact with TECs (Yokota et al. 2006; Harman et al. 2003; Hattori et al. 1996), definitive commitment takes place intrathymically, at least for some progenitors. Whether loss of B-cell potential precedes (Harman et al. 2005) or rapidly follows entry into the thymus compartment (Benz and Bleul 2005) is still a point of contention, but, because immature thymocytes retain a limited capacity to generate BLs, this question is not of primary importance for the understanding of early fetal T-cell development. The central issue of thymus colonization resides in the complex relationship between the selective capacity of thymus-colonizing cells to respond to signals emanating from the thymic parenchyma and the process of polarization or specification toward the T lineage (Rothenberg and Taghon 2005), both of which remain poorly understood.

A decade ago, Kawamoto et al. developed an improved clonal FTOC assay that allowed the simultaneous generation of TLs and BLs, as well as granulocytes (G) and macrophages (M), and searched for some extrathymic population of T-cell precursors in mouse embryos (Kawamoto et al. 1997). Analysis of FL $Lin^{-}c\text{-}kit^{+}$ mononuclear cells led to the findings that they segregated into two different populations based on IL-7R expression levels. Converse to the $IL\text{-}7R^{-}$ fraction that was enriched in cells with clonal T/B/M, T/M, B/M, or M potentials, $IL\text{-}7R^{+}$ cells were found to display either T- or B-restricted potentials, arguing thus for an early separation between the two lineages (Kawamoto et al. 2000). The balance between pre-T and pre-B populations rapidly evolved over time, since the pre-T–cell activity, which predominates between E11 and E12, drastically decreases thereafter to the point that, by day 15 of development, $IL\text{-}7R^{+}$ cells display an exclusive B-cell potential. This suggests either that, beyond day 12 of development, pre-T cells leave the FL to enter the blood circulation and reach the thymus, or that they are diluted by the rise of B lymphopoiesis. $Lin^{-}c\text{-}kit^{+}IL\text{-}7R^{+}$ cells circulate in the fetal blood, where they are detected based on paired immunoglobulin-like receptors (PIR) A and B expression (Masuda et al. 2005a). Also, as expected for prothymocyte, they progressively accumulate in the mesenchymal region surrounding the thymus anlage (Masuda et al. 2005b). For that reason, most current evidence supports the idea that FL-derived $Lin^{-}c\text{-}kit^{+}$ $IL\text{-}7R^{+}$ cells might represent the first wave of prothymocytes. Douagi et al. identified in the FL another population of $c\text{-}kit^{+}B220^{lo}Thy1^{-}$ $NK1.1^{-}$ T-cell precursors, which was referred to as "fraction e" and accounted for up to 70% of FL T-cell precursor activity (Douagi et al. 2002). These precursors express only low levels of IL-7R transcripts and are apparently devoid of B-lymphopoietic activity. But whether they are related to the $AA4.1^{+}Fc\gamma RII/III^{+}IL\text{-}7R^{-}$ T/B/M population described by Lacaud et al. or to previously mentioned LMPPs remains unclear (Lacaud et al. 1998; Mansson et al. 2007).

A third population of candidate T-cell precursors was identified in mouse fetal blood several years ago based on c-kit and Thy1/CD90 expression (Rodewald et al. 1994). By day 15.5, large numbers of $Thy1/CD90^{+}c\text{-}kit^{low}$ prothymocytes circulate in blood where they represent up to 8% of circulating $CD45^{+}$ blood cells.

When tested in vivo for T-cell differentiation potentials following direct intrathymic injection or intravenous transfer, the Thy1$^+$c-kit^{low} cells generate a single wave of thymocytes but fail to reconstitute the B-cell compartment. This seminal work was confirmed by Zuniga-Pflucker and colleagues who subsequently showed that, after reaching the thymus, Thy1/CD90$^+$c-kit^{low} prothymocytes rapidly acquire surface expression of NK1.1, thereby defining an intermediate bipotent T/NK-restricted developmental stage during progression along the T lineage (Carlyle et al. 1997; Carlyle and Zuniga-Pflucker 1998a). Interestingly, they also showed that thymus homing and interaction with TECs are not absolute prerequisites for acquisition of NK1.1 expression, since a significant number of circulating prothymocytes already express this marker (Carlyle and Zuniga-Pflucker 1998b). That Thy1/CD90$^+$c-kit^{low} NK1.1$^{+/-}$ cells are absent from the FL remains intriguing; this suggests either that they differentiate somewhere else, for example, in spleen or BM, or that Thy1/CD90 and NK1.1 expression is acquired only when FL T-cell precursors gain access to the bloodstream. Reconstitution of the ontogenic linkage between diverse population FL lymphoid progenitors and circulating prothymocytes therefore represents a difficult task. Although both populations express IL-7R, the relationship between the early FL pre-T cells described by Katsura and colleagues and circulating prothymocytes is still unclear. More interestingly, that Thy1/CD90$^+$c-kit^{low} prothymocytes also express FcγRII/III strongly suggests that they may correspond to downstream successors of previously described FL AA4.1$^+$FcγRII/III$^+$IL-7R$^-$ T/B/M lymphoid progenitors (Lacaud et al. 1998; Carlyle and Zuniga-Pflucker 1998b; Rodewald et al. 1992) that probably display partial overlap with Lin$^-$Sca1$^+$Kit$^+$CD34$^+$Flt3^{hi} LMPPs (Mansson et al. 2007). Finally, that a minority of FL cells actively transcribe CCR9 locus at day 11.5 of development further suggests that acquisition of CCR9 surface expression may confer to extrathymic T-cell precursors the capacity to respond selectively to thymus homing signals (Benz et al. 2004, 2008). Most importantly, Benz and Bleul (2005) have recently shown that Lin$^-$c-kit$^+$CCR9/EGFP$^+$ T-cell precursors from adults BM also express Thy1/CD90 and IL-7R marker, thus strengthening the developmental link established between T-lineage polarization and thymus colonization. Thus, although the possibility formally remains that a small number of multipotent progenitors also migrate into the thymus, most current evidence indicates that, during fetal life, thymus colonization is ensured by diverse already T-lineage–specified precursors that rapidly enter the circulation to reach the thymus rudiment. In this regard, one may hypothesize that the first wave of thymus colonization is ensured by liver-derived T-cell precursors. The origin of subsequent waves of thymus seeding cells remains questionable, but they may emerge in BM or spleen and reach the thymus parenchyma thereafter (Arcangeli et al. 2005; Gautreau et al. 2007). Finally, the phenotypic heterogeneity of fetal blood prothymocytes that split into two ontogenetically linked populations of Thy1/CD90$^+$c-kit^{low} NK1.1$^-$ and Thy1/CD90$^+$c-kit^{low} NK1.1$^+$ precursors, both of which correspond to thymus seeding cells, suggests that precursors at different stages of T-lineage polarization may synergetically seed the fetal thymus.

Dynamics of Human Fetal T-Cell Precursors

Historical background of human thymus colonization is somewhat paradoxical since, conversely to the mouse, the nature of thymus-colonizing cells has long remained an almost black hole. In humans, thymus colonization starts by development weeks 8–9 and mature intrathymic TLs are detected about 1 week later (Campana et al. 1989; Haynes et al. 1988; Lobach and Haynes 1987; Lobach et al. 1985). The thymus corticomedullary partitioning occurs by week 14, with definitive thymus architecture being established by gestation week 16 (Haar 1974; Norris 1938; von Gaudecker and Muller-Hermelink 1979). Twenty years ago Haynes and colleagues identified CD7 as the first T-lineage–affiliated marker to be expressed extrathymically (Lobach et al. 1985) and provided some evidence that CD7-expressing cells might accumulate at the vicinity of thymus rudiment as early as week 8.5 of development (Haynes et al. 1988). Although this population subsequently proved to correspond to contaminant TLs of maternal origin, this pioneering work paved the way for the characterization of human fetal prothymocytes. In the early 1990s, Spits and colleagues searched for T-cell precursors in early embryonic human FL and found that, although liver parenchyma supports the development of B and NK lineages, it does not contain T-lineage–specified precursors. Among all FL $CD34^{+}$ HPC populations, the T-cell differentiation potential was restricted to the most immature stem cell–like $CD34^{+}CD38^{-}$ fraction (Blom et al. 1997; Res et al. 1996). For that reason, until recently, it was thus commonly assumed that the human thymus is passively seeded by multipotent liver-derived HSCs (Rothenberg and Taghon 2005; Spits 2002). Ten years ago, Canque et al. identified in the umbilical cord blood (UCB) a population of $CD34^{hi}CD45RA^{hi}CD7^{+}$ precursors lacking erythroid potential, which comprises bipotent NK and dendritic cell progenitors and shares with immature postnatal $CD34^{+}CD1a^{-}$ thymocytes the capacity to differentiate into Langerhans cells via a TGF-β1-independent pathway (Canque et al. 2000). Combining clonal analysis and microarray transcriptional profiling, this population was subsequently shown to already express germline TCRγ and IL-7R transcripts and display a predominant T/NK cell differentiation potential (Haddad et al. 2004). Because $CD34^{hi}CD45RA^{hi}CD7^{+}$ UCB precursors fit the criterion for T-lineage–specified precursors, a developmental approach, based on precise in vivo tracking, was then devised to search for their possible contribution to fetal thymus colonization (Haddad et al. 2006). This led to the finding that $CD34^{hi}CD45RA^{hi}CD7^{+}$ precursors, which are absent from the early FL, actually differentiate in the BM where they are first detected by weeks 8–9 of gestation, at the very onset of thymus colonization. Later on, low numbers of $CD34^{hi}CD45RA^{hi}CD7^{+}$ precursors are also detected in the fetal spleen; given that the human spleen is devoid of hematopoietic activity, their presence at this level indicates that they circulate in the fetal blood. As expected for immediate upstream precursors of $CD34^{hi}IL\text{-}7R^{-}CD1a^{-}$ thymocytes, $CD34^{hi}CD45RA^{hi}CD7^{+}$ precursors undergo a selective thymic recruitment in an ex vivo xenogeneic thymus colonization assays, indicating that they actually correspond to fetal prothymocytes. Most interestingly, this represents evidence that this population is subjected to a precise developmental regulation. CD34^{hi}CD45RA^{hi}CD7^{+} precursors are detected in the fetal BM by gestation weeks 8–9, at the very beginning of its colonization by FL-derived HSCs, where they follow an original bell curve dynamics. This population was found to accumulate in the fetal BM until the end of the second trimester of development to represent up to 40% of total BM CD34^{+} hematopoietic cells. Later on, during the third trimester of gestation, they abruptly decline and return to basal levels around birth and persist at only trace levels in the postnatal BM. Interestingly, fetal BM CD34^{+}CD45RA^{hi}Lin^{-}CD10^{+} pre-pro-B cells follow grossly similar kinetics, except that they persist at significantly higher levels during postnatal life (Rossi et al. 2003). Although the precise mechanisms involved in this process remain elusive, it must be noted that the decline in lymphopoietic activity of the late fetal BM coincides with the cessation of FL hematopoiesis. For that reason, the most plausible explanation is that this phenomenon relates to major microenvironmental changes associated with establishment of an active myelopoiesis in the BM.

Mechanisms Regulating Thymus Colonization

The first phase of thymus colonization by FL-derived precursors occurs before vascular connections are established, thereby proceeding by direct transtissular migration. In mice, thymic immigrants start to accumulate at E10.5 in the mesenchyme surrounding the thymus rudiment and by day 14.5 most of them are completely surrounded by the epithelium (Yokota et al. 2006; Suniara et al. 1999). Upon contact with TECs, immature thymocytes are rapidly driven through the T-cell differentiation pathway. They acquire expression of CD3ε and RAG1 proteins and actively transcribe Notch target genes (PTCRA, Deltex1, HES1); expression of intracytoplasmic TCRβ^{+} chain is detected 24 hours after they have reached the thymic parenchyma (Harman et al. 2003, 2005; Suniara et al. 1999). The route of early immigrants to the thymus has been well explored in the zebrafish using the CD41/GFP transgenic model (Kissa et al. 2008). Thymus-colonizing cells migrate a long way through the mesenchyme to reach the thymus, coming from widely diverse locations (rostral and caudal), an indication that most of them traveled through the blood. That they extravasated at quite diverse sites to reach their final target strongly suggests the presence of a chemoattractant gradient that extends and can be perceived far from the source. Interestingly, these early immigrants keep moving in and out of the thymus, probably in the phase of resynthesis of chemokine receptor and/or the downstream intracellular signaling component. Indeed, these movements occur by 5-day postfecundation, 2 days after the first detection of *Rag1* expression, marking their commitment to a T-cell fate. Later on, when thymic vascularization develops, subsequent waves of blood-born T-cell precursors extravasate the capillaries to directly enter the thymus parenchyma and seed the so-called developmental niches (Kyewski 1987; Lind et al. 2001). In humans, early fetal thymus immigrants are scattered throughout the cortical zone from the subcapsular region deep into the inner cortex, where they mainly reside in the immediate vicinity of vascular structures (Haddad et al. 2006). Le Douarin et al. obtained similar results in the avian model, arguing thus for a multifocal seeding of the fetal thymus (Le Douarin et al. 1984). These observations also support the idea that T-cell precursors gain access to the thymus parenchyma through multiple routes during the phases of active colonization. This picture sharply contrasts with the postnatal period, at the steady state, when thymus entry seems to be confined to a narrow region located at the junction between cortical and medullary regions (Porritt et al. 2003). Although the concept of a thymic "developmental niche" remains poorly defined, there is some evidence that early thymic immigrants specifically interact with dendritic-like processes emanating from the cortical TECs, with which they form patchy structures resembling "lymphostromal synapses." This suggested that definitive commitment toward the T lineage may occur in specific ordered structures (Haddad et al. 2006).

The molecules guiding the recruitment of the prothymocytes and their role are still poorly characterized (Fontaine-Perus et al. 1981; Wilkinson et al. 1999). Direct migration of cells in response to a gradient of chemotactic factors may require the ability to interact with the extracellular matrix (ECM) proteins. Matrix metalloproteinases (MMP) are believed to play an important role in the degradation of the ECM, but their involvement has not been clearly established (Goetzl et al. 1996). From other studies, it was shown that the fetal thymus produces chemokines CCL21, CCL25, and CXCL12 (Bleul and Boehm 2000; Liu et al. 2005, 2006). In line with these observations, the CCL21-deficient mice (or in mice deficient for the receptor CCR7) harbor only a small number of fetal thymocytes until E14.5 (Liu et al. 2005). Similarly, in CCR9 (and CCL25) deficient mice, a threefold decrease in total thymocyte cellularity until E17.5 is observed (Liu et al. 2005; Wurbel et al. 2001). This suggested that fetal thymus colonization involves CCL21 and CCL25 and to a lesser extend CXCL12 (Ara et al. 2003), which is in accordance with a report that thymus seeding progenitors express CCR7, CCR9, and CXCR4. Locally produced chemokines may thus generate a systemic gradient attracting T-cell precursors from the blood and coordinate with adhesion molecules their specific accumulation in the tissue. But whether they act cooperatively remains unclear.

Whether thymus colonization occurs in "continuous" or "cyclical" timing is still a long-standing question. In the avian model where T-cell precursors emerge in the para-aortic mesoderm and BM, thymus colonization proceeds in three waves occurring by days 6, 12, and 18 of development (Dieterlen-Lievre et al. 1996; Vainio et al. 1996). Each wave of new thymic immigrants starts to divide 2–3 days after settlement in the thymic parenchyma to give rise to almost synchronous waves of developing thymocytes (Le Douarin and Jotereau 1975; Jotereau and Le Douarin 1982). Based on these pioneering observations, it was then proposed that such cyclical colonization is dictated by the thymus itself, which permits precursor entry only during short receptive periods once the thymus niches are emptied (Jotereau and Le Douarin 1982; Coltey et al. 1987). Although attractive, this appears somewhat more complex. There is convincing evidence that the level of circulating T-cell precursors is subjected to important fluctuations. Most interestingly, they also found that the decline in circulating T-cell precursors between the first waves of colonization coincides also with the transition between AGM and BM hematopoiesis. Inasmuch as only first-wave thymus-colonizing cells originate in the AGM, this suggests that, rather than to the thymus itself, their fluctuations actually relate to major developmental transitions (Dunon et al. 1999).

Attempts to extend to mammals the cyclical model of avian thymus colonization have not yielded convincing results. Jotereau and Le Douarin (1982) proposed that the first wave of mouse thymus immigrants ensures production of TLs throughout gestation, whereas those entering at later time points produce TLs only after week 1 of postnatal life, but their findings never received formal experimental demonstration (Jotereau et al. 1987). This topic has been extensively investigated in adult mice by Foss et al. They used parabiotic and adoptive transfer models to search for possible fluctuation in thymic receptivity and proposed that, as for its fetal avian counterpart, adult mouse thymus alternates phases of receptive (1 week) and refractory periods (3 weeks) allowing for the gated importation of T-cell precursors (Foss et al. 2001). Surprisingly, they also proposed that T-cell precursors might accumulate in the BM approximately 1 week before the opening of intrathymic niches, thus linking BM production and release of T-cell precursors to receptive thymic periods (Donskoy et al. 2003). It is worth noting that, although technically demanding, the gated importation model relies on a global assessment of BM thymopoietic activity and thymic chimerism, which is poorly sensitive; this topic needs to be reinvestigated in more recent genetic mouse models. Also, most if not all studies that have been performed in the field of thymus colonization have led to the same conclusion that extrathymic emergence of T-cell precursors is thymus independent, thus ruling out the hypothesis that their development might be driven by so-called thymus-derived hormones (Goldstein et al. 1970; Schlesinger et al. 1975), since T-cell precursors are detected in normal numbers in the FL, blood, or BM of athymic nu/nu mice (Douagi et al. 2002; Chatterjea-Matthes et al. 2003; Kawamoto et al. 1999). To what degree thymus entry should be regulated remains an opened question. One may consider that global thymic cellular content or the degree of niche occupancy may exert some modulatory effect on the recruitment of T-cell precursors through expression of putative endothelial thymus homing receptors and/or the release of chemotactic factors (Wilkinson et al. 1999). The observation that alymphoid fetal thymus lobes possess a greater capacity to attract T-cell precursors strongly supports this view.

Conclusion

The period of fetal development provides a unique opportunity to study the ontogenic relationship between diverse lymphoid populations and the developmental regulation of adaptive immunity. Fetal lymphoid development also differs in several aspects from the postnatal period that makes it original. Although not thoroughly discussed herein, these peculiarities also extend to the molecular level, since Ikaros null mice have been shown to display a selective defect in fetal TL development (Wang et al. 1996). On the other hand, deletion of IL-7 or IL-7R that blocks post-natal B-cell development does not affect the differentiation of fetal B-cell precursors that display a unique capacity to respond to thymic stromal lymphopoietin (Vosshenrich et al. 2003).

References

Adolfsson J, Borge OJ, Bryder D, Theilgaard-Monch K, Astrand-Grundstrom I, Sitnicka E, Sasaki Y, Jacobsen SE (2001) Upregulation of Flt3 expression within the bone marrow Lin(-)Sca1(+) c-kit(+) stem cell compartment is accompanied by loss of self-renewal capacity. Immunity 15:659–669

Adolfsson J, Mansson R, Buza-Vidas N, Hultquist A, Liuba K, Jensen CT, Bryder D, Yang L, Borge OJ, Thoren LA, Anderson K, Sitnicka E, Sasaki Y, Sigvardsson M, Jacobsen SE (2005) Identification of Flt3+ lympho-myeloid stem cells lacking erythro-megakaryocytic potential a revised road map for adult blood lineage commitment. Cell 121:295–306

Allman D, Sambandam A, Kim S, Miller JP, Pagan A, Well D, Meraz A, Bhandoola A (2003) Thymopoiesis independent of common lymphoid progenitors. Nat Immunol 4:168–174

Ara T, Itoi M, Kawabata K, Egawa T, Tokoyoda K, Sugiyama T, Fujii N, Amagai T, Nagasawa T (2003) A role of CXC chemokine ligand 12/stromal cell-derived factor-1/pre-B cell growth stimulating factor and its receptor CXCR4 in fetal and adult T cell development in vivo. J Immunol 170:4649–4655

Arcangeli ML, Lancrin C, Lambolez F, Cordier C, Schneider E, Rocha B, Ezine S (2005) Extrathymic hemopoietic progenitors committed to T cell differentiation in the adult mouse. J Immunol 174:1980–1988

Bell JJ, Bhandoola A (2008) The earliest thymic progenitors for T cells possess myeloid lineage potential. Nature 452:764–767

Benz C, Bleul CC (2005) A multipotent precursor in the thymus maps to the branching point of the T versus B lineage decision. J Exp Med 202:21–31

Benz C, Heinzel K, Bleul CC (2004) Homing of immature thymocytes to the subcapsular microenvironment within the thymus is not an absolute requirement for T cell development. Eur J Immunol 34:3652–3663

Benz C, Martins VC, Radtke F, Bleul CC (2008) The stream of precursors that colonizes the thymus proceeds selectively through the early T lineage precursor stage of T cell development. J Exp Med 205:1187–1199

Berardi AC, Meffre E, Pflumio F, Katz A, Vainchenker W, Schiff C, Coulombel L (1997) Individual CD34+CD38lowCD19–CD10– progenitor cells from human cord blood generate B lymphocytes and granulocytes. Blood 89:3554–3564

Blackburn CC, Manley NR (2004) Developing a new paradigm for thymus organogenesis. Nat Rev Immunol 4:278–289

Bleul CC, Boehm T (2000) Chemokines define distinct microenvironments in the developing thymus. Eur J Immunol 30:3371–3379

Bleul CC, Corbeaux T, Reuter A, Fisch P, Monting JS, Boehm T (2006) Formation of a functional thymus initiated by a postnatal epithelial progenitor cell. Nature 441:992–996

Blom B, Res P, Noteboom E, Weijer K, Spits H (1997) Prethymic CD34+ progenitors capable of developing into T cells are not committed to the T cell lineage. J Immunol 158:3571–3577

Bloom W, Bartelmez GW (1940) Hematopoiesis in young human embryos. Am J Anat 67:21–53

Boggs SS, Trevisan M, Patrene K, Geogopoulos K (1998) Lack of natural killer cell precursors in fetal liver of Ikaros knockout mutant mice. Nat Immunol 16:137–145

Campana D, Janossy G, Coustan-Smith E, Amlot PL, Tian WT, Ip S, Wong L (1989) The expression of T cell receptor-associated proteins during T cell ontogeny in man. J Immunol 142:57–66
Canque B, Camus S, Dalloul A, Kahn E, Yagello M, Dezutter-Dambuyant C, Schmitt D, Schmitt C, Gluckman JC (2000) Characterization of dendritic cell differentiation pathways from cord blood CD34(+)CD7(+)CD45RA(+) hematopoietic progenitor cells. Blood 96:3748–3756
Carlyle JR, Zuniga-Pflucker JC (1998a) Regulation of NK1.1 expression during lineage commitment of progenitor thymocytes. J Immunol 161:6544–6551
Carlyle JR, Zuniga-Pflucker JC (1998b) Requirement for the thymus in alpha beta T lymphocyte lineage commitment. Immunity 9:187–197
Carlyle JR, Michie AM, Furlonger C, Nakano T, Lenardo MJ, Paige CJ, Zuniga-Pflucker JC (1997) Identification of a novel developmental stage marking lineage commitment of progenitor thymocytes. J Exp Med 186:173–182
Chagraoui J, Lepage-Noll A, Anjo A, Uzan G, Charbord P (2003) Fetal liver stroma consists of cells in epithelial-to-mesenchymal transition. Blood 101:2973–2982
Chatterjea-Matthes D, Garcia-Ojeda ME, Dejbakhsh-Jones S, Jerabek L, Manz MG, Weissman IL, Strober S (2003) Early defect prethymic in bone marrow T cell progenitors in athymic nu/nu mice. J Immunol 171:1207–1215
Choi K, Kennedy M, Kazarov A, Papadimitriou JC, Keller G (1998) A common precursor for hematopoietic and endothelial cells. Development 125:725–732
Collins LS, Dorshkind K (1987) A stromal cell line from myeloid long-term bone marrow cultures can support myelopoiesis and B lymphopoiesis. J Immunol 138:1082–1087
Coltey M, Jotereau FV, Le Douarin NM (1987) Evidence for a cyclic renewal of lymphocyte precursor cells in the embryonic chick thymus. Cell Differ 22:71–82
Cumano A, Godin I (2007) Ontogeny of the hematopoietic system. Annu Rev Immunol 25:745–785
Cumano A, Paige CJ, Iscove NN, Brady G (1992) Bipotential precursors of B cells and macrophages in murine fetal liver. Nature 356:612–615
Cumano A, Ferraz JC, Klaine M, Di Santo JP, Godin I (2001) Intraembryonic, but not yolk sac hematopoietic precursors, isolated before circulation, provide long-term multilineage reconstitution. Immunity 15:477–485
Dahl CA (1980) Natural cytotoxicity in the mouse embryo. I. Evidence for the presence of natural cytotoxic function in the yolk sac of the 10-day mouse embryo. J Immunol 125:1924–1927
Dahl CA (1983) Natural cytotoxicity in the mouse embryo: characterization of yolk sac-associated natural cytotoxic cells and their activity. Eur J Immunol 13:747–755
Davidson WF, Pierce JH, Rudikoff S, Morse HC 3rd (1988) Relationships between B cell and myeloid differentiation. Studies with a B lymphocyte progenitor line, HAFTL-1. J Exp Med 168:389–407
de Bruijn MF, Speck NA, Peeters MC, Dzierzak E (2000) Definitive hematopoietic stem cells first develop within the major arterial regions of the mouse embryo. EMBO J 19:2465–2474
de Bruijn MF, Ma X, Robin C, Ottersbach K, Sanchez MJ, Dzierzak E (2002) Hematopoietic stem cells localize to the endothelial cell layer in the midgestation mouse aorta. Immunity 16:673–683
Dexter TM, Testa NG (1976) Differentiation and proliferation of hemopoietic cells in culture. Methods Cell Biol 14:387–405
Dieterlen-Lievre F, Godin I, Pardanaud L (1996) Ontogeny of hematopoiesis in the avian embryo: a general paradigm. Curr Top Microbiol Immunol 212:119–128
Donskoy E, Foss D, Goldschneider I (2003) Gated importation of prothymocytes by adult mouse thymus is coordinated with their periodic mobilization from bone marrow. J Immunol 171:3568–3575
Dorshkind K, Montecino-Rodriguez E (2007) Fetal B-cell lymphopoiesis and the emergence of B-1-cell potential. Nat Rev Immunol 7:213–219
Douagi I, Colucci F, Di Santo JP, Cumano A (2002) Identification of the earliest prethymic bipotent T/NK progenitor in murine fetal liver. Blood 99:463–471

Dunon D, Allioli N, Vainio O, Ody C, Imhof BA (1999) Quantification of T-cell progenitors during ontogeny: thymus colonization depends on blood delivery of progenitors. Blood 93:2234–2243

Dzierzak E, Speck NA (2008) Of lineage and legacy: the development of mammalian hematopoietic stem cells. Nat Immunol 9:129–136

Fontaine-Perus JC, Calman FM, Kaplan C, Le Douarin NM (1981) Seeding of the 10-day mouse embryo thymic rudiment by lymphocyte precursors in vitro. J Immunol 126:2310–2316

Foss DL, Donskoy E, Goldschneider I (2001) The importation of hematogenous precursors by the thymus is a gated phenomenon in normal adult mice. J Exp Med 193:365–374

Furuta C, Ema H, Takayanagi S, Ogaeri T, Okamura D, Matsui Y, Nakauchi H (2006) Discordant developmental waves of angioblasts and hemangioblasts in the early gastrulating mouse embryo. Development 133:2771–2779

Galy A, Travis M, Cen D, Chen B (1995) Human T, B, natural killer, and dendritic cells arise from a common bone marrow progenitor cell subset. Immunity 3:459–473

Gautreau L, Arcangeli ML, Pasqualetto V, Joret AM, Garcia-Cordier C, Megret J, Schneider E, Ezine S (2007) Identification of an IL-7-dependent pre-T committed population in the spleen. J Immunol 179:2925–2935

Gekas C, Dieterlen-Lievre F, Orkin SH, Mikkola HK (2005) The placenta is a niche for hematopoietic stem cells. Dev Cell 8:365–375

Goetzl EJ, Banda MJ, Leppert D (1996) Matrix metalloproteinases in immunity. J Immunol 156:1–4

Goldstein AL, Asanuma Y, White A (1970) The thymus as an endocrine gland: properties of thymosin, a new thymus hormone. Recent Prog Horm Res 26:505–538

Gordon J, Wilson VA, Blair NF, Sheridan J, Farley A, Wilson L, Manley NR, Blackburn CC (2004) Functional evidence for a single endodermal origin for the thymic epithelium. Nat Immunol 5:546–553

Gounari F, Aifantis I, Martin C, Fehling HJ, Hoeflinger S, Leder P, von Boehmer H, Reizis B (2002) Tracing lymphopoiesis with the aid of a pTalpha-controlled reporter gene. Nat Immunol 3:489–496

Gregoire C (1932) Contribution expérimentale à l'étude du thymus des mammifères. L'action comaprée des rayons X sur le thymus au cours de l'histogenèse et chez l'adulte, envisagé spécialement du point de vue de la radiosensibilité des petites cellules thymiques, de la nature du processus de phagocytose et du mécanisme de la régénération. Arch Intern Med 7:511–527

Haar JL (1974) Light and electron microscopy of the human fetal thymus. Anat Rec 179:463–476

Haar JL, Ackerman GA (1971) A phase and electron microscopic study of vasculogenesis and erythropoiesis in the yolk sac of the mouse. Anat Rec 170:199–223

Haddad R, Guardiola P, Izac B, Thibault C, Radich J, Delezoide AL, Baillou C, Lemoine FM, Gluckman JC, Pflumio F, Canque B (2004) Molecular characterization of early human T/NK and B-lymphoid progenitor cells in umbilical cord blood. Blood 104:3918–3926

Haddad R, Guimiot F, Six E, Jourquin F, Setterblad N, Kahn E, Yagello M, Schiffer C, Andre-Schmutz I, Cavazzana-Calvo M, Gluckman JC, Delezoide AL, Pflumio F, Canque B (2006) Dynamics of thymus-colonizing cells during human development. Immunity 24:217–230

Hammar JA (1905) Zur histogenese and Involution der thymusdrüsse. Anat Anz 27:23

Hardy RR, Hayakawa K (1991) A developmental switch in B lymphopoiesis. Proc Natl Acad Sci USA 88:11550–11554

Hardy RR, Hayakawa K (2001) B cell development pathways. Annu Rev Immunol 19:595–621

Harman BC, Jenkinson EJ, Anderson G (2003) Entry into the thymic microenvironment triggers Notch activation in the earliest migrant T cell progenitors. J Immunol 170:1299–1303

Harman BC, Jenkinson WE, Parnell SM, Rossi SW, Jenkinson EJ, Anderson G (2005) T/B lineage choice occurs prior to intrathymic Notch signaling. Blood 106:886–892

Hattori N, Kawamoto H, Fujimoto S, Kuno K, Katsura Y (1996) Involvement of transcription factors TCF-1 and GATA-3 in the initiation of the earliest step of T cell development in the thymus. J Exp Med 184:1137–1147

Hayakawa K, Hardy RR, Parks DR, Herzenberg LA (1983) The "Ly-1 B" cell subpopulation in normal immunodefective, and autoimmune mice. J Exp Med 157:202–218

Haynes BF, Martin ME, Kay HH, Kurtzberg J (1988) Early events in human T cell ontogeny. Phenotypic characterization and immunohistologic localization of T cell precursors in early human fetal tissues. J Exp Med 168:1061–1080

Herbomel P, Thisse B, Thisse C (1999) Ontogeny and behaviour of early macrophages in the zebrafish embryo. Development 126:3735–3745

Holmes KL, Pierce JH, Davidson WF, Morse HC 3rd (1986) Murine hematopoietic cells with pre-B or pre-B/myeloid characteristics are generated by in vitro transformation with retroviruses containing fes, ras, abl, and src oncogenes. J Exp Med 164:443–457

Huber TL, Kouskoff V, Fehling HJ, Palis J, Keller G (2004) Haemangioblast commitment is initiated in the primitive streak of the mouse embryo. Nature 432:625–630

Huyhn A, Dommergues M, Izac B, Croisille L, Katz A, Vainchenker W, Coulombel L (1995) Characterization of hematopoietic progenitors from human yolk sacs and embryos. Blood 86:4474–4485

Igarashi H, Gregory SC, Yokota T, Sakaguchi N, Kincade PW (2002) Transcription from the RAG1 locus marks the earliest lymphocyte progenitors in bone marrow. Immunity 17:117–130

Jotereau FV, Le Douarin NM (1982) Demonstration of a cyclic renewal of the lymphocyte precursor cells in the quail thymus during embryonic and perinatal life. J Immunol 129:1869–1877

Jotereau F, Heuze F, Salomon-Vie V, Gascan H (1987) Cell kinetics in the fetal mouse thymus: precursor cell input, proliferation, and emigration. J Immunol 138:1026–1030

Kantor AB, Stall AM, Adams S, Herzenberg LA, Herzenberg LA (1992) Differential development of progenitor activity for three B-cell lineages. Proc Natl Acad Sci USA 89:3320–3324

Katsura Y (2002) Redefinition of lymphoid progenitors. Nat Rev Immunol 2:127–132

Kawamoto H, Ohmura K, Katsura Y (1997) Direct evidence for the commitment of hematopoietic stem cells to T, B and myeloid lineages in murine fetal liver. Int Immunol 9:1011–1019

Kawamoto H, Ohmura K, Fujimoto S, Katsura Y (1999) Emergence of T cell progenitors without B cell or myeloid differentiation potential at the earliest stage of hematopoiesis in the murine fetal liver. J Immunol 162:2725–2731

Kawamoto H, Ikawa T, Ohmura K, Fujimoto S, Katsura Y (2000) T cell progenitors emerge earlier than B cell progenitors in the murine fetal liver. Immunity 12:441–450

Kingston R, Jenkinson EJ, Owen JJ (1985) A single stem cell can recolonize an embryonic thymus, producing phenotypically distinct T-cell populations. Nature 317:811–813

Kissa K, Murayama E, Zapata A, Cortes A, Perret E, Machu C, Herbomel P (2008) Live imaging of emerging hematopoietic stem cells and early thymus colonization. Blood 111:1147–1156

Kolliker A (1879) In entwicklungsgeschichte der menschen und der höheren tiere. Wilhelm Engelman, Leipzig, Germany

Kondo M, Weissman IL, Akashi K (1997) Identification of clonogenic common lymphoid progenitors in mouse bone marrow. Cell 91:661–672

Kumano K, Chiba S, Kunisato A, Sata M, Saito T, Nakagami-Yamaguchi E, Yamaguchi T, Masuda S, Shimizu K, Takahashi T, Ogawa S, Hamada Y, Hirai H (2003) Notch1 but not Notch2 is essential for generating hematopoietic stem cells from endothelial cells. Immunity 18:699–711

Kumaravelu P, Hook L, Morrison AM, Ure J, Zhao S, Zuyev S, Ansell J, Medvinsky A (2002) Quantitative developmental anatomy of definitive haematopoietic stem cells/long-term repopulating units (HSC/RUs): role of the aorta-gonad-mesonephros (AGM) region and the yolk sac in colonisation of the mouse embryonic liver. Development 129:4891–4899

Kyba M, Perlingeiro RC, Daley GQ (2002) HoxB4 confers definitive lymphoid-myeloid engraftment potential on embryonic stem cell and yolk sac hematopoietic progenitors. Cell 109:29–37

Kyewski BA (1987) Seeding of thymic microenvironments defined by distinct thymocyte-stromal cell interactions is developmentally controlled. J Exp Med 166:520–538

La Motte-Mohs RN, Herer E, Zuniga-Pflucker JC (2004) Induction of T cell development from human cord blood hematopoietic stem cells by Delta-like 1 in vitro. Blood 105: 1431-1439
Lacaud G, Carlsson L, Keller G (1998) Identification of a fetal hematopoietic precursor with B cell, T cell, and macrophage potential. Immunity 9:827–838
Laiosa CV, Stadtfeld M, Graf T (2006) Determinants of lymphoid-myeloid lineage diversification. Annu Rev Immunol 24:705–738
Le Douarin NM, Jotereau FV (1975) Tracing of cells of the avian thymus through embryonic life in interspecific chimeras. J Exp Med 142:17–40
Le Douarin NM, Dieterlen-Lievre F, Oliver PD (1984) Ontogeny of primary lymphoid organs and lymphoid stem cells. Am J Anat 170:261–299
Lind EF, Prockop SE, Porritt HE, Petrie HT (2001) Mapping precursor movement through the postnatal thymus reveals specific microenvironments supporting defined stages of early lymphoid development. J Exp Med 194:127–134
Liu C, Ueno T, Kuse S, Saito F, Nitta T, Piali L, Nakano H, Kakiuchi T, Lipp M, Hollander GA, Takahama Y (2005) The role of CCL21 in recruitment of T-precursor cells to fetal thymi. Blood 105:31–39
Liu C, Saito F, Liu Z, Lei Y, Uehara S, Love P, Lipp M, Kondo S, Manley N, Takahama Y (2006) Coordination between CCR7- and CCR9-mediated chemokine signals in prevascular fetal thymus colonization. Blood 108:2531–2539
Lobach DF, Haynes BF (1987) Ontogeny of the human thymus during fetal development. J Clin Immunol 7:81–97
Lobach DF, Hensley LL, Ho W, Haynes BF (1985) Human T cell antigen expression during the early stages of fetal thymic maturation. J Immunol 135:1752–1759
Luc S, Buza-Vidas N, Jacobsen SE (2007) Biological and molecular evidence for existence of lymphoid-primed multipotent progenitors. Ann N Y Acad Sci 1106:89–94
Lux CT, Yoshimoto M, McGrath K, Conway SJ, Palis J, Yoder MC (2008) All primitive and definitive hematopoietic progenitor cells emerging before E10 in the mouse embryo are products of the yolk sac. Blood 111:3435–3438
Mansson R, Hultquist A, Luc S, Yang L, Anderson K, Kharazi S, Al-Hashmi S, Liuba K, Thoren L, Adolfsson J, Buza-Vidas N, Qian H, Soneji S, Enver T, Sigvardsson M, Jacobsen SE (2007) Molecular evidence for hierarchical transcriptional lineage priming in fetal and adult stem cells and multipotent progenitors. Immunity 26:407–419
Manz MG, Miyamoto T, Akashi K, Weissman IL (2002) Prospective isolation of human clonogenic common myeloid progenitors. Proc Natl Acad Sci USA 99:11872–11877
Masuda K, Kubagawa H, Ikawa T, Chen CC, Kakugawa K, Hattori M, Kageyama R, Cooper MD, Minato N, Katsura Y, Kawamoto H (2005a) Prethymic T-cell development defined by the expression of paired immunoglobulin-like receptors. EMBO J 24:4052–4060
Masuda K, Itoi M, Amagai T, Minato N, Katsura Y, Kawamoto H (2005b) Thymic anlage is colonized by progenitors restricted to T, NK, and dendritic cell lineages. J Immunol 174:2525–2532
Mebius RE, Miyamoto T, Christensen J, Domen J, Cupedo T, Weissman IL, Akashi K (2001) The fetal liver counterpart of adult common lymphoid progenitors gives rise to all lymphoid lineages, CD45+CD4+CD3– cells, as well as macrophages. J Immunol 166:6593–6601
Migliaccio G, Migliaccio AR, Petti S, Mavilio F, Russo G, Lazzaro D, Testa U, Marinucci M, Peschle C (1986) Human embryonic hemopoiesis. Kinetics of progenitors and precursors underlying the yolk sac—liver transition. J Clin Invest 78:51–60
Montecino-Rodriguez E, Leathers H, Dorshkind K (2006) Identification of a B-1 B cell-specified progenitor. Nat Immunol 7:293–301
Moore MA, Metcalf D (1970) Ontogeny of the haemopoietic system: Yolk sac origin of in vivo and in vitro colony forming cells in the developing mouse embryo. Br J Haematol 18:279–296
Moore MA, Owen JJ (1967) Chromosome marker studies in the irradiated chick embryo. Nature 215:1081–1082
Morrison SJ, Hemmati HD, Wandycz AM, Weissman IL (1995) The purification and characterization of fetal liver hematopoietic stem cells. Proc Natl Acad Sci USA 92:10302–10306

Muller AM, Medvinsky A, Strouboulis J, Grosveld F, Dzierzak E (1994) Development of hematopoietic stem cell activity in the mouse embryo. Immunity 1:291–301
Nakano T, Kodama H, Honjo T (1994) Generation of lymphohematopoietic cells from embryonic stem cells in culture. Science 265:1098–1101
Norris EH (1938) The morphogenesis and histogenesis of the thymus gland in man in which the origin of the Hassall's corpuscles of the human thymus is discovered. Contrib Embryol Carnegie Inst 27:193–215
Ohmura K, Kawamoto H, Fujimoto S, Ozaki S, Nakao K, Katsura Y (1999) Emergence of T, B, and myeloid lineage-committed as well as multipotent hemopoietic progenitors in the aorta-gonad-mesonephros region of day 10 fetuses of the mouse. J Immunol 163:4788–4795
Orkin SH, Zon LI (2008) Hematopoiesis: an evolving paradigm for stem cell biology. Cell 132:631–644
Ottersbach K, Dzierzak E (2005) The murine placenta contains hematopoietic stem cells within the vascular labyrinth region. Dev Cell 8:377–387
Palis J, Robertson S, Kennedy M, Wall C, Keller G (1999) Development of erythroid and myeloid progenitors in the yolk sac and embryo proper of the mouse. Development 126:5073–5084
Peault B, Khazaal I, Weissman IL (1994) In vitro development of B cells and macrophages from early mouse fetal thymocytes. Eur J Immunol 24:781–784
Pevny L, Simon MC, Robertson E, Klein WH, Tsai SF, D'Agati V, Orkin SH, Costantini F (1991) Erythroid differentiation in chimaeric mice blocked by a targeted mutation in the gene for transcription factor GATA-1. Nature 349:257–260
Plum J, De Smedt M, Defresne MP, Leclercq G, Vandekerckhove B (1994) Human CD34+ fetal liver stem cells differentiate to T cells in a mouse thymic microenvironment. Blood 84:1587–1593
Porritt HE, Gordon K, Petrie HT (2003) Kinetics of steady-state differentiation and mapping of intrathymic-signaling environments by stem cell transplantation in nonirradiated mice. J Exp Med 198:957–962
Pouget C, Gautier R, Teillet MA, Jaffredo T (2006) Somite-derived cells replace ventral aortic hemangioblasts and provide aortic smooth muscle cells of the trunk. Development 133:1013–1022
Raff MC, Megson M, Owen JJ, Cooper MD (1976) Early production of intracellular IgM by B-lymphocyte precursors in mouse. Nature 259:224–226
Res P, Martinez-Caceres E, Cristina Jaleco A, Staal F, Noteboom E, Weijer K, Spits H (1996) CD34+CD38dim cells in the human thymus can differentiate into T, natural killer, and dendritic cells but are distinct from pluripotent stem cells. Blood 87:5196–5206
Rhodes KE, Gekas C, Wang Y, Lux CT, Francis CS, Chan DN, Conway S, Orkin SH, Yoder MC, Mikkola HK (2008) The emergence of hematopoietic stem cells is initiated in the placental vasculature in the absence of circulation. Cell Stem Cell 2:252–263
Rodewald HR, Moingeon P, Lucich JL, Dosiou C, Lopez P, Reinherz EL (1992) A population of early fetal thymocytes expressing Fc gamma RII/III contains precursors of T lymphocytes and natural killer cells. Cell 69:139–150
Rodewald HR, Kretzschmar K, Takeda S, Hohl C, Dessing M (1994) Identification of pro-thymocytes in murine fetal blood: T lineage commitment can precede thymus colonization. EMBO J 13:4229–4240
Rossi MI, Yokota T, Medina KL, Garrett KP, Comp PC, Schipul AH Jr, Kincade PW (2003) B lymphopoiesis is active throughout human life, but there are developmental age-related changes. Blood 101:576–584
Rossi SW, Jenkinson WE, Anderson G, Jenkinson EJ (2006) Clonal analysis reveals a common progenitor for thymic cortical and medullary epithelium. Nature 441:988–991
Rothenberg EV, Taghon T (2005) Molecular genetics of T cell development. Annu Rev Immunol 23:601–649
Rumfelt LL, Zhou Y, Rowley BM, Shinton SA, Hardy RR (2006) Lineage specification and plasticity in CD19– early B cell precursors. J Exp Med 203:675–687

Sabin FR (1920) Studies on the origin of blood vessels and of red corpuscles as seen in the living blastoderm of the chick during the second day of incubation. Contrib Embryol Carnegie Inst 9:213–262

Samokhvalov IM, Samokhvalova NI, Nishikawa S (2007) Cell tracing shows the contribution of the yolk sac to adult haematopoiesis. Nature 446:1056–1061

Schlesinger DH, Goldstein G, Scheid MP, Boyse EA (1975) Chemical synthesis of a peptide fragment of thymopoietin II that induces selective T cell differentiation. Cell 5:367–370

Schmitt TM, Zuniga-Pflucker JC (2002) Induction of T cell development from hematopoietic progenitor cells by delta-like-1 in vitro. Immunity 17:749–756

Scott EW, Simon MC, Anastasi J, Singh H (1994) Requirement of transcription factor PU.1 in the development of multiple hematopoietic lineages. Science 265:1573–1577

Scott EW, Fisher RC, Olson MC, Kehrli EW, Simon MC, Singh H (1997) PU.1 functions in a cell-autonomous manner to control the differentiation of multipotential lymphoid-myeloid progenitors. Immunity 6:437–447

Shivdasani RA, Fujiwara Y, McDevitt MA, Orkin SH (1997) A lineage-selective knockout establishes the critical role of transcription factor GATA-1 in megakaryocyte growth and platelet development. EMBO J 16:3965–3973

Spits H (2002) Development of alphabeta T cells in the human thymus. Nat Rev Immunol 2:760–772

Strasser A, Rolink A, Melchers F (1989) One synchronous wave of B cell development in mouse fetal liver changes at day 16 of gestation from dependence to independence of a stromal cell environment. J Exp Med 170:1973–1986

Suniara RK, Jenkinson EJ, Owen JJ (1999) Studies on the phenotype of migrant thymic stem cells. Eur J Immunol 29:75–80

Tavian M, Hallais MF, Peault B (1999) Emergence of intraembryonic hematopoietic precursors in the pre-liver human embryo. Development 126:793–803

Tavian M, Robin C, Coulombel L, Peault B (2001) The human embryo, but not its yolk sac, generates lympho-myeloid stem cells: mapping multipotent hematopoietic cell fate in intraembryonic mesoderm. Immunity 15:487–495

Terszowski G, Muller SM, Bleul CC, Blum C, Schirmbeck R, Reimann J, Pasquier LD, Amagai T, Boehm T, Rodewald HR (2006) Evidence for a functional second thymus in mice. Science 312:284–287

Tober J, Koniski A, McGrath KE, Vemishetti R, Emerson R, de Mesy-Bentley KK, Waugh R, Palis J (2007) The megakaryocyte lineage originates from hemangioblast precursors and is an integral component both of primitive and of definitive hematopoiesis. Blood 109:1433–1441

Vainio O, Dunon D, Aissi F, Dangy JP, McNagny KM, Imhof BA (1996) HEMCAM, an adhesion molecule expressed by c-kit+hemopoietic progenitors. J Cell Biol 135:1655–1668

von Gaudecker B, Muller-Hermelink HK (1979) Ontogenetic differentiation of epithelial and nonepithelial cells in the human thymus. Adv Exp Med Biol 114:19–23

Vosshenrich CA, Cumano A, Muller W, Di Santo JP, Vieira P (2003) Thymic stromal-derived lymphopoietin distinguishes fetal from adult B cell development. Nat Immunol 4:773–779

Wada H, Masuda K, Satoh R, Kakugawa K, Ikawa T, Katsura Y, Kawamoto H (2008) Adult T-cell progenitors retain myeloid potential. Nature 452:768–772

Wang JH, Nichogiannopoulou A, Wu L, Sun L, Sharpe AH, Bigby M, Georgopoulos K (1996) Selective defects in the development of the fetal and adult lymphoid system in mice with an Ikaros null mutation. Immunity 5:537–549

Whitlock CA, Witte ON (1982) Long-term culture of B lymphocytes and their precursors from murine bone marrow. Proc Natl Acad Sci USA 79:3608–3612

Whitlock CA, Tidmarsh GF, Muller-Sieburg C, Weissman IL (1987) Bone marrow stromal cell lines with lymphopoietic activity express high levels of a pre-B neoplasia-associated molecule. Cell 48:1009–1021

Wilkinson B, Owen JJ, Jenkinson EJ (1999) Factors regulating stem cell recruitment to the fetal thymus. J Immunol 162:3873–3881

Witt CM, Hurez V, Swindle CS, Hamada Y, Klug CA (2003) Activated Notch2 potentiates CD8 lineage maturation and promotes the selective development of B1 B cells. Mol Cell Biol 23:8637–8650

Wurbel MA, Malissen M, Guy-Grand D, Meffre E, Nussenzweig MC, Richelme M, Carrier A, Malissen B (2001) Mice lacking the CCR9 CC-chemokine receptor show a mild impairment of early T- and B-cell development and a reduction in T-cell receptor gammadelta(+) gut intra-epithelial lymphocytes. Blood 98:2626–2632

Xie H, Ye M, Feng R, Graf T (2004) Stepwise reprogramming of B cells into macrophages. Cell 117:663–676

Ye M, Ermakova O, Graf T (2005) PU.1 is not strictly required for B cell development and its absence induces a B-2 to B-1 cell switch. J Exp Med 202:1411–1422

Yoder MC, Hiatt K, Dutt P, Mukherjee P, Bodine DM, Orlic D (1997) Characterization of definitive lymphohematopoietic stem cells in the day 9 murine yolk sac. Immunity 7:335–344

Yokota T, Kouro T, Hirose J, Igarashi H, Garrett KP, Gregory SC, Sakaguchi N, Owen JJ, Kincade PW (2003) Unique properties of fetal lymphoid progenitors identified according to RAG1 gene expression. Immunity 19:365–375

Yokota T, Huang J, Tavian M, Nagai Y, Hirose J, Zuniga-Pflucker JC, Peault B, Kincade PW (2006) Tracing the first waves of lymphopoiesis in mice. Development 133:2041–2051

Yoshida T, Ng SY, Zuniga-Pflucker JC, Georgopoulos K (2006) Early hematopoietic lineage restrictions directed by Ikaros. Nat Immunol 7:382–391

Zon LI (2008) Intrinsic and extrinsic control of haematopoietic stem-cell self-renewal. Nature 453:306–313

GATA1 and GATA2 Function in Hematopoietic Differentiation

Takashi Moriguchi, Mikiko Suzuki, James Douglas Engel, and Masayuki Yamamoto

Abstract Transcription factors GATA1 and GATA2 play essential roles in gene regulation during hematopoietic differentiation. GATA1 is essential for the survival and terminal differentiation of erythroid, megakaryocytic, eosinophilic, and mast cell precursors, whereas GATA2 primarily regulates the proliferation and maintenance of immature progenitors in hematopoietic cells. The activities of GATA1 and GATA2 are regulated in a cell-type–specific manner, and their expression is strictly controlled during hematopoietic development and differentiation. A series of recent investigations highlight the strong correlation between GATA1 misexpression and human hematopoietic disorders, including inherited anemia, thrombocytopenia, and leukemia. This review summarizes current progress in functional analyses of GATA1 and GATA2, their relationship to blood diseases, and regulatory mechanisms that control their gene expression.

Introduction

The mechanisms specifying lineage-specific blood cell differentiation have been carefully evaluated to determine how the activity of a particular transcription factor influences these processes. Once hematopoietic stem cells (HSCs) commit to undergo lineage differentiation, their potential becomes progressively restricted. At that time a hematopoietic lineage-specific gene expression program is initiated by a cascade of transcription factors whose differential expression in hematopoietic

T. Moriguchi, M. Suzuki and M. Yamamoto (✉)
Department of Medical Biochemistry, Tohoku University Graduate School of Medicine, 2-1 Seiryo-cho, Aoba-ku, Sendai 980-8575, Japan
e-mail: masi@mail.tains.tohoku.ac.jp

J.D. Engel
Cell and Developmental Biology, University of Michigan Medical School, Ann Arbor, MI 48109-2220, USA

M. Kondo (ed.), *Hematopoietic Stem Cell Biology,* Stem Cell Biology and Regenerative Medicine, DOI 10.1007/978-1-60327-347-3_5,

progenitor cells promotes a diversity of functions that dictates oxygen-carrying capacity in one cell or elaborates roles in innate and acquired immunity in another related cell.

GATA1 is a prototypical transcription factor that promotes hematopoietic differentiation in the erythroid, eosinophilic, megakaryocytic, and mast cell lineages. GATA2, another member of the GATA transcription factor family, is preferentially expressed in hematopoietic stem and progenitor cells and regulates their proliferation and maintenance. The function of these two GATA factors during hematopoiesis has been extensively examined over the past two decades. A variety of experimental approaches have been adopted to evaluate the activity of *cis*-acting regulatory elements, as well as to examine domain structure-function relationships of the proteins in a more integrated approach. The activity of both GATA1 and GATA2 is regulated at multiple levels, from transcriptional through posttranslational modification. These two GATA factors interact with a variety of other regulatory cofactors to generate active transcriptional complexes. In this chapter, we summarize the recent advances that address the physiological function(s) of the GATA1 and GATA2 proteins as well as the transcriptional regulation of *Gata1* and *Gata2* genes during erythroid and megakaryocytic cell proliferation, differentiation, and homeostasis.

The GATA Family of Transcription Factors

The GATA family of transcription factors is composed of six members (GATA1 through GATA6) in mammals (reviewed in Laverriere et al. 1994; Ohneda and Yamamoto 2002; Orkin 1992; Yamamoto et al. 1990). GATA proteins bind most avidly to the consensus motif (T/A)GATA(A/G) through two characteristic C_4(Cys-X_2-Cys-X_{17}-Cys-X_2-Cys) zinc finger motifs specific to the GATA family (Ko and Engel 1993), and this DNA recognition motif is most highly conserved between the six paralogous members that constitute this small vertebrate multigene family (Patient and McGhee 2002; Yamamoto et al. 1990). GATA1, GATA2, and GATA3 constitute the "hematopoietic GATA" subfamily because of their prominent expression in hematopoietic cells, whereas GATA4, GATA5, and GATA6 constitute a second subfamily of "endodermal" GATA factors that are expressed predominantly in internal organs (Weiss and Orkin 1995). The *Gata* genes contain two alternative first exons and additional five translated exons (Fig. 1). The first exons are utilized in a tissue-restricted manner and contain only 5′-noncoding sequence in the mature mRNAs. Differential utilization of the alternative first exons has been most vigorously explored in the hematopoietic GATA factors. Evidence for differential, tissue-specific first exon utilization was first described for *Gata1* gene (Ito et al. 1993; Fig. 1) and subsequently reported for both *Gata2* (Minegishi et al. 1998; Pan et al. 2000) and *Gata3* genes (Asnagli et al. 2002).

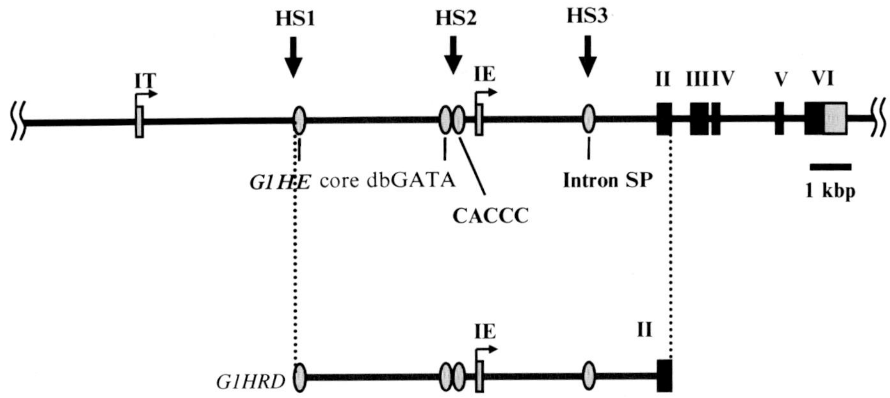

Fig. 1 Mouse *Gata1* gene. The mouse *Gata1* gene harbors two tissue-specific alternative first exons (IT, testis-specific exon 1; IE, hematopoietic-specific exon 1) and five coding exons (II–VI). The positions of three DNase I hypersensitive (HS) sites (HS1, HS2, and HS3) as well as the correspondent four regulatory modules (*G1HE* core, dbGATA, CACCC, and intron SP) referred to in the text are indicated. *G1HRD* (*Gata1* hematopoietic regulatory domain) bears 3.9 kbp of 5' flanking sequence plus the 4.2 kbp first intron, and includes the four *Gata1* regulatory modules

Expression Profiles of GATA1 and GATA2

GATA1 expression is highly restricted. GATA1 is expressed in erythroid cells, megakaryocytic cells, eosinophils, dendritic cells, and mast cells of hematopoietic lineage, whereas GATA2 is expressed in many multilineage progenitors and HSCs (Leonard et al. 1993; Minegishi et al. 1998; Nagai et al. 1994; Suzuki et al. 2006). Within the erythroid differentiation cascade, GATA1 expression can be detected as early as in common myeloid progenitors (CMPs), and subsequently its expression level dramatically increases when those progenitors differentiate into proerythroblasts (Kuhl et al. 2005; Suzuki et al. 2003). From the proerythroblast stage onward, GATA1 expression levels decrease as the red blood cell matures (Suzuki et al. 2003; Fig. 2). Throughout the erythroid differentiation process, GATA1 orchestrates a series of gene expression changes, which promote essential steps required for the proliferation and differentiation of erythroid progenitors. As erythroid cells enter terminal erythroid differentiation, GATA1 directly activates erythroid lineage-affiliated genes (i.e., β-globin, *Alas2*, and *Gata1* itself), while at the same time repressing *Gata2*, *c-Kit*, *c-Myb*, and *c-Myc*, all of which are essential for progenitor proliferation in earlier stages of hematopoiesis (Ferreira et al. 2005). Through gene ablation experiments, we learned that GATA1 is essential for erythroid differentiation *in vivo* (Fujiwara et al. 1996; Takahashi et al. 1997). Closer examination of both *Gata1*-deficient mice and *Gata1*-deficient embryonic stem-derived (ES-derived) cell lines revealed that erythroid cells lacking GATA1 fail to maturate beyond the proerythroblast stage (Pan et al. 2005; Pevny et al. 1995; Weiss et al. 1994).

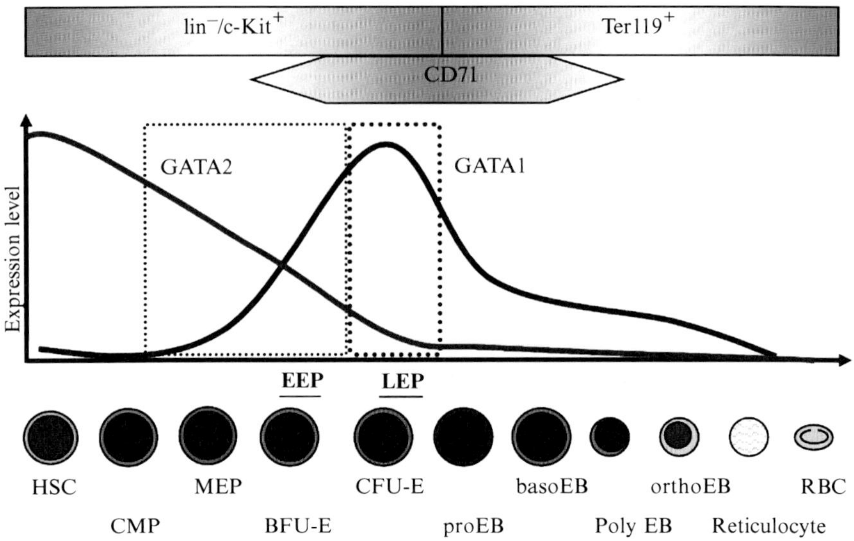

Fig. 2 Dynamic changes in GATA1 and GATA2 abundance during erythroid cell differentiation. GATA2 is expressed at high level in early erythroid progenitors (EEP). Once GATA1 expression is activated, GATA2 expression decreases as the GATA1 level increases. The lineage marker-negative (Lin^-), c-Kit, and CD71 double-positive fraction of *G1HRD*-GFP bone marrow cells is designated as late erythroid progenitors (LEP) and contains abundant colony-forming units-erythroid (CFU-E), while the c-Kit^+/$CD71^{low\ or\ -}$ fraction is referred to as the early erythroid progenitors (EEP) fraction and primarily contains burst forming unit-erythroid (BFU-E) (Suzuki et al. 2003)

GATA2 is preferentially expressed in HSCs and early hematopoietic progenitor cells and regulates development and proliferation of HSCs and multipotent hematopoietic precursors (Tsai et al. 1994; Tsai and Orkin 1997). GATA2 haploinsufficiency impairs the quality of both embryonic and adult HSCs, thus eventually causing a reduction in the number of early HSCs (Rodrigues et al. 2005). However, during erythroid differentiation GATA2 expression declines coincident with GATA1 induction around the later stages of c-Kit^+ erythroid progenitor differentiation (Fig. 2; Cheng et al. 1996; Grass et al. 2003; Weiss et al. 1994; Yamamoto et al. 1990).

Assuming that we might encounter difficulties in establishing mice that were mutant in the X chromosome–linked *Gata1* gene (in ES cells of X/Y karyotype), we instead generated a mutant mouse line bearing a promoter-specific *Gata1* knockdown allele (Takahashi et al. 1997). This mutant allele bears a neomycin resistance cassette (neo) inserted immediately 5′ to the *Gata1* IE exon, thus separating three known upstream *cis*-acting elements (the *Gata1* hematopoietic enhancer [G1HE], double GATA site, and CACCC site) from the hematopoietic transcription start site, and was originally created anticipating that the insertion would interfere with the transcription of *Gata1* (Takahashi et al. 1997). In fact, *Gata1* mRNA is suppressed to approximately 5% of the wild-type level in this mutant allele, so we designated this allele *Gata1.05*. Hemizygous ($X^{G1.05}$/Y) male embryos harboring this knockdown allele die at around embryonic day 10.5 (E10.5) due to hematopoietic deficiency, suggesting that 5% of GATA1 production is insufficient for primitive erythropoiesis. In contrast, although heterozygous ($X^{G1.05}$/X) female embryos were born at a lower frequency than expected from Mendelian statistics, they exhibited various degrees of anemia during the perinatal stage, presumably due to random X chromosome inactivation (Takahashi et al. 1998).

These surviving $X^{G1.05}$/X female mice recovered from anemia before adolescence and were fertile. Therefore, this mutant allele can be transmitted from mother to daughter in a stable fashion, but not to male offspring. Taking advantage of this inheritance profile, we initiated a transgenic complementation rescue analysis of *Gata1.05* mutant embryos by crossing male transgenic mice expressing GATA1, GATA2, or GATA3 cDNAs under the control of the *Gata1* gene hematopoietic regulatory domain (*G1HRD*; 3.9 kbp 5′ region plus 4.2 kbp containing the *Gata1* first exon and intron; Figs. 1 and 5), which recapitulates the endogenous GATA1 expression profile. We found that transgenic GATA1, GATA2, or GATA3 expression rescued *Gata1.05* knockdown mutant mice from embryonic lethality (Takahashi et al. 2000). These data led to the conclusion that the stage-specific dynamic regulation of GATA factor expression during erythroid differentiation is more important than the identities of individual GATA transcription family members (Ferreira et al. 2007).

Importantly, forced GATA1 expression in the terminal erythroid differentiation stage leads to maturation arrest, indicating that aberrant GATA1 accumulation inhibits terminal differentiation (Ferreira et al. 2007; Whyatt et al. 2000), suggesting that turning off GATA1 expression could be as important as turning it on. However, the specific regulatory mechanisms directing GATA1 reduction during terminal erythroid maturation are still elusive. Although the mechanisms that regulate how persistent, high-level GATA1 expression affects terminal maturation of erythroid cells are poorly understood, glimpses into mechanisms are slowly coming to light. For example, it was reported recently that caspase 3 induces GATA1 protein degradation in erythroblasts, which is normally suppressed by erythropoietin signaling (De Maria et al. 1999; Ribeil et al. 2007). Thus, posttranslational modifications or protein degradation might also contribute to the dynamic changes in GATA1 protein levels during erythroid differentiation and maturation.

The partially overlapping expression pattern of GATA1 and GATA2 in some hematopoietic cells implies a functional redundancy between these two closely related proteins, and the very severe hematopoietic deficiencies observed in *Gata1/Gata2* compound knockout mice support this notion. *Gata1*$^{+/-}$:*Gata2*$^{+/-}$ compound mutant embryos die at midgestation from hematopoietic deficiency, and yet single heterozygotes for either gene live to adulthood and are fertile (Fujiwara et al. 2004). However, several reports have also identified distinct, presumably nonoverlapping functions for these two GATA family members. This is highlighted by the fact that *in vitro* differentiated GATA1-deficient ES cells (*G1E* cells) induce compensatory 50-fold GATA2 expression, yet fail to circumvent the differentiation arrest and apoptotic cell death of GATA1-deficient cells (Weiss and Orkin 2005). Whereas transgenic GATA2 and GATA3 expression rescue the embryonic lethality of the *Gata1.05* knockdown mutant, these transgenic complementation mutants reproducibly exhibited hematopoietic disorders after reaching maturity (Takahashi et al. 2000). The latter observations revealed inherent biochemical differences between GATA1 and GATA2 and suggested that the balance of expression between these GATA factors could be quite important in the maintenance of hematopoiesis.

Transcriptional Regulation of the *Gata1* Gene

The GATA1 Hematopoietic Regulatory Domain (G1HRD)

Mouse and rat harbor a *Gata1* gene that contains two noncoding first exons, termed IT and IE, and five coding exons (Fig. 1; Ito et al. 1993; Tsai et al. 1991). The distal, IT promoter and first exon primarily direct *Gata1* expression in testicular sertoli cells, while the proximal IE promoter and first exon mediates expression of *Gata1* in hematopoietic cells (Ito et al. 1993). Both testicular- and erythroid-specific promoters contain GATA binding sites that are required for the proper function of these tissue-specific promoters, suggesting a possible autoregulatory loop in both cell types (Ito et al. 1993; Onodera et al. 1997; Vyas et al. 1999; Tsai et al. 1991). Recently, two additional minor alternative first exons were identified in the first intron: IE_B is located 3.7 kbp downstream of the authentic IE exon, and IE_C is found 42 bp 3′ to exon $1E_B$ (Dyer et al. 2007). Specific function of these minor transcriptional start sites is under investigation.

The 8 kbp region spanning from 3.9 kbp upstream of exon IE to the second exon contains sufficient regulatory information to confer tissue-specific expression of green fluorescent protein (GFP) or β-galactosidase reporter gene in both yolk sac–derived primitive and fetal liver–derived definitive hematopoietic cells in transgenic mouse assays (Onodera et al. 1997; Box 1). We originally designated this region as *G1HRD* (Motohashi et al. 2000; Fig. 1). A GATA1 cDNA expressed under the direction of *G1HRD* in transgenic mice restored hematopoiesis to GATA1-deficient mice, rescuing them from embryonic lethality. This observation demonstrates that *G1HRD* harbors sufficient regulatory information to support physiologically essential hematopoiesis in the mouse (Takahashi et al. 2000).

Sequences and Mechanisms that Regulate Gata1 Transcription

Much research has been conducted on *Gata1* transcriptional regulation, focusing on several *cis*-regulatory elements located near the 5′-flanking region of the gene. DNase I hypersensitive (HS) site mapping of erythroid cell lines revealed the three major HS sites in the *Gata1* locus. HS1 is located 3.7 kbp 5′ to exon IE, HS2 is located in the IE promoter region and HS3 is located within the first intron (Fig. 1;

McDevitt et al. 1997; Valverde-Garduno et al. 2004). Earlier studies employing transfection assays into erythroid cell lines (e.g., K562 or mouse erythroleukemia [MEL] cells) revealed that these elements bore regulatory activity (Nicolis et al. 1991). Studies from our group and others revealed the physiological functions of those *Gata1* regulatory regions *in vivo* utilizing a *G1HRD*-based β-galactosidase reporter transgenic system (Nishimura et al. 2000; Onodera et al. 1997; Vyas et al. 1999). In the following sections, we consider recent results that enlighten the precise role of each of the *Gata1* transcriptional regulatory modules.

***Gata1* Hematopoietic Enhancer**

G1HE is located in the 5′ end of the *G1HRD* (Fig. 1). In transgenic mouse assays, deletion of 1.3 kbp (thus including *G1HE*) from the 5′ end of *G1HRD* significantly suppressed reporter gene expression both in the yolk sac and fetal liver hematopoietic cells (Nishimura et al. 2000; Onodera et al. 1997). We concluded that this 1.3 kbp region was responsible for *G1HE* activity. *G1HE* coincides with an erythroid-specific DNase I HS site (HS1), and histone H3/H4 hyperacetylation has been found in the *G1HE* region of MEL cells (McDevitt et al. 1997; Valverde-Garduno et al. 2004).

Further dissection of *G1HE* revealed that a 235 bp region at the most 5′ end of *G1HE* is highly conserved among species. This core element plays an essential role in enhancer activity and was thus aptly named as the *G1HE* core region (Nishimura et al. 2000; Ohneda et al. 2002). The *G1HE* core region contains a highly conserved GATA binding site that is separated by 10 bp from an adjacent E-box. This sequence arrangement binds a hematopoietic-specific transcription factor complex that includes GATA1, SCL/tal-1, E2A, Lmo2, and Ldb-1 (Wadman et al. 1997). Substitution mutations within the GATA site completely abolished complex formation, whereas mutation of the E-box motif barely affected complex formation, suggesting an essential requirement of the GATA site in generating a stable E-box-GATA complex (Nishimura et al. 2000; Vyas et al. 1999).

More precise analysis of the *G1HE* core region demonstrate that the 5′-most 149 bp, including the conserved GATA site, is indispensable for erythroid-specific reporter gene expression, whereas the 3′-most 86 bp of the *G1HE* core is required for megakaryocyte-specific expression (Nishimura et al. 2000). Extensive transgenic analysis revealed a critical 25 bp element within this 86 bp 3′ core region that is required for megakaryocytic expression (Guyot et al. 2006). Guyot et al. proposed that the Kruppel-related zinc finger transcription factor ZBP89 was responsible for the binding and transcriptional activity derived from this megakaryocyte-specific regulatory element. Furthermore, three ZBP89-immunoprecipitated binding sites in a megakaryocyte cell line correspond to the same three major DNase I HS sites that contain GATA binding sites in the *Gata1* locus (–3.1 kbp, IE promoter and +3.5 kbp; Guyot et al. 2006). Consistently, Woo et al. revealed that ZBP89 forms a transcriptional complex with GATA1 in megakaryocytic cell lines, further supporting that a GATA1-ZBP89 complex regulates megakaryocytic GATA1 expression through the identical GATA sites that control *Gata1* erythroid-specific expression (Woo et al. 2008).

CACCC Box

CACCC box was initially recognized as a DNase I HS site HS2 located within the IE promoter (Fig. 1). This site contributes significantly to *Gata1* transcription activity in transfection assays (Tsai et al. 1991) and is occupied by acetylated histones in the endogenous *Gata1* locus in erythroid cell lines. Transgenic reporter experiments demonstrated that the CACCC box was essential for the initiation of *Gata1* transcription, as mutation or deletion of this element completely disrupted *G1HRD*-based transgene promoter activity (Ohneda et al. 2002). One of the most plausible *trans*-activating factors that binds to this element is erythroid Krüppel-like factor (EKLF or KLF1), although no direct evidence for this assertion has been reported. Recently it was proposed that SP1/SP3 binds to this *Gata1* CACCC site, allowing its transcriptional activation in K562 cells (Hou et al. 2008). Further investigation will be required to clarify the physiological significance of SP1/SP3 versus KLF binding in *Gata1* transcriptional activation.

Double GATA Site

A palindromic GATA binding site (dbGATA), which has eight-fold higher binding affinity than a single GATA site, is located 700 bp 5′ to the IE transcriptional start site (Trainor et al. 1996; 2000). This regulatory element was originally discovered in MEL cells and K562 cells (Nicolis et al. 1991; Tsai et al. 1991). This dbGATA site is essential for GATA1 expression in fetal liver hematopoietic cells (Ohneda et al. 2002). Germ line mutation of the dbGATA site in mice leads to selective loss of the eosinophil lineage, and this mutant mouse (ΔdbGATA) exhibits resistance to the pathological changes that occur during chronic bronchial asthma (Humbles et al. 2004; Yu et al. 2002a). Because ΔdbGATA mutant mice exhibit only mild suppression of erythrocyte differentiation, the significance of the dbGATA site in *Gata1* gene expression in erythroid progenitor cells has largely been unidentified (Yu et al. 2002a).

CP2, a ubiquitously expressed transcription factor belonging to the *Drosophila* grainyhead-like gene family, is known to coordinate globin-gene expression. CP2 binding sites are usually located adjacent to GATA binding sites in the promoter and enhancer of erythroid genes (e.g., GATA1, EKLF, and p45 NF-E2). Indeed, a CP2-binding site lies adjacent to the dbGATA site in the *Gata1* regulatory region and presumably forms a transcriptional complex with GATA1 in K562 human erythroid cells (Bosè et al. 2006). Direct interactions between CP2 and GATA1 might therefore be responsible for the erythroid-specific function(s) of the ubiquitously expressed CP2 factor. Elucidation of CP2 function in *Gata1* expression *in vivo* is currently under investigation.

Intron SP Element

The third major DNase I HS site is known to be located within the first intron in erythroid cell lines (Nicolis et al. 1991). We initially identified that sequences within the first intron are required for definitive hematopoietic cell-specific *G1HRD* transgene expression, whereas this region was dispensable for yolk sac primitive erythroid cell–specific expression (Onodera et al. 1997). In the transgenic reporter analysis, we further identified that *Gata1* expression in definitive erythroid cells and megakaryocytes requires a 320 bp sequence (intron SP element, where S and P stand for the *Sac*I and *Pst*I sites in the *Gata1* first intron, respectively) containing GATA and AP1 repeats (Ohneda et al. 2002). A highly conserved $[GATA]_7$ motif, as well as two new transcription start sites (IE_B and IE_C), were also recently identified in this region (Dyer et al. 2007; Seshasayee et al. 2000; Ohneda et al. 2002). A broader understanding of the physiological function of this intron SP element in a genomic context is a current ongoing goal.

Differences Between Human GATA1 Gene and Mouse Gata1 Gene Transcription

Recently it was suggested that species-specific sequences flanking the *Gata1* gene are responsible for the differences in murine and human GATA1 expression. Comparison of DNA sequences by DNase I HS site-mapping, along with analysis of the acetylation status of histones H3 and H4 over a 120 kbp flanking region of the human *GATA1* and mouse *Gata1* genes revealed that the chromatin structures of the two orthologous loci are significantly different. There are two novel erythroid cell-specific GATA binding sites, one in each species (at –25 kbp in the mouse genome and +15 kbp in the human genome), which are not conserved either in position or in surrounding sequences. These findings indicate that some *cis*-acting elements regulate the human and mouse *Gata1* genes differently. The human-specific regulatory mechanisms might be an interesting avenue for further investigation, assuming that these sequences have the potential to cause or affect hematopoietic disorders (Valverde-Garduno et al. 2004).

Use of G1HRD and Gata1 BACs for Characterization of Erythroid Progenitors

One of the first applications we performed exploiting the *Gata1* gene regulatory domains was to establish a new cell sorting strategy for early erythroid progenitors (EEP) based on the expression of *G1HRD*. Although the separation of erythroblasts at different erythroid differentiation stages has usually been conducted using Ter119 and CD71 (transferrin receptor) antibodies (Socolovsky et al. 2001), no practical method was developed to decipher more immature erythroid progenitors. The most immature erythroid-committed progenitor is designated as a burst forming unit-erythroid (BFU-E; Gregory and Eaves 1977; 1978). The BFU-E gives rise to more differentiated erythroid progenitor cells with less proliferative potential, the colony-forming units-erythroid (CFU-E). Subsequently, CFU-E cells differentiate into erythroblasts, which later on extrude their nuclei (as reticulocytes) and then become terminally mature circulating erythrocytes. Thus, the two most immature erythroid progenitors (BFU-E and CFU-E) could only be identified retrospectively by colony-forming assays (Stephenson et al. 1971).

In order to isolate these two erythroid progenitor fractions prospectively, cells were sorted using a protocol that utilizes mouse bone marrow from transgenic lines expressing the GFP reporter transgene under the control of *G1HRD*. Two putative erythroid progenitor fractions were identified as GFP⁺ cells from the bone marrow preparations of *G1HRD*-GFP transgenic mice (Suzuki et al. 2003). The c-Kit and CD71 double-positive fraction, designated as late erythroid progenitors (LEP), contains primarily CFU-E, whereas the c-Kit⁺/CD71-low or negative fraction (called EEP) contains primarily BFU-E (Fig. 2; Suzuki et al. 2003).

Although the *G1HRD* transgenic mouse system provides significant new insights into the *Gata1* gene regulation, there are several limitations to this approach. For example, as often happens in other transgenic mouse experiments, the *G1HRD* activity is subject to position-effect variegation (Onodera et al. 1997; Suzuki et al. 2003). A more serious problem is that *G1HRD* appears insufficient to recapitulate fully the *Gata1* gene expression profile, especially in c-Kit⁺ EEPs (Suzuki et al. 2003).

To circumvent this shortcoming, we began to explore the potential of *Gata1* bacteria artificial chromosome (BAC) transgenic mice. We found that one BAC (RP23-443E19) harbors an approximately 196 kbp genomic fragment including the *Gata1* gene (Fig. 3), providing significantly more potential regulatory capacity than the plasmid-based transgene, as *G1HRD* is approximately 8 kbp in length. Thus, the *Gata1* BAC clone may contain a more comprehensive set of *Gata1* gene regulatory

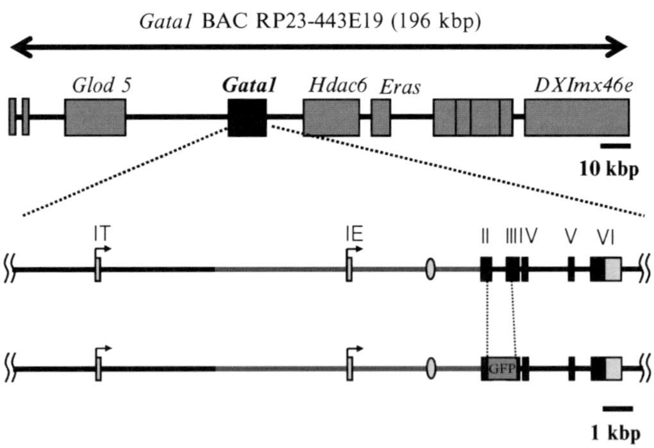

Fig. 3 Genes in the vicinity of *Gata1* in bacteria artificial chromosome (BAC) clone RP23-443E19. This BAC covers a 196 kbp genomic region including and surrounding *Gata1* gene. A green fluorescent protein (GFP) expression cassette was inserted at the translation start site in the second exon (bottom line) using bacterially based homologous recombination (Suzuki et al., 2009)

elements that is able to more accurately mediate proper spatiotemporal gene expression. Our most recent studies indicate that these transgenic BAC mouse lines expressing GFP under the regulatory influence of the 196 kbp of mouse *Gata1* gene plus flanking sequences recapitulate endogenous *Gata1* gene expression more faithfully than the *G1HRD* transgenic plasmid (Suzuki et al., 2009).

Transcription Factors Interacting with GATA1 in Higher Transcriptional Complexes

FOG1 (Friend of GATA1)

Friend of GATA1 (FOG1) is a 998 amino acid protein that was identified by yeast two-hybrid assays as a GATA1 interacting molecule (Tsang et al. 1998). FOG1 contains nine predicted zinc finger domains, which were specifically reported to interact with the N-fingers of GATA1, but primarily through interactions with the sixth zinc finger (Fig. 4A; Fox et al. 1998). Other fingers, such as the first, fifth, and ninth fingers, also interact with GATA1 (Fox et al. 1999). FOG1 is abundantly expressed in erythroid and megakaryocytic lineages. *Fog1*$^{-/-}$ mice die in utero during midgestation (E10.5–E11.5) due to defective primitive and definitive erythropoiesis. The anemia that develops is quite similar to that observed in *Gata1*-deficient embryos, suggesting that FOG1 cooperatively aids GATA1 in mediating hematopoiesis (Tsang et al. 1998).

Analysis of GATA1 point mutations identified which amino acids in N-finger (Fig. 4B) lead to defective physical interactions with FOG1 and proved that these interactions are required for GATA1 function during erythroid differentiation (Crispino et al. 1999). Such mutations have been identified in patients with X-linked anemia and thrombocytopenia (Table 1; Nichols et al. 2000; Freson et al. 2001, 2002; Mehaffey et al. 2001; Yu et al. 2002b; Phillips et al. 2007). One example is the germ line V205M mutation (Nichols et al. 2000). Amino acid 205 is located within the N-finger of GATA1, and mutation of this residue in mice results in embryonic lethality around E11.5 due to fetal anemia (Chang et al. 2002). Complementation of the *Gata1.05* knockdown mutant with a GATA1 V205G transgene also failed to fully prevent this anemia, suggesting that amino acid 205 is essential for the interaction between GATA1 and FOG1 in erythropoiesis (Shimizu et al. 2004).

PU.1

PU.1 is a member of the Ets family transcription factors and is essential for the development of granulocytic, monocytic, and lymphoid cells (Scott et al. 1994; McKercher et al. 1996; Hromas et al. 1993). PU.1 induces the expression of the

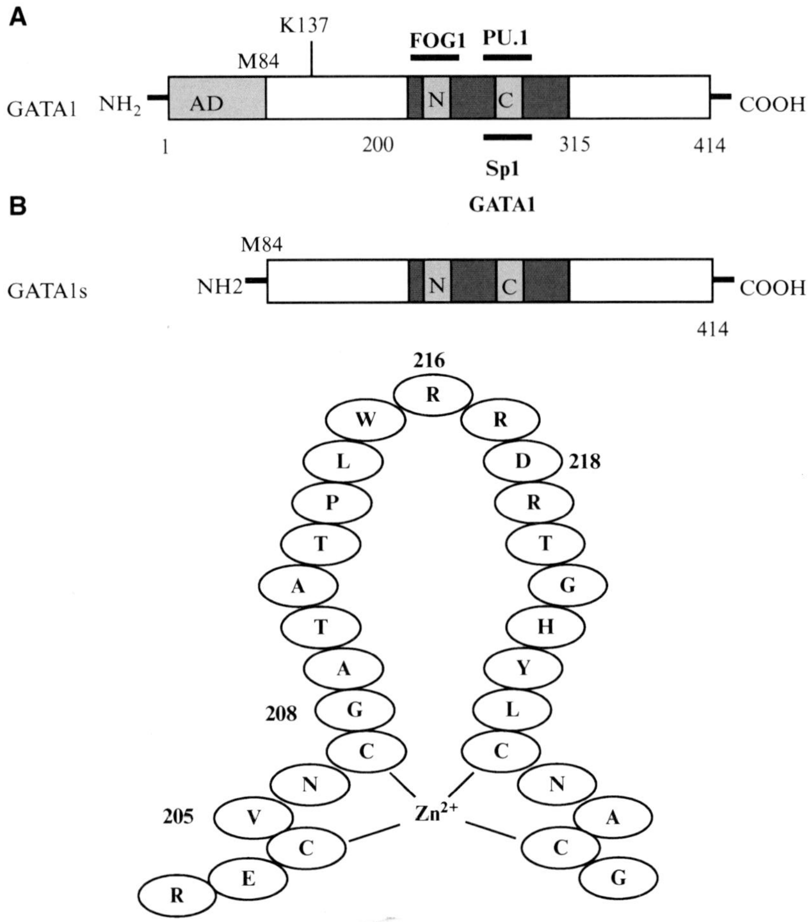

Fig. 4 Positions of GATA1 mutations. **(A)** Schematic representation of wild-type GATA1. The amino-terminal (putative transcriptional activation) domain, as well as the amino (N) and carboxyl (C) zinc finger domains, is indicated. Friend of GATA1 (FOG1) interacts with the N-finger, whereas PU.1 and Sp1 associate with the C-finger. The sumoylation site (K137) is indicated (*top*). An N-terminal truncated isoform of GATA1 is generated in transient myeloproliferative disorder (TMD) and acute megakaryocytic leukemia (AMKL) associated with Down syndrome (DS) (GATA1s; *below*). **(B)** Amino acid sequences of the N-finger of human GATA1. Substitution mutations at V205, G208, R216, and D218 have been identified in human hematopoietic disorders (see Table 1)

granulocyte macrophage colony-stimulating factor (GM-CSF), granulocyte colony-stimulating factor (G-CSF), and macrophage colony-stimulating factor (M-CSF) receptors, and controls the responsiveness of myeloid progenitors to those myeloid-specific cytokines. Transgenic expression of PU.1 in the murine erythroid lineage results in erythroleukemia at higher rates than wild-type controls, presumably because ectopic expression of PU.1 in erythroid cells suppresses erythroid

Table 1 GATA1 substitution mutations involved in human hematopoietic disorder

	Symptom		Reduced affinty to	
Mutatiaon	Thrombocytopenia	Dyserythropoiesis	FOG1	DNA
V205M	Severe	Severe	Severe	None
G208S	Severe	None	Mild	None
R216Q	Mild	Thalassemia (Gray platelet syndrome)	None	Severe[a]
R216W	Severe	Thalassemia HbF Porphyria	N/A	N/A
D218G	Severe	Mild	Mild	None
D218Y	Severe	Severe	Severe	None

N/A, not assessed. HbF, fetal hemoglobin. [a]Reduced binding affinity for palindromic GATA sites, but normal association with single GATA site

maturation (Moreau-Gachelin et al. 1996). Meanwhile, forced expression of GATA1 in myelomonocytic cells transforms them into erythroid, megakaryocytic, and eosinophilic cells (Kulessa et al. 1995; Visvader et al. 1992). These observations, together with the observed mutually exclusive expression patterns of PU.1 and GATA1, originally suggested an antagonistic effect between these two transcription factors, possibly through direct physical interactions (Fig. 4A; Nerlov et al. 2000; Rekhtman et al. 1999; Zhang et al. 1999, 2000). It was found that the PU.1 and GATA1 interaction is mediated via the DNA binding domain of both proteins (Rekhtman et al. 1999; Zhang et al. 2000), but those two transcription factors antagonize each other through quite distinctive mechanisms: GATA1 inhibits PU.1 by competing for its interaction with co-activator c-Jun (Zhang et al. 1999), while PU.1 suppresses the DNA binding of GATA1 by masking the C-finger DNA binding domain (Fig. 4A; Zhang et al. 2000).

NLI/Ldb1, LMO2, SCL/TAL1-E2A Complex

GATA1 is essential for the formation of a DNA binding complex composed of NLI/Ldb1, LMO2 and a SCL/TAL1-E2A heterodimer in murine erythroid cell extracts (Wadman et al. 1997). This complex binds both a GATA site and an adjacent E-box (usually separated by nine nucleotides), creating a bridge between the two. GATA1 and SCL, in the presence of Lmo2, synergistically activate transcription of GATA-E-box reporter genes (Osada et al. 1997). The GATA-SCL complex functions as a positive regulator of several erythroid genes, including EKLF, β-globin, and glycophorin A, all by mediating complex formation at the GATA-E-box in their promoters (Anderson et al. 1998; Cohen-Kaminsky et al. 1998; Lahlil et al. 2004; Lécuyer et al. 2002; Wadman et al. 1997; Xu et al. 2003). It has been reported that approximately 10% of the SCL-immunoprecipitated DNA fragments from erythroleukemia cell lines contain adjacent GATA binding sites (Cohen-Kaminsky et al. 1998). The E-box-GATA complex is also observed within the *G1HE* core region of the 5' *Gata1* gene enhancer, and this composite element is essential for positive autoregulation of *Gata1* transcription (Nishimura et al. 2000; Valverde-Garduno et al. 2004). This GATA1-SCL complex also binds within the β-globin locus control region (LCR) and functions as a potential mediator of long-range interactions between the β-globin promoter and the LCR (Song et al. 2007).

Domain Function Analyzes of GATA1

Three domains within the GATA1 protein have been shown to be responsible for discrete functions (Fig. 4A): an N-terminal activation domain and two highly conserved zinc finger domains ("N-finger" and "C-finger"). The C-finger is essential for the recognition of the GATA box and binding to DNA (Yang and Evans 1992; Martin and Orkin 1990). The function of the N-finger is to stabilize GATA1 DNA binding to palindromic GATA sites (Newton et al. 2001; Pedone et al. 1997; Trainor et al. 1996, 2000). Another important property of the N-finger is to promote complex formation with FOG1 (Fox et al. 1998). It has been demonstrated that GATA1 self-association occurs via the C-finger, as well as its interactions with Sp1 and EKLF (Crossley et al. 1995; Gregory et al. 1996; Merika and Orkin 1995). GATA1 homodimerization is essential for positive autoregulation and maintenance of GATA1 abundance in zebrafish and in mice (Kobayashi et al. 2001; Nishikawa et al. 2003; Shimizu et al. 2007).

In vitro erythroid differentiation studies using G1E cells (derived from immortalized GATA1 deficient ES cells) demonstrated that, while both N-finger and C-finger domains are essential for erythropoiesis, the N-terminal transactivation domain is dispensable for erythroid and megakaryocytic differentiation (Weiss et al. 1997). By contrast, our *in vivo* transgenic complementation rescue analysis revealed that the N-terminal domain possesses distinct functions that are required for hematopoiesis (Shimizu et al. 2001). In the complementation analysis, $X^{G1.05}/X$ female mice were crossed with transgenic male mice expressing various GATA1 domain deletion mutants driven by *G1HRD* and the specific role of each functional domain was determined by the hematopoietic-deficient phenotype (Fig. 5). Consistent with the *in vitro* G1E analysis, the C-finger deletion of GATA1 (ΔCF) failed to support either primitive or definitive embryonic erythropoiesis. Interestingly, whereas the N-finger deletion mutant (ΔNF) was able to support primitive erythropoiesis, it did not support definitive erythropoiesis.

We found that mice expressing GATA1 without N-terminal transactivation domain (ΔNT) at wild-type abundance failed to reconstitute definitive erythropoiesis, whereas mice expressing more abundant ΔNT transgene maintained both primitive and definitive erythropoiesis. Based on these observations, we conclude that all three GATA1 domains are essential for definitive erythropoiesis, while the GATA1 N-terminal transactivation domain and the N-finger are both dispensable for primitive erythropoiesis (Shimizu et al. 2001). These results imply that

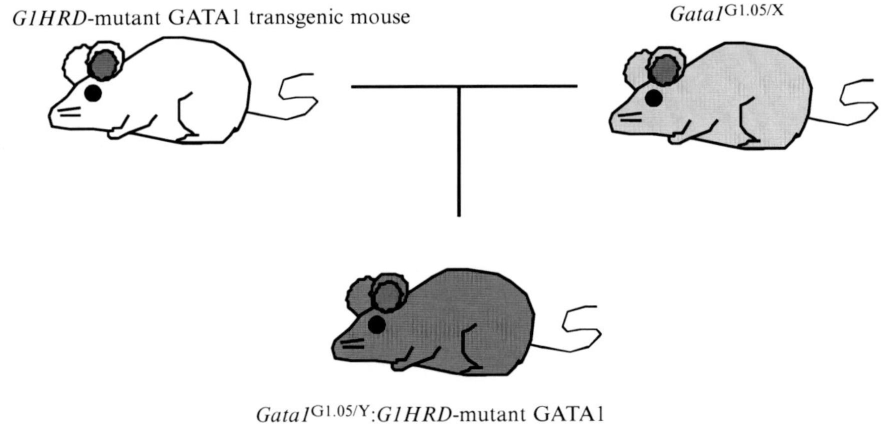

Fig. 5 Schematic depiction of the transgenic complementation rescue analysis. The breeding of $X^{G1.05}$/X mutant mice with transgenic mice expressing various mutant GATA1 alleles or mutated cDNAs results in transgenic-rescued $X^{G1.05}$/Y progeny. Hematopoietic deficiencies detected in the mutant GATA1-rescued $X^{G1.05}$/Y progeny directly reveal the physiological function of each domain or amino acid residue of GATA1

primitive and definitive erythropoiesis require distinctly different modes of domain utilization to mediate GATA1 function, and further that the transgenic complementation approach provides a powerful tool to address the functional contributions of specific GATA1 domains to hematopoiesis.

To analyze the function of each amino acid residue within the GATA1 N-finger domain, the transgenic complementation approach was also utilized. As described above, the first GATA1 structural mutation linked to human disease was identified in a patient with hereditary thrombocytopenia (Nichols et al. 2000). The patient carries a substitution of valine 205 in the NF domain to methionine and presents clinically with severe dyserythropoietic anemia and thrombocytopenia (Table 1). This is likely because the V205M mutation abrogates the GATA1-FOG1 interaction. Since that first report, several other mutations in the GATA1 N-finger domain have been identified in human hereditary diseases (Table 1 and Fig. 4B). During this time, a mouse model harboring the V205G GATA1 mutation was generated, and these mice succumb to dyserythropoietic death around midgestational age, making it difficult to obtain adult mutant animals in which the contributions of the V205G GATA1 mutation to adult hematopoiesis can be analyzed (Chang et al. 2002).

Consistent with this report, we found that *Gata1.05* mice complemented with a transgenic V205G GATA1 mutant failed to be rescued from embryonic lethality. By contrast, however, more abundant expression of the same GATA1 V205G mutant directed by the *G1HRD* rescued GATA1 knockdown ($X^{G1.05}$/Y) male mice from the embryonic lethality (Shimizu et al. 2004). The rescued mutant mice were fertile and thrombocytopenia was the only noted phenotypic abnormality under unstressed conditions. Thus, the GATA1-FOG1 association is required critically for megakaryopoiesis. One interesting feature of the rescued mice is that upon challenge of anemic stresses, recovery from the anemia is substantially delayed in the mice, suggesting that erythropoiesis is also recovered partially even by the abundant expression of the GATA1 V205G mutant (Shimizu et al. 2004).

Posttranslational Modification of GATA1

In the erythroid differentiation process, the physiological function of GATA1 is regulated by a variety of posttranslational modifications. For example, GATA1 is acetylated at four major sites (K245, K246, K252, and K312) and seven more minor sites (K233, K287, K299, K308, K314, K315, and K316). These modifications are thought to augment GATA1 DNA binding and to enhance transcription from a GATA1-dependent promoter (Boyes et al. 1998; Hung et al. 1999). Mutation of the main acetylation sites eliminates the ability of GATA1 to induce erythroid differentiation in G1E cells, suggesting that GATA1 acetylation is an essential requirement for proper hematopoiesis (Lamonica et al. 2006).

Similarly, GATA1 is phosphorylated constitutively at six serine residues within the amino terminus (S26, S49, S72, S142, S178, and S187). Phosphorylation of a seventh residue (S310) near the carboxyl boundary of the DNA binding domain occurs following induction of erythroid differentiation (Crossley and Orkin 1994) or in response to erythropoietin signaling via the PI3-kinase/AKT signaling pathway, thus enhancing GATA1 transcriptional activity (Zhao et al. 2006). Phosphatase inhibition increases GATA1 binding to target sequences in the human erythroid cell line K562 (Partington and Patient 1999).

Although GATA1 is mainly known as a transcriptional activator, a number of downstream genes have been identified as negatively regulated target genes of GATA1 during hematopoiesis. GATA1 induced cell cycle arrest of differentiating erythrocytes by suppressing two genes, c-Kit and c-Myc, through direct binding to regulatory regions on both genes (Munugalavadla et al. 2005; Rylski et al. 2003). Nearing the completion of terminal differentiation of the erythroblasts, *Gata2* transcription is suppressed directly by GATA1 that binds to the GATA sites at –2.8 and –1.8 kbp relative to the *Gata2* IS hematopoietic promoter (Grass et al. 2003; Martowicz et al. 2005; Ohneda and Yamamoto 2002). The molecular mechanism by which GATA1 mediates transcriptional suppression remains largely unexplored.

Recently it was reported that K137 residue near the transactivation domain of mouse GATA1 was sumoylated (Fig. 4A; Collavin et al. 2004). SUMO (small ubiquitin-like modifier) covalently links to substrates by forming an isopeptide bond with lysine residues through a multistep process analogous to ubiquitination. Sumoylation is carried out via the consecutive actions of three enzymes: an activating enzyme (E1; Aos1/Uba2), a conjugating enzyme (E2; Ubc9), and an E3 ligase (i.e., PIAS family proteins; Sachdev et al. 2001). SUMO conjugation requires the consensus sequence ΨKxE (Ψ; large hydrophobic residue; x, any amino acid) around the target lysine, and K137 of mouse GATA1 fits this profile (Johnson 2004). SUMO modification of transcription factors is often associated with transcriptional repression. A series of experiments has been performed to clarify the physiological significance of SUMO modification of GATA1 using either erythroid or nonerythroid cell lines. However, those analyzes encountered difficulties in revealing the functional differences between sumoylated- and nonsumoylated-GATA1 (Collavin et al. 2004). Our recent analysis using the transgenic complementation strategy suggests that sumoylation of GATA1 actually contributes to the suppression of a subset of negatively regulated target genes (Ohmori et al., unpublished observation).

GATA1-Related Leukemias

There is an alternative translational initiation site located at GATA1 M84 (Calligaris et al. 1995). This alternative start codon produces a truncated 40 kDa protein, GATA1s or ΔNT-GATA1, which lacks the first 83 amino acids of GATA1 or the N-terminal transactivation domain (Fig. 4A). ΔNT-GATA1 can be detected in MEL and K562 cells as well as in mouse tissues, but expression levels are far lower than full-length GATA1. The ΔNT-GATA1 protein shows normal DNA binding activity but reduced transactivation potential, consistent with the transactivation activity ascribed to the NT domain (Martin and Orkin 1990). Unless expressed at high levels, GATA1 mutants lacking the NT domain fail to rescue the *GATA1.05* knockdown phenotype (Shimizu et al. 2001), suggesting that ΔNT-GATA1 cannot compensate in driving terminal erythroid or megakaryocytic differentiation when it is expressed at only wild-type abundance.

Although GATA1 mutations have been found to cause anemia, thrombocytopenia, thalassemia, and porphyria, direct evidence for a relationship between leukemogenesis and GATA1 were unknown until the recent discovery of human GATA1-related acute megakaryocytic leukemia. It was originally reported that acquired GATA1 mutations are associated with acute megakaryoblastic leukemia (AMKL) in children with Down syndrome (DS; Wechsler et al. 2002). Since then, frequent GATA1 mutations have been reported in patients with AMKL and transient myeloproliferative disorder (TMD), a preleukemic condition for DS-associated AMKL (Ahmed et al. 2004; Groet et al. 2003; Hitzler et al. 2003; Mundschau et al. 2003). These mutations occur in the second exon of GATA1 and are either small insertions or deletions that (1) introduced premature stop codon, (2) disrupted splicing, or (3) altered the start codon of GATA1 (Hirose et al. 2003; Rainis et al. 2003; Wechsler et al. 2002; Xu et al. 2003). These mutations all obviate full-length GATA1 translation and result in shortened GATA1 isoforms (GATA1s or ΔNT-GATA1), the translation of which all appear to initiate at M84 (Fig. 4A).

DS children, all of whom bear an extra copy of chromosome 21, develop TMD, AMKL, AML, and acute lymphoblastic leukemia much more frequently (10- to 20-fold) than unpredisposed children. AMKL (FAB M7) is a rare disease for non-DS patients, but prevalent in DS children. Most AMKL patients frequently present with a history of TMD (Zipursky et al. 1992). TMD cases associated with pancytopenia, hepatosplenomegaly, and circulating immature white blood cells are observed in DS patients with a high incidence of approximately 10%. In most of the cases, the TMD undergoes spontaneous regression within 2–3 months after birth, but approximately 20% of DS-TMD children develop AMKL in later life. The finding that identical GATA1 mutations in DS-AMKL blasts are observed antecedent to TMD implies a causal relationship of these disorders (Shimizu et al. 2008).

Roles GATA2 Plays in Hematopoietic Stem Cells

HSCs possess the defining properties of self-renewal and pluripotency and are located at the top of the extensive blood cell differentiation hierarchy. GATA2 is most abundantly expressed in HSCs as well as in immature progenitors in hematopoietic lineages (Akashi et al. 2000; Tsai and Orkin 1997). We previously demonstrated that mouse *Gata2* has two alternative first exons (Minegishi et al. 1998). The distal IS (hematopoietic tissue-specific) exon directs GATA2 transcription exclusively in hematopoietic and neural cells, whereas the IG (general) exon is active in almost all tissues where endogenous GATA2 is expressed.

The germ line loss-of-function studies demonstrated that GATA2 is essential for multilineage hematopoietic progenitor proliferation (Tsai et al. 1994; Tsai and Orkin 1997). GATA2-null mutant mice die during gestation at E10–11 of severe hematopoietic defects. Moreover, in the *Gata2*$^{+/-}$ heterozygous mutant, the size of functional stem cell pool is reduced due to altered cell survival caused by decreased Bcl-xL expression (Rodrigues et al. 2005).

We hypothesized that HSCs can be isolated efficiently from bone marrow cells using *Gata2*-directed GFP fluorescence in combination with Sca1 antibody staining (Suzuki et al. 2006). To this end, the GFP gene was inserted into the *Gata2* IS exon by targeted homologous recombination (EIS-KI mice). In EIS-KI mouse bone marrow cells, a significant fraction of the GFP^{+}Sca1^{+} cells (25%) are dormant CD34^{-}Lin^{-}Sca1^{+}c-Kit^{+} HSCs. Moreover, EIS-KI enabled the visualization of the movement of HSCs in situ in living bone marrow cells (movie in Suzuki et al. 2006). Immunohistochemical analysis of the bone marrow employing an anti-GFP antibody further revealed that all those GFP^{+} HSCs are directly contacting with osteoblasts. These results suggest that the endogenous HSC niche is located close to the endosteum and includes direct contact with osteoblasts.

GATA Switching

We demonstrated in transgenic mouse assays that *Gata2* IS promoter and the upstream region (7 kbp) can direct reporter gene expression in an early hematopoietic progenitor compartment (Kobayashi-Osaki et al. 2005; Minegishi et al. 2003). Furthermore, IS-mediated transcription is preferentially activated in human CD34^{+}

hematopoietic progenitor cells (Pan et al. 2000). Meanwhile, we also reported that murine *Gata2* contains numerous hematopoietic regulatory elements scattered over more than 250 kbp of the locus, so that a 250 kbp yeast artificial chromosome (YAC) transgene was capable of rescuing the hematopoietic lethal deficiency in *Gata2*-null mutant embryos (Zhou et al. 1998). The latter observation suggests that *Gata2* gene expression is determined by multiple regulatory elements spread over a relatively large expanse of the mouse genome.

Interestingly, although most of the *cis*-acting regulatory motifs for the *Gata2* gene reside within the aforementioned YAC, the regulatory region required for *Gata2* gene expression in the urogenital system was found to be located beyond the YAC regions (Khandekar et al. 2004; Brandt et al. 2008), further suggesting that the size of the *Gata2* locus is massive.

Although HSCs and the progenitor population contain GATA2 abundantly, this expression is suppressed concomitantly with an increase in GATA1 abundance during terminal erythroid differentiation. This phenomenon is referred to as GATA switching (Grass et al. 2003; Ohneda and Yamamoto 2002). This biphasic regulatory mechanism is thought to be first governed by the initial binding of GATA2 to a GATA site in the *Gata2* promoter region at early stages of erythroid differentiation, facilitating a positive autoregulatory loop. Later on, during terminal differentiation of erythroblasts, GATA1 displaces GATA2 and binds various GATA switching sites at –3.9, –2.8, and –1.8 kbp relative to the *Gata2* IS exon, thus suppressing the expression of *Gata2* (Grass et al. 2003; Martowicz et al. 2005; Ohneda and Yamamoto 2002). Recently, "ChIP-on-chip" array data examining from –100 to +30 kbp of the *Gata2* locus revealed additional GATA switching site at –77 kbp (Grass et al. 2006).

Perspectives and Closing Remarks

We described here recent observations suggesting that each of the *Gata1* gene regulatory regions possesses distinct spatiotemporal regulatory functions. *Gata1* BAC transgenic mice, coupled with homologous recombination based mutation analysis, should be an outstanding approach that leads to the identification and dissection of these regulatory domains, as well as to detailed understanding of their activities. As mutations or regulatory region SNPs present in human diseases become better interpreted, it is possible that these mutations could be responsible for variations in the expression of human GATA1 and lead to specific hematopoietic disorder(s). Meanwhile, the identification of *Gata1* target genes, along with genome-wide chromatin immunoprecipitation analysis, will help to clarify the function of GATA1 at various target loci during erythroid differentiation. By deciphering those two aspects of the GATA1 regulatory network, we would be able to have a first glimpse of the complexity of regulatory control underlying erythroid differentiation. There are still many unresolved issues in our understanding of GATA biology and function, but fortunately the field is rapidly advancing.

Box 1 Transcriptional Control of Hematopoiesis

In early hematopoietic development, the first wave of embryonic "primitive" erythropoiesis arises from the blood islands in the visceral yolk sac, generating nucleated erythrocytes that express the embryonic εγ and βH1-globin genes in E6.5–7 mouse embryos. Subsequently, the site of hematopoiesis moves into the aorta-gonad-mesonephros (AGM) region of the embryo, and then eventually the fetal liver becomes the main embryonic hematopoietic tissue, generating "definitive" enucleated erythrocytes that express adult β-globin. Finally, in the later stages of embryogenesis, hematopoiesis finally switches to the bone marrow (and spleen in mice), which continuously generates blood throughout life.

Unraveling of the molecular regulatory strategies taken by GATA1 and the effects caused by its mutation give hope to the potential future pharmacologic or genetic intervention in preventing hematopoietic disorders caused by GATA deficiencies.

Acknowledgments We thank Dr. Jon Maher for help and critical reading of the manuscript. This work is supported in part by grants-in-aid from the Ministry of Education, Culture, Sports, Science and Technology, Japan Science and Technology Corporation, National Institutes of Health, the Naito Foundation, and Takeda Science Foundation.

References

Ahmed M, Sternberg A, Hall G, Thomas A, Smith O, O'Marcaigh A, Wynn R, Stevens R, Addison M, King D, Stewart B, Gibson B, Roberts I, Vyas P (2004) Natural history of GATA-1 mutations in Down syndrome. Blood 103:2480–2489

Akashi K, Traver D, Miyamoto T, Weissman IL (2000) A clonogenic common myeloid progenitor that gives rise to all myeloid lineages. Nature 404:193–197

Anderson KP, Crable SC, Lingrel JB (1998) Multiple proteins binding to a GATA-E box-GATA motif regulate the erythroid Krüppel-like factor (EKLF) gene. J Biol Chem 273:14347–14354

Asnagli H, Afkarian M, Murphy KM (2002) Cutting edge: Identification of an alternative GATA-3 promoter directing tissue-specific gene expression in mouse and human. J Immunol 168:4268–4271

Bosè F, Fugazza C, Casalgrandi M, Capelli A, Cunningham JM, Zhao Q, Jane SM, Ottolenghi S, Ronchi A (2006) Functional interaction of CP2 with GATA-1 in the regulation of erythroid promoters. Mol Cell Biol 26:3942–3954

Boyes J, Byfield P, Nakatani Y, Ogryzko V (1998) Regulation of activity of the transcription factor GATA-1 by acetylation. Nature 396:594–598

Brandt W, Khandekar M, Suzuki N, Yamamoto M, Lim KC, Engel JD (2008) Defining the functional boundaries of the Gata2 locus by rescue with a linked bacterial artificial chromosome transgene. J Biol Chem 283:8976–8983

Calligaris R, Bottardi S, Cogoi S, Apezteguia I, Santoro C (1995) Alternative translation initiation site usage results in two functionally distinct forms of the GATA-1 transcription factor. Proc Natl Acad Sci USA 92:11598–11602

Chang AN, Cantor AB, Fujiwara Y, Lodish MB, Droho S, Crispino JD, Orkin SH (2002) GATA-factor dependence of the multitype zinc-finger protein FOG-1 for its essential role in megakaryopoiesis. Proc Natl Acad Sci USA 14:9237–9242

Cheng T, Shen H, Giokas D, Gere J, Tenen DG, Scadden DT (1996) Temporal mapping of gene expression levels during the differentiation of individual primary hematopoietic cells. Proc Natl Acad Sci USA 93:13158–13163

Cohen-Kaminsky S, Maouche-Chrétien L, Vitelli L, Vinit MA, Blanchard I, Yamamoto M, Peschle C, Roméo PH (1998) Chromatin immunoselection defines a TAL-1 target gene. EMBO J 17:5151–5160

Collavin L, Gostissa M, Avolio F, Secco P, Ronchi A, Santoro C, Del Sal G (2004) Modification of the erythroid transcription factor GATA-1 by SUMO-1. Proc Natl Acad Sci USA 101:8870–8875

Crispino JD, Lodish MB, MacKay JP, Orkin SH (1999) Use of altered specificity mutants to probe a specific protein-protein interaction in differentiation: the GATA-1:FOG complex. Mol Cell 3:219–228

Crossley M, Orkin SH (1994) Phosphorylation of the erythroid transcription factor GATA-1. J Biol Chem 269:16589–16596

Crossley M, Merika M, Orkin SH (1995) Self-association of the erythroid transcription factor GATA-1 mediated by its zinc finger domains. Mol Cell Biol 15:2448–2456

De Maria R, Zeuner A, Eramo A, Domenichelli C, Bonci D, Grignani F, Srinivasula SM, Alnemri ES, Testa U, Peschle C (1999) Negative regulation of erythropoiesis by caspase mediated cleavage of GATA-1. Nature 401:489–493

Dyer KD, Czapiga M, Foster B, Foster PS, Kang EM, Lappas CM, Moser JM, Naumann N, Percopo CM, Siegel SJ, Swartz JM, Ting-De Ravin S, Rosenberg HF (2007) Eosinophils from lineage-ablated Delta dblGATA bone marrow progenitors: the dblGATA enhancer in the promoter of GATA-1 is not essential for differentiation ex vivo. J Immunol 179:1693–1699

Ferreira R, Ohneda K, Yamamoto M, Philipsen S (2005) GATA1 function, a paradigm for transcription factors in hematopoiesis. Mol Cell Biol 25:1215–1227

Ferreira R, Wai A, Shimizu R, Gillemans N, Rottier R, von Lindern M, Ohneda K, Grosveld F, Yamamoto M, Philipsen S (2007) Dynamic regulation of Gata factor levels is more important than their identity. Blood 109:5481–5490

Fox AH, Kowalski K, King GF, Mackay JP, Crossley M (1998) Key residues characteristic of GATA N-fingers are recognized by FOG. J Biol Chem 273:33595–33603

Fox AH, Liew C, Holmes M, Kowalski K, Mackay J, Crossley M (1999) Transcriptional cofactors of the FOG family interact with GATA proteins by means of multiple zinc fingers. EMBO J 18:2812–2822

Freson K, Devriendt K, Matthijs G, Van Hoof A, De Vos R, Thys C, Minner K, Hoylaerts MF, Vermylen J, Van Geet C (2001) Platelet characteristics in patients with X-linked macrothrombocytopenia because of a novel GATA1 mutation. Blood 98:85–92

Freson K, Matthijs G, Thys C, Mariën P, Hoylaerts MF, Vermylen J, Van Geet C (2002) Different substitutions at residue D218 of the X-linked transcription factor GATA1 lead to altered clinical severity of macrothrombocytopenia and anemia and are associated with variable skewed X inactivation. Hum Mol Genet 11:147–152

Fujiwara Y, Browne CP, Cunniff K, Goff SC, Orkin SH (1996) Arrested development of embryonic red cell precursors in mouse embryos lacking transcription factor GATA-1. Proc Natl Acad Sci USA 93:12355–12358

Fujiwara Y, Chang AN, Williams AM, Orkin SH (2004) Functional overlap of GATA-1 and GATA-2 in primitive hematopoietic development. Blood 103:583–585

Grass JA, Boyer ME, Pal S, Wu J, Weiss MJ, Bresnick EH (2003) GATA-1-dependent transcriptional repression of GATA-2 via disruption of positive autoregulation and domain-wide chromatin remodeling. Proc Natl Acad Sci USA 100:8811–8816

Grass JA, Jing H, Kim SI, Martowicz ML, Pal S, Blobel GA, Bresnick EH (2006) Distinct functions of dispersed GATA factor complexes at an endogenous gene locus. Mol Cell Biol 26:7056–7067

Gregory CJ, Eaves AC (1977) Human marrow cells capable of erythropoietic differentiation in vitro: definition of three erythroid colony responses. Blood 49:855–864

Gregory CJ, Eaves AC (1978) Three stages of erythropoietic progenitor cell differentiation distinguished by a number of physical and biologic properties. Blood 51:527–537

Gregory RC, Taxman DJ, Seshasayee D, Kensinger MH, Bieker JJ, Wojchowski DM (1996) Functional interaction of GATA1 with erythroid Krüppel-like factor and Sp1 at defined erythroid promoters. Blood 87:1793–1801

Groet J, McElwaine S, Spinelli M, Rinaldi A, Burtscher I, Mulligan C, Mensah A, Cavani S, Dagna-Bricarelli F, Basso G, Cotter FE, Nizetic D (2003) Acquired mutations in GATA1 in neonates with Down's syndrome with transient myeloid disorder. Lancet 361:1617–1620

Guyot B, Murai K, Fujiwara Y, Valverde-Garduno V, Hammett M, Wells S, Dear N, Orkin SH, Porcher C, Vyas P (2006) Characterization of a megakaryocyte-specific enhancer of the key hemopoietic transcription factor GATA1. J Biol Chem 281:13733–13742

Hirose Y, Kudo K, Kiyoi H, Hayashi Y, Naoe T, Kojima S (2003) Comprehensive analysis of gene alterations in acute megakaryoblastic leukemia of Down's syndrome. Leukemia 17:2250–2252

Hitzler JK, Cheung J, Li Y, Scherer SW, Zipursky A (2003) GATA1 mutations in transient leukemia and acute megakaryoblastic leukemia of Down syndrome. Blood 101:4301–4304

Hou CH, Huang J, He QY, Zhang CN, Zhang XJ, Qian RL (2008) Involvement of Sp1/Sp3 in the activation of the GATA-1 erythroid promoter in K562 cells. Cell Res 18:302–310

Hromas R, Orazi A, Neiman RS, Maki R, Van Beveran C, Moore J, Klemsz M (1993) Hematopoietic lineage- and stage-restricted expression of the ETS oncogene family member PU.1. Blood 82:2998–3004

Humbles AA, Lloyd CM, McMillan SJ, Friend DS, Xanthou G, McKenna E, Ghiran S, Gerard NP, Yu C, Orkin SH, Gerard CA (2004) Critical role for eosinophils in allergic airways remodeling. Science 305:1776–1779

Hung HL, Lau J, Kim AY, Weiss MJ, Blobel GA (1999) CREB-binding protein acetylates hematopoietic transcription factor GATA-1 at functionally important sites. Mol Cell Biol 19:3496–3505

Ito E, Toki T, Ishihara H, Ohtani H, Gu L, Yokoyama M, Engel JD, Yamamoto M (1993) Erythroid transcription factor GATA-1 is abundantly transcribed in mouse testis. Nature 362:466–468

Johnson ES (2004) Protein modification by SUMO. Annu Rev Biochem 73:355–382

Khandekar M, Suzuki N, Lewton J, Yamamoto M, Engel JD (2004) Multiple, distant Gata2 enhancers specify temporally and tissue-specific patterning in the developing urogenital system. Mol Cell Biol 24:10263–10276

Ko LJ, Engel JD (1993) DNA-binding specificities of the GATA transcription factor family. Mol Cell Biol 13:4011–4022

Kobayashi M, Nishikawa K, Yamamoto M (2001) Hematopoietic regulatory domain of gata1 gene is positively regulated by GATA1 protein in zebrafish embryos. Development 128:2341–2350

Kobayashi-Osaki M, Ohneda O, Suzuki N, Minegishi N, Yokomizo T, Takahashi S, Lim KC, Engel JD, Yamamoto M (2005) TA motifs regulate early hematopoietic lineage-specific expression of the Gata2 gene. Mol Cell Biol 25:7005–7020

Kuhl C, Atzberger A, Iborra F, Nieswandt B, Porcher C, Vyas P (2005) GATA1-mediated megakaryocyte differentiation and growth control can be uncoupled and mapped to different domains in GATA1. Mol Cell Biol 25:8592–8606

Kulessa H, Frampton J, Graf T (1995) GATA-1 reprograms avian myelomonocytic cell lines into eosinophils, thromboblasts, and erythroblasts. Genes Dev 9:1250–1262

Lahlil R, Lécuyer E, Herblot S, Hoang T (2004) SCL assembles a multifactorial complex that determines glycophorin A expression. Mol Cell Biol 24:1439–1452

Lamonica JM, Vakoc CR, Blobel GA (2006) Acetylation of GATA-1 is required for chromatin occupancy. Blood 12:37368–37380

Laverriere AC, MacNeill C, Mueller C, Poelmann RE, Burch JB, Evans T (1994) GATA-4/5/6, a subfamily of three transcription factors transcribed in developing heart and gut. J Biol Chem 269:23177–23184

Lécuyer E, Herblot S, Saint-Denis M, Martin R, Begley CG, Porcher C, Orkin SH, Hoang T (2002) The SCL complex regulates c-kit expression in hematopoietic cells through functional interaction with Sp1. Blood 100:2430–2440

Leonard M, Brice M, Engel JD, Papayannopoulou T (1993) Dynamics of GATA transcription factor expression during erythroid differentiation. Blood 82:1071–1079

Martin DI, Orkin SH (1990) Transcriptional activation and DNA binding by the erythroid factor GF-1/NF-E1/Eryf 1. Genes Dev 4:1886–1898

Martowicz ML, Grass JA, Boyer ME, Guend H, Bresnick EH (2005) Dynamic GATA factor interplay at a multicomponent regulatory region of the GATA-2 locus. J Biol Chem 280:1724–1732

McDevitt MA, Shivdasani RA, Fujiwara Y, Yang H, Orkin SH (1997) An upstream, DNase I hypersensitive region of the hematopoietic-expressed transcription factor GATA-1 gene confers developmental specificity in transgenic mice. Proc Natl Acad Sci USA 94:7976–7981

McKercher SR, Torbett BE, Anderson KL, Henkel GW, Vestal DJ, Baribault H, Klemsz M, Feeney AJ, Wu GE, Paige CJ, Maki RA (1996) Targeted disruption of the PU.1 gene results in multiple hematopoietic abnormalities. EMBO J 15:5647–5658

Mehaffey MG, Newton AL, Gandhi MJ, Crossley M, Drachman JG (2001) X-linked thrombocytopenia caused by a novel mutation of GATA-1. Blood 98:2681–2688

Merika M, Orkin SH (1995) Functional synergy and physical interactions of the erythroid transcription factor GATA-1 with the Krüppel family proteins Sp1 and EKLF. Mol Cell Biol 15:2437–2447

Minegishi N, Ohta J, Suwabe N, Nakauchi H, Ishihara H, Hayashi N, Yamamoto M (1998) Alternative promoters regulate transcription of the mouse GATA-2 gene. J Biol Chem 273:3625–3634

Minegishi N, Suzuki N, Yokomizo T, Pan X, Fujimoto T, Takahashi S, Hara T, Miyajima A, Nishikawa S, Yamamoto M (2003) Expression and domain-specific function of GATA-2 during differentiation of the hematopoietic precursor cells in midgestation mouse embryos. Blood 102:896–905

Moreau-Gachelin F, Wendling F, Molina T, Denis N, Titeux M, Grimber G, Briand P, Vainchenker W, Tavitian A (1996) Spi-1/PU.1 transgenic mice develop multistep erythroleukemias. Mol Cell Biol 16:2453–2463

Motohashi H, Katsuoka F, Shavit JA, Engel JD, Yamamoto M (2000) Positive or negative MARE-dependent transcriptional regulation is determined by the abundance of small Maf proteins. Cell 103:865–875

Mundschau G, Gurbuxani S, Gamis AS, Greene ME, Arceci RJ, Crispino JD (2003) Mutagenesis of GATA1 is an initiating event in Down syndrome leukemogenesis. Blood 101:4298–4300

Munugalavadla V, Dore LC, Tan BL, Hong L, Vishnu M, Weiss MJ, Kapur R (2005) Repression of c-kit and its downstream substrates by GATA-1 inhibits cell proliferation during erythroid maturation. Mol Cell Biol 25:6747–6759

Nagai T, Harigae H, Ishihara H, Motohashi H, Minegishi N, Tsuchiya S, Hayashi N, Gu L, Andres B, Engel JD (1994) Transcription factor GATA-2 is expressed in erythroid, early myeloid, and CD34+ human leukemia-derived cell lines. Blood 84:1074–1084

Nerlov C, Querfurth E, Kulessa H, Graf T (2000) GATA-1 interacts with the myeloid PU.1 transcription factor and represses PU.1-dependent transcription. Blood 95:2543–2551

Newton A, Mackay J, Crossley M (2001) The N-terminal zinc finger of the erythroid transcription factor GATA-1 binds GATC motifs in DNA. J Biol Chem 276:35794–35801

Nichols KE, Crispino JD, Poncz M, White JG, Orkin SH, Maris JM, Weiss MJ (2000) Familial dyserythropoietic anaemia and thrombocytopenia due to an inherited mutation in GATA1. Nat Genet 24:266–270

Nicolis S, Bertini C, Ronchi A, Crotta S, Lanfranco L, Moroni E, Giglioni B, Ottolenghi S (1991) An erythroid specific enhancer upstream to the gene encoding the cell-type specific transcription factor GATA-1. Nucleic Acids Res 19:5285–5291

Nishikawa K, Kobayashi M, Masumi A, Lyons SE, Weinstein BM, Liu PP, Yamamoto M (2003) Self-association of GATA-1 enhances transcriptional activity in vivo in zebra fish embryos. Mol Cell Biol 23:8295–8305

Nishimura S, Takahashi S, Kuroha T, Suwabe N, Nagasawa T, Trainor C, Yamamoto M (2000) A GATA box in the GATA-1 gene hematopoietic enhancer is a critical element in the network of GATA factors and sites that regulate this gene. Mol Cell Biol 20:713–723

Ohneda K, Yamamoto M (2002) Roles of hematopoietic transcription factors GATA-1 and GATA-2 in the development of red blood cell lineage. Acta Haematol 108:237–245

Ohneda K, Shimizu R, Nishimura S, Muraosa Y, Takahashi S, Engel JD, Yamamoto M (2002) A minigene containing four discrete *cis*-elements recapitulates GATA-1 gene expression in vivo. Genes Cells 7:1243–1254

Onodera K, Takahashi S, Nishimura S, Ohta J, Motohashi H, Yomogida K, Hayashi N, Engel JD, Yamamoto M (1997) GATA-1 transcription is controlled by distinct regulatory mechanisms during primitive and definitive erythropoiesis. Proc Natl Acad Sci USA 94:4487–4492

Orkin SH (1992) GATA-binding transcription factors in hematopoietic cells. Blood 80:575–581

Osada H, Grutz GG, Axelson H, Forster A, Rabbitts TH (1997) LIM-only protein Lmo2 forms a protein complex with erythroid transcription factor GATA-1. Leukemia Suppl 3:307–312

Pan X, Minegishi N, Harigae H, Yamagiwa H, Minegishi M, Akine Y, Yamamoto M (2000) Identification of human GATA-2 gene distal IS exon and its expression in hematopoietic stem cell fractions. J Biochem 127:105–112

Pan X, Ohneda O, Ohneda K, Lindeboom F, Iwata F, Shimizu R, Nagano M, Suwabe N, Philipsen S, Lim KC, Engel JD, Yamamoto M (2005) Graded levels of GATA-1 expression modulate survival, proliferation, and differentiation of erythroid progenitors. J Biol Chem 280:22385–22394

Partington GA, Patient RK (1999) Phosphorylation of GATA-1 increases its DNA-binding affinity and is correlated with induction of human K562 erythroleukaemia cells. Nucleic Acids Res 27:1168–1175

Patient RK, McGhee JD (2002) The GATA family (vertebrates and invertebrates). Curr Opin Genet Dev 12:416–422

Pedone PV, Omichinski JG, Nony P, Trainor C, Gronenborn AM, Clore GM, Felsenfeld G (1997) The N-terminal fingers of chicken GATA-2 and GATA-3 are independent sequence-specific DNA binding domains. EMBO J 16:2874–2882

Pevny L, Lin CS, D'Agati V, Simon MC, Orkin SH, Costantini F (1995) Development of hematopoietic cells lacking transcription factor GATA-1. Development 121:163–172

Phillips JD, Steensma DP, Pulsipher MA, Spangrude GJ, Kushner JP (2007) Congenital erythropoietic porphyria due to a mutation in GATA1: the first trans-acting mutation causative for a human porphyria. Blood 109:2618–2621

Rainis L, Bercovich D, Strehl S, Teigler-Schlegel A, Stark B, Trka J, Amariglio N, Biondi A, Muler I, Rechavi G, Kempski H, Haas OA, Izraeli S (2003) Mutations in exon 2 of GATA1 are early events in megakaryocytic malignancies associated with trisomy 21. Blood 102:981–986

Rekhtman N, Radparvar F, Evans T, Skoultchi AI (1999) Direct interaction of hematopoietic transcription factors PU.1 and GATA-1: functional antagonism in erythroid cells. Genes Dev 13:1398–1411

Ribeil JA, Zermati Y, Vandekerckhove J, Cathelin S, Kersual J, Dussiot M, Coulon S, Moura IC, Zeuner A, Kirkegaard-Sørensen T, Varet B, Solary E, Garrido C, Hermine O (2007) Hsp70 regulates erythropoiesis by preventing caspase-3-mediated cleavage of GATA-1. Nature 445:102–105

Rodrigues NP, Janzen V, Forkert R, Dombkowski DM, Boyd AS, Orkin SH, Enver T, Vyas P, Scadden DT (2005) Haploinsufficiency of GATA-2 perturbs adult hematopoietic stem-cell homeostasis. Blood 106:477–484

Rylski M, Welch JJ, Chen YY, Letting DL, Diehl JA, Chodosh LA, Blobel GA, Weiss MJ (2003) GATA-1-mediated proliferation arrest during erythroid maturation. Mol Cell Biol 23:5031–5042

Sachdev S, Bruhn L, Sieber H, Pichler A, Melchior F, Grosschedl R (2001) PIASy, a nuclear matrix-associated SUMO E3 ligase, represses LEF1 activity by sequestration into nuclear bodies. Genes Dev 15:3088–3103

Scott EW, Simon MC, Anastasi J, Singh H (1994) Requirement of transcription factor PU.1 in the development of multiple hematopoietic lineages. Science 265:1573–1577

Seshasayee D, Geiger JN, Gaines P, Wojchowski DM (2000) Intron 1 elements promote erythroid-specific GATA-1 gene expression. J Biol Chem 275:22969–22977
Shimizu R, Takahashi S, Ohneda K, Engel JD, Yamamoto M (2001) In vivo requirements for GATA-1 functional domains during primitive and definitive erythropoiesis. EMBO J 20:5250–5260
Shimizu R, Ohneda K, Engel JD, Trainor CD, Yamamoto M (2004) Transgenic rescue of GATA-1-deficient mice with GATA-1 lacking a FOG-1 association site phenocopies patients with X-linked thrombocytopenia. Blood 103:2560–2567
Shimizu R, Trainor CD, Nishikawa K, Kobayashi M, Ohneda K, Yamamoto M (2007) GATA-1 self-association controls erythroid development in vivo. J Biol Chem 282:15862–15871
Shimizu R, Engel JD, Yamamoto M (2008) GATA1-related leukaemias. Nat Rev Cancer 8:279–287
Socolovsky M, Nam H, Fleming MD, Haase VH, Brugnara C, Lodish HF (2001) Ineffective erythropoiesis in Stat5a(–/–)5b(–/–) mice due to decreased survival of early erythroblasts. Blood 98:3261–3273
Song SH, Hou C, Dean A (2007) A positive role for NLI/Ldb1 in long-range beta-globin locus control region function. Mol Cell 28:810–822
Stephenson JR, Axelrad AA, McLeod DL, Shreeve MM (1971) Induction of colonies of hemoglobin-synthesizing cells by erythropoietin in vitro. Proc Natl Acad Sci USA 7:1542–1546
Suzuki M, Moriguchi T, Ohneda K, Yamamoto M (2009) Differential contribution of the Gata1 gene hematopoietic enhancer to erythroid differentiation. Mol Cell Biol 29:1163–1175
Suzuki N, Suwabe N, Ohneda O, Obara N, Imagawa S, Pan X, Motohashi H, Yamamoto M (2003) Identification and characterization of 2 types of erythroid progenitors that express GATA-1 at distinct levels. Blood 102:3575–3583
Suzuki N, Ohneda O, Minegishi N, Nishikawa M, Ohta T, Takahashi S, Engel JD, Yamamoto M (2006) Combinatorial Gata2 and Sca1 expression defines hematopoietic stem cells in the bone marrow niche. Proc Natl Acad Sci USA 103:2202–2207
Takahashi S, Onodera K, Motohashi H, Suwabe N, Hayashi N, Yanai N, Nabesima Y, Yamamoto M (1997) Arrest in primitive erythroid cell development caused by promoter-specific disruption of the GATA-1 gene. J Biol Chem 272:12611–12615
Takahashi S, Komeno T, Suwabe N, Yoh K, Nakajima O, Nishimura S, Kuroha T, Nagasawa T, Yamamoto M (1998) Role of GATA-1 in proliferation and differentiation of definitive erythroid and megakaryocytic cells in vivo. Blood 92:434–442
Takahashi S, Shimizu R, Suwabe N, Kuroha T, Yoh K, Ohta J, Nishimura S, Lim KC, Engel JD, Yamamoto M (2000) GATA factor transgenes under GATA-1 locus control rescue germline GATA-1 mutant deficiencies. Blood 96:910–916
Trainor CD, Omichinski JG, Vandergon TL, Gronenborn AM, Clore GM, Felsenfeld G (1996) A palindromic regulatory site within vertebrate GATA-1 promoters requires both zinc fingers of the GATA-1 DNA-binding domain for high-affinity interaction. Mol Cell Biol 16:2238–2247
Trainor CD, Ghirlando R, Simpson MA (2000) GATA zinc finger interactions modulate DNA binding and transactivation. J Biol Chem 275:28157–28166
Tsai FY, Orkin SH (1997) Transcription factor GATA-2 is required for proliferation/survival of early hematopoietic cells and mast cell formation, but not for erythroid and myeloid terminal differentiation. Blood 89:3636–3643
Tsai SF, Strauss E, Orkin SH (1991) Functional analysis and in vivo footprinting implicate the erythroid transcription factor GATA-1 as a positive regulator of its own promoter. Genes Dev 5:919–931
Tsai FY, Keller G, Kuo FC, Weiss M, Chen J, Rosenblatt M, Alt FW, Orkin SH (1994) An early haematopoietic defect in mice lacking the transcription factor GATA-2. Nature 371:221–226
Tsang AP, Fujiwara Y, Hom DB, Orkin SH (1998) Failure of megakaryopoiesis and arrested erythropoiesis in mice lacking the GATA-1 transcriptional cofactor FOG. Genes Dev 12:1176–1188
Valverde-Garduno V, Guyot B, Anguita E, Hamlett I, Porcher C, Vyas P (2004) Differences in the chromatin structure and cis-element organization of the human and mouse GATA1 loci: implications for cis-element identification. Blood 104:3106–3116

Visvader JE, Elefanty AG, Strasser A, Adams JM (1992) GATA-1 but not SCL induces megakaryocytic differentiation in an early myeloid line. EMBO J 11:4557–4564

Vyas P, McDevitt MA, Cantor AB, Katz SG, Fujiwara Y, Orkin SH (1999) Different sequence requirements for expression in erythroid and megakaryocytic cells within a regulatory element upstream of the GATA-1 gene. Development 126:2799–2811

Wadman IA, Osada H, Grütz GG, Agulnick AD, Westphal H, Forster A, Rabbitts TH (1997) The LIM-only protein Lmo2 is a bridging molecule assembling an erythroid, DNA-binding complex which includes the TAL1, E47, GATA-1 and Ldb1/NLI proteins. EMBO J 16:3145–3157

Wechsler J, Greene M, McDevitt MA, Anastasi J, Karp JE, Le Beau MM, Crispino JD (2002) Acquired mutations in GATA1 in the megakaryoblastic leukemia of Down syndrome. Nat Genet 32:148–152

Weiss MJ, Orkin SH (1995) GATA transcription factors: key regulators of hematopoiesis. Exp Hematol 23:99–107

Weiss MJ, Keller G, Orkin SH (1994) Novel insights into erythroid development revealed through in vitro differentiation of GATA-1 embryonic stem cells. Genes Dev 8:1184–1197

Weiss MJ, Yu C, Orkin SH (1997) Erythroid-cell-specific properties of transcription factor GATA-1 revealed by phenotypic rescue of a gene-targeted cell line. Mol Cell Biol 17:1642–1651

Whyatt D, Lindeboom F, Karis A, Ferreira R, Milot E, Hendriks R, de Bruijn M, Langeveld A, Gribnau J, Grosveld F, Philipsen S (2000) An intrinsic but cell-nonautonomous defect in GATA-1-overexpressing mouse erythroid cells. Nature 406:519–524

Woo AJ, Moran TB, Schindler YL, Choe SK, Langer NB, Sullivan MR, Fujiwara Y, Paw BH, Cantor AB (2008) Identification of ZBP-89 as a novel GATA-1-associated transcription factor involved in megakaryocytic and erythroid development. Mol Cell Biol 28:2675–2689

Xu G, Nagano M, Kanezaki R, Toki T, Hayashi Y, Taketani T, Taki T, Mitui T, Koike K, Kato K, Imaizumi M, Sekine I, Ikeda Y, Hanada R, Sako M, Kudo K, Kojima S, Ohneda O, Yamamoto M, Ito E (2003) Frequent mutations in the GATA-1 gene in the transient myeloproliferative disorder of Down syndrome. Blood 102:2960–2968

Yamamoto M, Ko LJ, Leonard MW, Beug H, Orkin SH, Engel JD (1990) Activity and tissue-specific expression of the transcription factor NFE1 multigene family. Genes Dev 4:1650–1662

Yang HY, Evans T (1992) Distinct roles for the two cGATA-1 finger domains. Mol Cell Biol 12:4562–4570

Yu C, Cantor AB, Yang H, Browne C, Wells RA, Fujiwara Y, Orkin SH (2002a) Targeted deletion of a high-affinity GATA-binding site in the GATA-1 promoter leads to selective loss of the eosinophil lineage in vivo. J Exp Med 195:1387–1395

Yu C, Niakan KK, Matsushita M, Stamatoyannopoulos G, Orkin SH, Raskind WH (2002b) X-linked thrombocytopenia with thalassemia from a mutation in the amino finger of GATA-1 affecting DNA binding rather than FOG-1 interaction. Blood 100:2040–2045

Zhang P, Behre G, Pan J, Iwama A, Wara-Aswapati N, Radomska HS, Auron PE, Tenen DG, Sun Z (1999) Negative cross-talk between hematopoietic regulators: GATA proteins repress PU.1. Proc Natl Acad Sci USA 96:8705–8710

Zhang P, Zhang X, Iwama A, Yu C, Smith KA, Mueller BU, Narravula S, Torbett BE, Orkin SH, Tenen DG (2000) PU.1 inhibits GATA-1 function and erythroid differentiation by blocking GATA-1 DNA binding. Blood 96:2641–2648

Zhao W, Kitidis C, Fleming MD, Lodish HF, Ghaffari S (2006) Erythropoietin stimulates phosphorylation and activation of GATA-1 via the PI3-kinase/AKT signaling pathway. Blood 107:907–915

Zhou Y, Lim KC, Onodera K, Takahashi S, Ohta J, Minegishi N, Tsai FY, Orkin SH, Yamamoto M, Engel JD (1998) Rescue of the embryonic lethal hematopoietic defect reveals a critical role for GATA-2 in urogenital development. EMBO J 17:6689–6700

Zipursky A, Poon A, Doyle J (1992) Leukemia in Down syndrome: A review. Pediatr Hematol Oncol 9:139–149

Role of the IL-7 Receptor in γδ T-Cell Development from Hematopoietic Stem Cells

Koichi Ikuta and Shizue Tani-ichi

Abstract Interleukin 7 (IL-7) is an essential cytokine for early lymphocyte development. The IL-7 receptor (IL-7R) transmits at least two signals in lymphocyte progenitors. One is for survival and proliferation of early and mature lymphocytes. The other is to promote V(D)J recombination of the immunoglobulin heavy (IgH) and T-cell receptor γ (TCRγ) loci. γδ T cells with a specific variable (V) γ-chain appear in successive waves in the thymus, which is controlled at the levels of hematopoietic stem cells and V(D)J recombination. The IL-7R and signal transducer and activator of transcription 5 (STAT5) control the accessibility of the TCRγ locus by interacting with STAT consensus motifs in the Jγ germline promoters and the Eγ and HsA elements. STAT5 binding to the Jγ germline promoters induces germline transcription, histone acetylation, and chromatin remodeling that make the chromatin accessible for DNA recombinase. In addition, STAT5 is recruited to the consensus motifs in the Eγ elements by cytokine stimulation and augments Eγ activity in collaboration with Runx and c-Myb. Furthermore, STAT5 directly binds to and controls the HsA element, but STAT5 indirectly activates the Vγ5 promoter. STAT5 and the TCRγ locus is an excellent model system to understand the control mechanisms of V(D)J recombination in lymphocyte antigen receptor genes.

IL-7 and the IL-7R

Interleukin 7 (IL-7) is a cytokine for the growth and survival of early lymphocytes and mature T cells. It was identified as a growth factor that stimulates proliferation of pre-B cells in Whitlock-Witte bone marrow culture (Namen et al. 1988). Later, IL-7 was shown to support the growth and survival of early T cells (Suda and Zlotnik 1991). IL-7 is produced by thymic and bone marrow stromal cells, and skin,

K. Ikuta (✉) and S. Tani-ichi
Laboratory of Biological Protection, Department of Biological Responses, Institute for Virus Research, Kyoto University, Kyoto, 606-8507, Japan
e-mail: ikuta@virus.kyoto-u.ac.jp.

M. Kondo (ed.), *Hematopoietic Stem Cell Biology,* Stem Cell Biology and Regenerative Medicine, DOI 10.1007/978-1-60327-347-3_6,

thymic, and intestinal epithelial cells. IL-7 production is augmented by cell adhesion of B-cell precursors to stromal cells (Sudo et al. 1989). A cytokine with similar biological activities to IL-7, thymic stromal lymphopoietin (TSLP), was identified from a thymic stromal cell line (Ray et al. 1996).

The interleukin 7 receptor (IL-7R) consists of two polypeptides, a unique α-chain (IL-7Rα), and a common γ (γc)-chain (Fig. 1) (Noguchi et al. 1993; Kondo et al. 1994). The IL-7R shares the γc with the receptors for IL-2, IL-4, IL-9, IL-15, and IL-21. The IL-7Rα chain is a type I transmembrane protein consisting of 459 amino acids including a signal peptide (Goodwin et al. 1990). The extracellular domain of the IL-7Rα contains four conserved cysteine residues and a WSXWS motif and belongs to the cytokine receptor superfamily. The cytoplasmic domain of IL-7Rα contains three tyrosine residues conserved between mouse and human (Y401, Y449, and Y456). The membrane-proximal region of the cytoplasmic domain contains a box 1 motif conserved among cytokine receptors. JAK1 is expected to interact with IL-7Rα through this region (Leonard and O'Shea 1998). In contrast, JAK3 interacts with the γc-chain. The receptor for TSLP consists of the

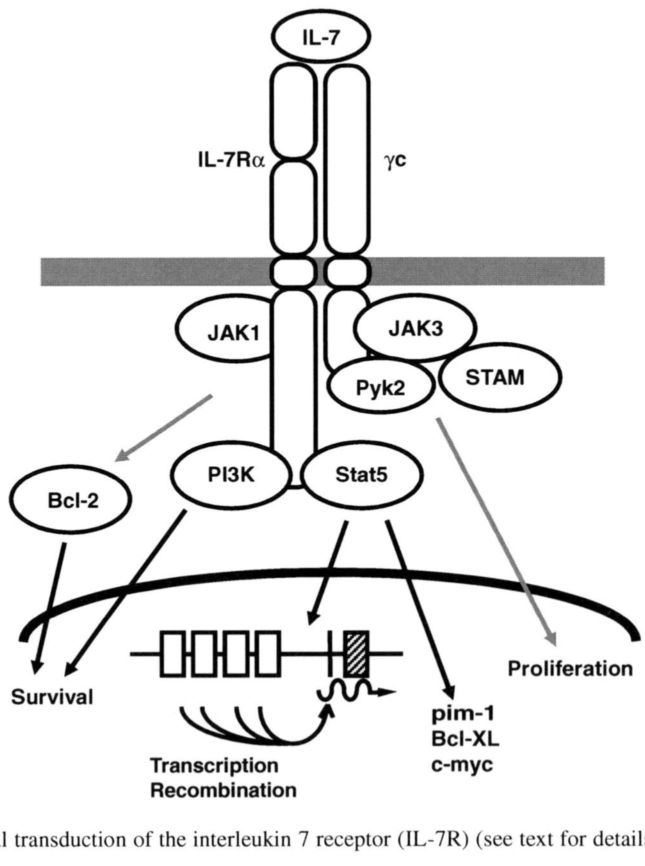

Fig. 1 Signal transduction of the interleukin 7 receptor (IL-7R) (see text for details)

IL-7Rα chain and a unique TSLP-R chain (Fujio et al. 2000; Pandey et al. 2000). The TSLP-R chain has homology with the γc-chain.

IL-7 binding to the IL-7R triggers the phosphorylation and activation of receptor-associated Jak1 and Jak3 tyrosine kinases (Leonard and O'Shea 1998). Following their activation, the JAK kinases phosphorylate the Y449 residue of the IL-7Rα. The signal transducer and activator of transcription (STAT) proteins as well as phosphatidyl inositol 3 (PI3) kinase are recruited to the tyrosine residue and subsequently phosphorylated and activated by the JAK kinases. The IL-7R mainly activates STAT5a and STAT5b, and to a lesser extent, STAT1 and STAT3. The phosphorylated STAT5 proteins then form homo- and heterodimers, bind to a consensus motif (TTCNNNGAA), and activate the transcription of various target genes, including TCRγ genes. PI3 kinase phosphorylates phosphatidylinositol and activates Akt. Akt protein in turn inactivates Bad, caspase-9, and IKKα proteins by phosphorylation and transmits antiapoptotic signals. The JAK kinases also activate STAM and Pyk2 proteins. This signaling pathway then activates the MAP kinase cascade and results in cell proliferation. The IL-7R also induces Bcl-2 expression in early T cells by an unidentified pathway (von Freeden-Jeffry et al. 1997). IL-2Rβ mobilizes a similar set of signaling molecules to the IL-7Rα. The TSLP receptor also activates STAT5 (Isaksen et al. 1999).

Role of the IL-7R in Early Lymphocyte Development

IL-7 is an essential cytokine for early T- and B-cell development when V(D)J recombination takes place. The IL-7R transmits at least two signals in T and B progenitors (Mazzucchelli and Durum 2007). One signal is for survival and proliferation. For instance, the IL-7R induces the expression of Bcl-2 in T-cell precursors (von Freeden-Jeffry et al. 1997), and introduction of a *bcl-2* transgene restores αβ T-cell development in IL-7R-deficient mice (Akashi et al. 1997; Maraskovsky et al. 1997). The IL-7R also promotes the proliferation of lymphocyte precursors through the activation of PI3 kinase (Corcoran et al. 1996; Pallard et al. 1999). The second signal from the IL-7R is to promote V(D)J recombination in the immunoglobulin heavy (IgH) and T-cell receptor γ (TCRγ) loci. For example, IL-7R signaling induces germline transcription and DNA rearrangement in D-distal V_H segments in pro-B cells (Corcoran et al. 1998; Bertolino et al. 2005). The V-J recombination of TCRγ genes is also severely impaired in IL-7Rα-deficient mice (Maki et al. 1996a; Candeias et al. 1997). We will describe the roles for the IL-7R in γδ T-cell development in detail.

In fetal thymic organ culture, addition of IL-7 promotes the expansion of mature γδ T cells but prevents the generation of mature αβ T cells (Watanabe et al. 1991; Plum et al. 1993). The epithelial cells in the skin and the gut produce IL-7 (Matsue et al. 1993; Watanabe et al. 1995), and Thy-1$^+$ dendritic epidermal T cells (DETC) proliferate in response to IL-7 (Matsue et al. 1993). Collectively, these results suggested that IL-7 might be involved in the development and maintenance of γδ T cells in the thymus and periphery.

Injection of neutralizing antibodies to IL-7 or IL-7Rα leads to a blockade of lymphocyte development (Grabstein et al. 1993; Sudo et al. 1993). After a brief period of anti-IL-7 antibody treatment, most of the pro-B cells and all of the pre-B and immature B cells are depleted from the bone marrow. After extended periods of antibody administration, thymic cellularity is profoundly reduced. $CD3^-CD4^-CD8^-$ precursors are depleted with interruption at the $CD44^+CD25^+$ stage. Addition of an anti–IL–7Rα antibody to Whitlock-Witte culture inhibits proliferation of B-lineage cells. Continuous injection of the anti–IL–7Rα antibody results in a decrease of B-precursor cells and also of thymocytes. These results indicated that IL-7 is an essential cytokine for the generation of B and T cells in the bone marrow and thymus, respectively.

Gene inactivation studies have been carried out to elucidate the in vivo function of IL-7 and IL-7R. IL-7-deficient mice are highly lymphopenic in the peripheral blood and lymphoid organs (von Freeden-Jeffry et al. 1995). Bone marrow B lymphopoiesis is partially blocked at the transition from pro-B to pre-B cells. Thymic cellularity is reduced 20-fold, but their thymocytes retain the normal CD4/CD8 profile. IL-7Rα-deficient mice have impaired lymphocyte development at the earliest stages (Peschon et al. 1994). B lymphopoiesis tends to be blocked at the HSA^{lo} $BP\text{-}1^-$ stage in the pro–B-cell fraction in IL-7Rα-deficient mice. Similarly, thymocyte development is hampered at the $CD44^+$ $CD25^-$ stage in $CD3^-CD4^-CD8^-$ fraction. Although IL-7Rα-deficient mice have small numbers of B cells and αβ T cells in periphery, they totally lack γδ T cells (Maki et al. 1996b; He and Malek 1996). γδ T cells are absent from fetal and adult thymus, spleen, liver, and the epithelium of skin and intestine in IL-7Rα-deficient mice. On the other hand, natural killer (NK) cell development is normal. IL-2Rβ-deficient mice have decreased numbers of γδ intraepithelial lymphocytes (IEL) and mature $V\gamma3^+$ fetal thymocytes and lack DETC in the skin (Suzuki et al. 1997; Kawai et al. 1998). These results clearly demonstrated that the signal from the IL-7R is indispensable for γδ T-cell development in both thymic and extrathymic pathways. Additionally, it has been shown that IL-15 supports the expansion of γδ T cells in the epithelium of skin and small intestine and is required for NK-cell development (Lodolce et al. 1998).

The IL-7R Controls the Recombination in the TCRγ Locus

γδ T cells have unique features in development (Ikuta et al. 1992; Haas et al. 1993; Xiong and Raulet 2007). γδ T cells expressing specific Vγ chain appear as several successive waves in the developing thymus (Fig. 2). In fetal thymus, $V\gamma3^+$ T cells develop as the first wave, followed by $V\gamma4^+$ T cells as the second wave. In neonatal and adult, $V\gamma2^+$ T cells become the major γδ T-cell population in the thymus, while $V\gamma5^+$ T cells develop in the small intestine. Each wave of γδ T cells shows specific tissue distribution in the adult mouse. For example, $V\gamma3^+$ T cells reside as DETC in

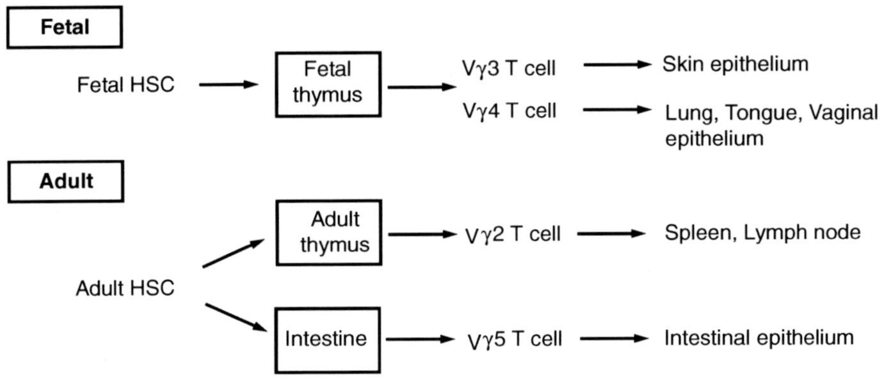

Fig. 2 γδ T-cell development during mouse ontogeny. The γδ T cells expressing specific Vγ proteins develop as successive waves during mouse development. They become distributed to specific organs in the adult

the skin, while Vγ4$^+$ T cells exist in the epithelium of the lung, tongue, and vagina. Vγ2$^+$ T cells become the major γδ T-cell population in the spleen and lymph nodes, while Vγ5$^+$ T cells are distributed in the gut as IEL. The development of γδ T cells is determined at least in part at the level of hematopoietic stem cells (HSC) (Ikuta et al. 1990; Ikuta and Weissman 1991). Fetal HSC give rise to Vγ3$^+$ and Vγ4$^+$ fetal type γδ T cells in the fetal thymic organ culture, while adult HSC do not.

IL-7 was implicated to promote rearrangements of the TCRγ locus. It was reported that IL-7 supports embryonic day 14 fetal liver cells in inducing TCRγ transcripts (Appasamy 1992). IL-7 induces rearrangement of Vγ2 and Vγ4 but not Vγ3 and Vγ5 genes and sustains expression of *RAG-1* and *RAG-2* genes (Appasamy et al. 1993). It was reported that IL-7 supports survival and growth of early thymocytes and promotes rearrangement of TCRβ chain in the thymus (Muegge et al. 1993). In addition, IL-7 also supports D-J but not V-DJ rearrangement of TCRβ gene in fetal liver cells (Tsuda et al. 1996). Although these results implicate that IL-7 might be involved in V(D)J recombination of the TCRβ and TCRγ loci, it was difficult to discriminate whether IL-7 directly induces the recombination or supports survival of early T cells and the transcription of rearranged TCR genes.

The IL-7R controls the recombination of the TCRγ locus. To elucidate the reason why IL-7Rα-deficient mice completely lack γδ T cells, we analyzed the rearrangements of TCR genes in IL-7Rα-deficient thymocytes (Maki et al. 1996a). Southern blot analysis with a Jγ1 probe revealed that more than 70% of Jγ1 and Jγ2 alleles are recombined to form distinct Vγ1.2-Jγ2 and Vγ2-Jγ1 rearranged fragments in wild-type thymocytes. In contrast, no such recombination is detected in IL-7Rα-deficient thymocytes. Because γδ T cells are less than 1% of wild-type thymocytes, this result demonstrated that the TCRγ locus is rearranged in the majority of αβ T cells and that the rearrangement of the TCRγ locus is reduced in αβ T cells of IL-7Rα-deficient mice. Polymerase chain reaction (PCR) analysis indicated that the V-J recombination of all the Vγ genes is severely reduced in

IL-7Rα-deficient mice. The rearrangements in the TCRα, β, and δ loci are comparable between control and mutant mice, suggesting that this block is specific for the TCRγ locus. The mRNA of *RAG-1*, *RAG-2*, *Ku-80*, and terminal deoxynucleotidyl transferase genes are equally detected between control and mutant thymus, suggesting that expression of the recombination machinery is not affected. These results demonstrated that V-J recombination of the TCRγ locus is specifically blocked in IL-7Rα-deficient mice and suggested the presence of highly specific regulation for TCRγ gene rearrangement. This conclusion was also supported by Southern analyses of IL-7Rα-deficient Bcl-2 transgenic mice with the Jγ1 probe (Ye et al. 1999) and of IL-7Rα-deficient mice with a Cγ probe (Kang et al. 1999). These mice have far lower levels of TCRγ gene rearrangement than normal mice.

The IL-7R induces the transcription of the TCRγ genes as well as the rearrangement. The notion that IL-7R-mediated signals are necessary for normal expression of rearranged TCRγ genes was supported by examining transcription of a rearranged TCRγ transgene in IL-7Rα-deficient mice (Kang et al. 1999). In addition, mature αβ T cells in the periphery are induced to transcribe endogenous rearranged TCRγ genes by IL-7 stimulation. Consistent with these results, defective germline transcription was observed in the TCRγ locus of IL-7R-deficient thymocytes (Durum et al. 1998). Recently, we demonstrated that STAT5 is recruited to the Eγ elements by cytokine stimulation and augments Eγ activity (Tani-ichi et al. 2009). Because the Eγ element was shown to enhance the transcription of minimal promoters in γδ T cells (Spencer et al. 1991; Kappes et al. 1991), these results implied that IL-7R induces the transcription of the TCRγ locus by activating the transcriptional enhancers.

STAT5 Controls the Accessibility of Jγ Gene Segments

Rearrangements of TCR and Ig genes are mediated by conserved recombinational signal sequences and V(D)J recombinases. The recombinational accessibility model postulates that in developing T and B cells specific molecular mechanisms should exist that make the appropriate Ig or TCR loci accessible to the common recombinase activity in a lineage- and stage-specific manner (Alt et al. 1987; Hempel et al. 1998; Sleckman et al. 1998; Krangel 2003). Two kinds of *cis*-control elements are involved in regulating the accessibility. One is the promoter for germline transcription that controls the local accessibility in V(D)J recombination. Where V(D)J recombination takes place, the germline transcription is associated with an open chromatin configuration (Villey et al. 1996; Whitehurst et al. 1999). For example, T early α (TEA) germline transcription takes place in the 5′ region of the Jα cluster in immature thymocytes. Targeted deletion of the TEA promoter results in severe impairment of the rearrangement of the 5′-most Jα segments (Villey et al. 1996). In another case, targeted deletion of the germline promoter of the Dβ1 gene segment abolishes Dβ germline transcription and reduces the rearrangement of the Dβ1 gene segment (Whitehurst et al. 1999). These results suggested a critical role for germline transcription in V(D)J recombination of the antigen receptor genes. The other is the enhancer elements that govern locus-wide accessibility in V(D)J recombination. Targeted deletion of the respective enhancers abolishes rearrangement of the TCRβ locus and greatly reduces rearrangement of the IgH, Igκ, and TCRα loci (Bories et al. 1996; Bouvier et al. 1996; Sleckman et al. 1997; Inlay et al. 2002; Perlot et al. 2005). These results suggested a central role for the enhancers in V(D)J recombination of the antigen receptor genes.

The accessibility of the TCRγ locus is reduced in IL-7Rα-deficient mice. This was analyzed by two different approaches (Schlissel et al. 2000). One is ligation-mediated (LM)-PCR that detects signal broken ends. The signal ends of the TCRγ locus are not observed in IL-7Rα-deficient fetal thymocytes, whereas the recombination intermediates at the TCRβ locus were easily detected. The other is in vitro recombination signal sequence cleavage assay in which intact nuclei are incubated in vitro with nuclear extract containing RAG-1 and RAG-2, and the high molecular weight DNA isolated from the nuclei is subjected to LM-PCR. Chromatin accessibility for *RAG*-mediated cleavage is reduced in IL-7Rα-deficient thymocytes compared with wild-type thymocytes. These results collectively demonstrated that the IL-7R controls V-J recombination at the TCRγ locus by regulating locus accessibility.

To identify the control mechanism by the IL-7R, we analyzed the germline transcription in the TCRγ locus (Ye et al. 1999). It was reported that germline transcripts of the TCRγ locus are induced in IL-3–dependent hematopoietic progenitor cell lines (Weinstein et al. 1989). We showed that the germline transcripts are induced in IL-3–dependent Ba/F3 cells and IL-7–dependent pre–B-cell lines by cytokine stimulation. Furthermore, the germline transcripts are detected in Ba/F3 cells transfected with a constitutively active STAT5 expression vector and cultured without cytokine, suggesting that STAT5 has potential to induce the germline transcription of the TCRγ locus. By examining the promoter region of Jγ-Cγ germline transcripts, we found that STAT consensus motifs (TTCNNNGAA) are conserved at approximately 100 bp upstream of the transcription initiation sites in all Jγ germline promoters. By gel shift and reporter assays, we showed that STAT5 binds to the consensus motifs and activates the activity of the Jγ germline promoters. It was also found that the constitutively active STAT5 restores the germline transcription and V-J recombination of TCRγ genes and partially rescues T-cell development from IL-7Rα–deficient T-cell precursors, especially in favor of γδ T cells. Thus, these results revealed a potential role for STAT5 in T-cell development and imply that STAT5 might control the accessibility of the TCRγ locus by inducing the germline transcription.

To identify the molecular mechanism that links STAT5-induced germline transcription to the accessibility of the TCRγ locus, we characterized the role of transcriptional coactivators and histone acetylation (Ye et al. 2001). Our results demonstrated that transcriptional coactivators augment STAT5-induced germline transcription of the TCRγ locus by their histone acetyltransferase activity. It was also found that histones are hyperacetylated at transcriptionally active Jγ segments in normal thymocyte precursors and Ba/F3 cells and that the levels of histone acetylation are reduced in IL-7Rα–deficient thymocytes. Finally, forced expression of *RAG-1* and *RAG-2* induces cleavage at the Jγ segment in Ba/F3 cells, suggesting that the chromatin region where germline transcription and histone hyperacetylation take place is accessible to the recombinational machinery. Thus, these results revealed a role for histone acetylation in controlling the accessibility of the TCRγ locus through IL-7R signaling and provided the first evidence identifying all the signaling molecules from cell surface to chromatin involved in the regulation of V(D)J recombination (Fig. 3). According to this model, STAT5 activated by IL-7R signaling interacts with the conserved motifs in the Jγ germline promoters and recruits transcriptional coactivators. The coactivators acetylate the histones at nearby nucleosomes and render the chromatin region accessible to the recombinational and the transcriptional machinery.

Chromatin remodeling is considered to play an important role in chromatin accessibility (Kwon et al. 2000). It was reported that SWI-SNF chromatin remodeling complexes including BRG1 and BRM are recruited to the Dβ germline promoter in early thymocytes and that forced recruitment of BRG1 to a

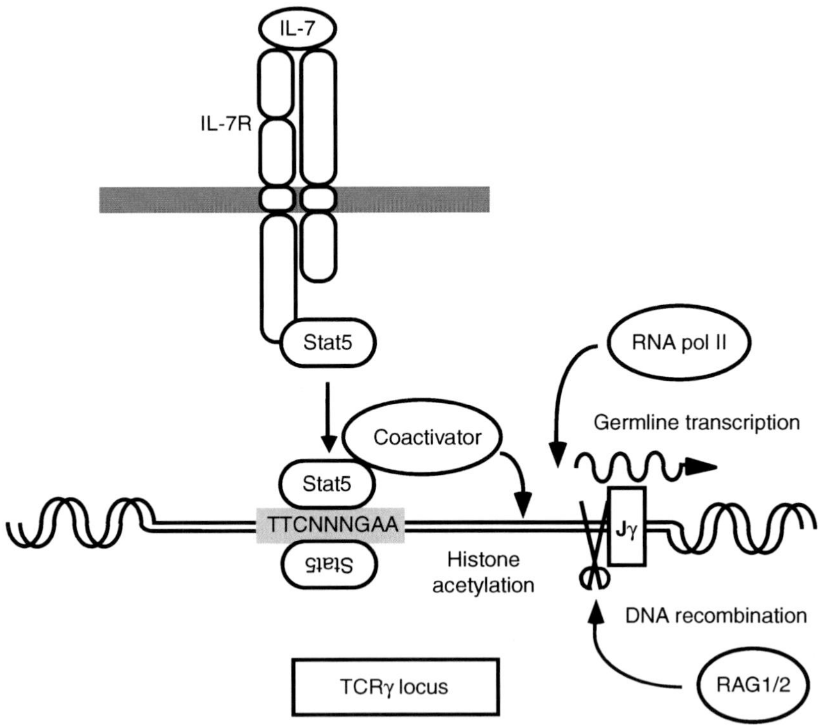

Fig. 3 Signal transducer and activator of transcription 5 (STAT5) controls the accessibility of the Jγ gene segment by transcriptional coactivators and histone acetylation. STAT5 activated by interleukin 7 receptor (IL-7R) signaling interacts with the conserved motifs in the Jγ germline promoters and recruits transcriptional coactivators. The coactivators acetylate the histones at nearby nucleosomes and render the chromatin region accessible to the recombinational and the transcriptional machinery

Dβ promoter-less chromatin restores Dβ-Jβ recombination (Osipovich et al. 2007). Recently, we found that BRG1 is recruited to the Jγ germline promoter in *RAG-2*-deficient thymocytes and augments the activity of the Jγ germline promoter (Sumi, Tani-ichi, and Ikuta, unpublished results). These results suggested that BRG1 is involved in chromatin remodeling of the TCRγ locus that makes the Jγ gene segment completely accessible for transcriptional and recombinational machinery.

STAT5 controls the Jγ germline promoters in different TCRγ clusters. The mouse TCRγ locus consists of four clusters, each containing Vγ, Jγ, and Cγ gene segments as well as 3′ enhancer (Eγ) elements (Fig. 4). The γ1, γ2, and γ3 clusters are highly conserved with one another, while the γ4 cluster is less homologous to the others. Although STAT motifs are present in the Jγ4 germline promoter and the Eγ4 element in the γ4 cluster, it was unknown whether STAT5 regulates the transcription of the Jγ4 promoter. Recently, we reported that cytokine stimulation induces Jγ4-Cγ4 germline transcripts in a pre–T-cell line, Scid.adh, and a hematopoietic cell line, Ba/F3 (Masui et al. 2008; Maki and Ikuta 2008). We found that STAT5 binds to the STAT motif of the Jγ4 germline promoter in vitro by gel shift assay. In addition, by chromatin immunoprecipitation assay we detected that STAT5 is recruited to the endogenous Jγ4 chromatin in Ba/F3 cells after cytokine stimulation. Finally, using reporter assay, we showed that the Jγ4 germline promoter is activated by STAT5 and that a mutation in the STAT motif abrogates the activity. Furthermore, transcriptional coactivators CBP and p300 augment this transactivation. Collectively, these results demonstrated that STAT5 binds to the STAT motif in the Jγ4 promoter and induces germline transcription. Thus, the IL-7R/STAT5 signal controls the transcription and accessibility of different clusters in the mouse TCRγ locus.

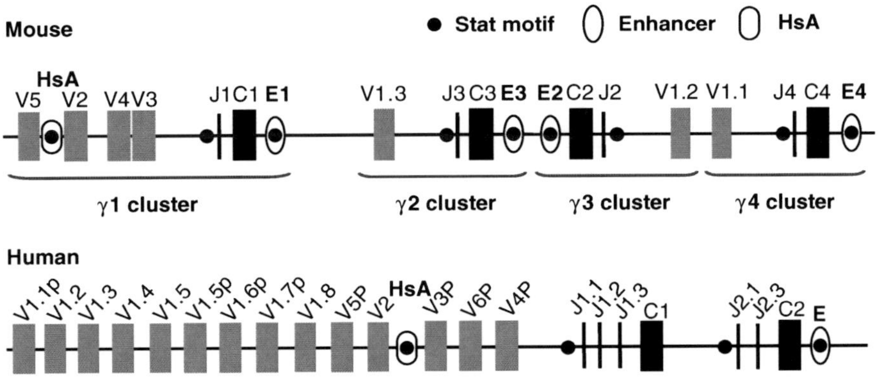

Fig. 4 Conserved signal transducer and activator of transcription (STAT) motifs in the mouse and human T-cell receptor γ (TCRγ) loci. The mouse TCRγ locus consists of four clusters, each containing Vγ, Jγ, and Cγ gene segments as well as the Eγ element. In the human TCRγ locus, the V gene segments with "P" indicate pseudogenes. STAT consensus motifs are conserved in the Jγ germline promoters, 3′ enhancers (E), and HsA elements between mouse and human

STAT5 and the TCRγ Locus

The recombinational accessibility of the TCRγ locus might be regulated by several steps. First, the localization in the nucleus might control the global activity of the TCRγ locus. Second, it is probable that the enhancers of the TCRγ locus (Eγ) control locus-wide accessibility of the TCRγ locus, as has been described for other TCR and Ig loci (Bories et al. 1996; Bouvier et al. 1996; Sleckman et al. 1997; Inlay et al. 2002; Perlot et al. 2005). A locus control region-like element between Vγ5 and Vγ2 gene segments, HsA, might also contribute to the accessibility of the Vγ region (Baker et al. 1999). Finally, STAT5 induces germline transcription and regulates the local accessibility near Jγ gene segments by histone acetylation and chromatin remodeling (Ye et al. 1999; Ye et al. 2001). STAT5 might be involved in all of these steps.

In the mouse and human TCRγ loci, STAT consensus motifs are conserved in the Jγ germline promoters and the Eγ and HsA elements (Lee et al. 2001) (Fig. 4). This conservation is very strict in that the motifs are present exactly at the same portion within each control element. In contrast, the STAT motifs are not conserved in cis-controlling elements of the TCRα, β, and δ loci. Thus, it is highly probable that STAT5 interacts with these STAT motifs and contributes to the accessibility control of the TCRγ locus. Indeed, the Eγ element is histone hyperacetylated in normal thymocyte precursors and the acetylation levels are reduced in IL-7Rα–deficient mice (Ye et al. 2001). This result is consistent with the previous report that the histone acetylation of enhancers is tightly correlated with the accessibility of the TCRα/δ loci and is proposed as a mechanism for coupling enhancer activity to accessibility (McMurry and Krangel 2000). Recently we found that the bovine TCRγ locus contains conserved STAT consensus motifs in the Jγ germline promoters and the Eγ elements (Wada and Ikuta, unpublished results), suggesting that the control of the TCRγ locus by STAT5 is probably conserved through mammals.

STAT5 activates the Eγ elements in the TCRγ locus. Although STAT consensus motifs are conserved not only in the Jγ promoters but also in the Eγ elements, little has been known about the function of STAT5 on Eγ activity. Recently, we found that the Eγ elements are substantially histone hyperacetylated regardless of cytokine stimulation in cell lines and $Rag2^{-/-}$ thymocytes, and that STAT5 has the potential to elevate histone acetylation at these elements (Tani-ichi et al. 2009). We also found that STAT5 is recruited to the STAT consensus motifs in Eγ elements after cytokine stimulation, and that transcription factors Runx and c-Myb are constitutively recruited to Eγ. Furthermore, we showed that STAT5 synergistically augments the enhancer activity of Eγ in cooperation with Runx and c-Myb that were reported to activate the Eγ elements. These results demonstrated that STAT5 is recruited to the consensus motifs in the Eγ elements by cytokine stimulation and augments Eγ activity in collaboration with Runx and c-Myb. Therefore, STAT5 might control locus-wide accessibility of the TCRγ locus by activating Eγ elements.

The conservation of the STAT motifs indicates that the TCRγ locus is under the strong influence of IL-7R signaling. After entry into the thymus, T-cell precursors first proliferate by stimuli from c-kit and the IL-7R (Rodewald et al. 1997). At this stage they receive a signal from the IL-7R to induce the rearrangement of the TCRγ locus. This will help them to commit and maintain themselves to the γδ T-cell lineage. In αβ T-cell development, however, this signal for γδ T cells has to be shut off. We speculate that pre-TCR signaling might cancel this IL-7R signal to maintain commitment to the αβ T-cell lineage. During the evolution of immune system, γδ T cells might have emerged first with the simple mechanism that IL-7 produced in epithelial cells of the skin and intestine induces the V(D)J recombination and cell expansion. αβ T cells probably evolved later with a more sophisticated system where pre-TCR signaling invalidates the IL-7R signal, enabling positive and negative selection to operate on the basis of interactions between αβ TCR and self–major histocompatibility complex.

Control of Vγ Recombination by the IL-7R

The V-J recombination of the mouse TCRγ1 cluster takes place in a developmentally regulated manner. The Vγ region contains four Vγ genes, Vγ5, Vγ2, Vγ4, and Vγ3, and a control element, HsA, between Vγ5 and Vγ2 (Fig. 4). The γδ T cells expressing specific Vγ proteins develop as successive waves during mouse development (Ikuta et al. 1992; Haas et al. 1993; Xiong and Raulet 2007). Interestingly, the four Vγ genes are arranged sequentially from Jγ1 gene segment in the order of appearance in the thymus and intestine. The unique development of γδ T cells is controlled at the level of V-J recombination: J-proximal Vγ3 and Vγ4 are rearranged in the fetus, while J-distal Vγ2 and Vγ5 are rearranged in the neonatal and adult (Itohara et al. 1993). Histone acetylation is well correlated with the chromatin accessibility of each Vγ gene. Mice with targeted deletions suggested that Eγ and HsA have partially overlapping effects on Vγ-Jγ recombination (Xiong et al. 2002). Deficiency in E2A transcription factor resulted in persistence of Vγ3^{+} T cells in adult thymus, suggesting that E2A is involved in fetal to adult switch of Vγ recombination (Bain et al. 1999). Finally, it has been reported that IL-15R/STAT5 signal induces Vγ5-Jγ1 recombination in IL-15 transgenic mice (Zhao et al. 2005). In spite of these studies, the precise mechanism controlling the Vγ recombination is largely unknown.

The IL-7R is essential for chromatin accessibility of not only Jγ but also Vγ gene segments. It was reported that germline transcription of Vγ genes is severely reduced in *IL-7* × *RAG-2* double deficient thymocytes compared with *RAG-2*–deficient thymocytes (Huang et al. 2001). Histone acetylation of Vγ5 and Vγ2 genes is also decreased in *IL-7* × *RAG-2* double deficient thymocytes (Tani-ichi and Ikuta, unpublished results). These results indicated that the IL-7R controls the chromatin accessibility of Vγ genes. It was reported that IL-15R/STAT5 signal induces Vγ5-Jγ1 recombination in IL-15 transgenic mice (Zhao et al. 2005), suggesting that STAT5 is involved in opening Vγ5 chromatin. Recently, we found that that the histone acetylation and germline transcription in Vγ5 gene and the HsA element are induced by cytokine signal and STAT5 in a hematopoietic cell line BaF3 and a pre–T-cell line Scid.adh (Tani-ichi and Ikuta, unpublished results). Next, we found that STAT5 is recruited to the STAT consensus motif in the HsA element after cytokine stimulation. In contrast, only an atypical STAT motif is found in the Vγ5 promoter, and STAT5 is not recruited to the motif. Furthermore, we showed that the Vγ5 promoter is activated by cytokine stimulation even with a mutation in the atypical motif. Finally, we showed a bimodal pattern of histone acetylation at Vγ5 and HsA, suggesting that the histone acetylation in the Vγ5 promoter is independent of that in the HsA element. These results suggested that STAT5 directly binds to and controls the HsA element but that STAT5 indirectly activates the Vγ5 promoter.

It is largely unknown what the control mechanism of Vγ gene recombination is during mouse development. It was reported that a basic helix-loop-helix protein, E2A, regulates the fetal (Vγ3 and Vγ4) to adult (Vγ2 and Vγ5) switch of preferential recombination of the TCRγ locus (Bain et al. 1999). Although Vγ3-Jγ1 rearrangement is not detected in the adult thymus of wild-type mice, E2A-deficient mice show sustained Vγ3-Jγ1 rearrangement in the adult. Although it is clear that E2A is involved in the developmental switch, the precise mechanism of Vγ gene recombination during development is yet to be elucidated. In addition, we found that Runx transcription factor is constitutively recruited to the HsA element in *RAG-2*–deficient thymocytes (Tani-ichi and Ikuta, unpublished results). However, it is still unclear whether Runx is involved in Vγ gene recombination.

Conclusion

IL-7 is an essential cytokine for early T- and B-cell development when V(D)J recombination takes place. The IL-7R and STAT5 control the TCRγ locus by interacting with STAT consensus motifs in the Jγ germline promoters and the Eγ and HsA elements (Fig. 5). STAT5 binding induces germline transcription, histone acetylation, and chromatin remodeling that make the chromatin accessible for DNA recombinases. A question yet to be resolved is why the TCRγ locus is rearranged only in T cells, even though the IL-7R signal operates in both T and B cells. There should exist some T-cell–specific transcription factors that make the locus active in cooperation with STAT5. It is also of interest whether the localization of the TCRγ locus in the nucleus is regulated by STAT5 and other transcription factors. STAT5 and the TCRγ locus will remain an excellent model system to understand the control mechanisms of V(D)J recombination in lymphocyte antigen receptor genes.

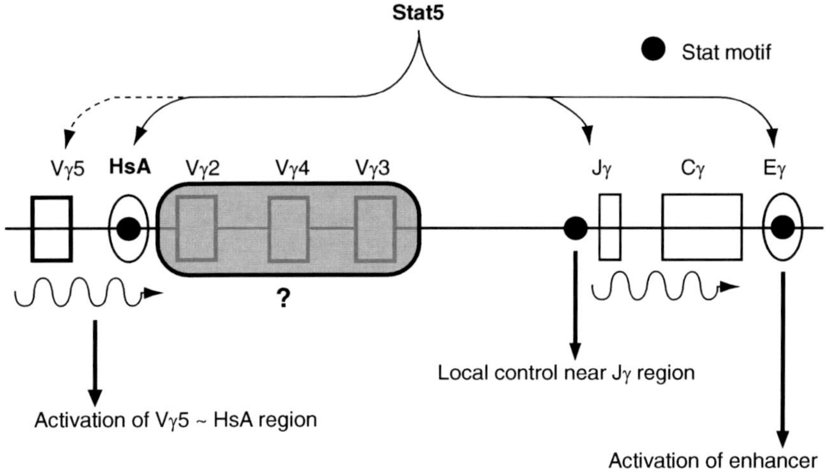

Fig. 5 Control of the T-cell receptor γ (TCRγ) locus by signal transducer and activator of transcription 5 (STAT5). STAT5 controls the TCRγ locus by interacting with STAT consensus motifs in the Jγ germline promoters and the Eγ and HsA elements. STAT5 binding to the Jγ germline promoters induces germline transcription, histone acetylation, and chromatin remodeling that make the chromatin accessible for DNA recombinases. In addition, STAT5 is recruited to the consensus motifs in the Eγ elements by cytokine stimulation and augments Eγ activity. Furthermore, STAT5 directly binds to and controls the HsA element but that STAT5 indirectly activates the Vγ5 promoter

Acknowledgments We thank H. Miyachi and S. Kitano for study mice, S. Hayashi, M. Kishida, and S. Kamioka for technical assistance, and members of the Ikuta lab for helpful discussions. Published and unpublished works from this laboratory were supported by Grants-in-Aid from the Ministry of Education, Culture, Sports, Science, and Technology of Japan and by grants provided by Takeda Science Foundation, Naito Foundation, Sapporo Bioscience Foundation, Astellas Foundation for Research on Metabolic Disorders, Sumitomo Foundation, Ichiro Kanehara Foundation, Uehara Memorial Foundation, Novartis Foundation (Japan) for the Promotion of Science, Shimadzu Science Foundation, TERUMO Life Science Foundation, Mitsubishi Pharma Research Foundation, and Inoue Foundation for Science. S. Tani-ichi was supported by the 21st Century COE Program of the Ministry of Education, Culture, Sports, Science and Technology to the Graduate School of Bio Studies and the Institute for Virus Research, Kyoto University.

References

Akashi K, Kondo M, von Freeden-Jeffry U, Murray R, Weissman IL (1997) Bcl-2 rescues T lymphopoiesis in interleukin-7 receptor-deficient mice. Cell 89:1033–1041

Alt FW, Blackwell TK, Yancopoulos GD (1987) Development of the primary antibody repertoire. Science 238:1079–1087

Appasamy PM (1992) IL 7-induced T cell receptor-γ gene expression by pre-T cells in murine fetal liver cultures. J Immunol 149:1649–1656

Appasamy PM, Kenniston TW Jr, Weng Y, Holt EC, Kost J, Chambers WH (1993) Interleukin 7-induced expression of specific T cell receptor γ variable region genes in murine fetal liver cultures. J Exp Med 178:2201–2206

Bain G, Romanow WJ, Albers K, Havran WL, Murre C (1999) Positive and negative regulation of V(D)J recombination by the E2A proteins. J Exp Med 189:289–300

Baker JE, Kang J, Xiong N, Chen T, Cado D, Raulet DH (1999) A novel element upstream of the Vγ2 gene in the murine T cell receptor γ locus cooperates with the 3′ enhancer to act as a locus control region. J Exp Med 190:669–779

Bertolino E, Reddy K, Medina KL, Parganas E, Ihle J, Singh H (2005) Regulation of interleukin 7-dependent immunoglobulin heavy-chain variable gene rearrangements by transcription factor STAT5. Nat Immunol 6:836–843

Bories JC, Demengeot J, Davidson L, Alt FW (1996) Gene-targeted deletion and replacement mutations of the T-cell receptor β-chain enhancer: The role of enhancer elements in controlling V(D)J recombination accessibility. Proc Natl Acad Sci USA 93:7871–7876

Bouvier G, Watrin F, Naspetti M, Verthuy C, Naquet P, Ferrier P (1996) Deletion of the mouse T-cell receptor β gene enhancer blocks αβ T-cell development. Proc Natl Acad Sci USA 93:7877–7881

Candeias S, Peschon JJ, Muegge K, Durum SK (1997) Defective T-cell receptor γ gene rearrangement in interleukin-7 receptor knockout mice. Immunol Lett 57:9–14

Corcoran AE, Smart FM, Cowling RJ, Crompton T, Owen MJ, Venkitaraman AR (1996) The interleukin-7 receptor α chain transmits distinct signals for proliferation and differentiation during B lymphopoiesis. EMBO J 15:1924–1932

Corcoran AE, Riddell A, Krooshoop D, Venkitaraman AR (1998) Impaired immunoglobulin gene rearrangement in mice lacking the IL-7 receptor. Nature 391:904–907

Durum SK, Candeias S, Nakajima H, Leonard WJ, Baird AM, Berg LJ, Muegge K (1998) Interleukin 7 receptor control of T cell receptor γ gene rearrangement: Role of receptor-associated chains and locus accessibility. J Exp Med 188:2233–2241

Fujio K, Nosaka T, Kojima T, Kawashima T, Yahata T, Copeland NG, Gilbert DJ, Jenkins NA, Yamamoto K, Nishimura T, Kitamura T (2000) Molecular cloning of a novel type 1 cytokine receptor similar to the common gamma chain. Blood 95:2204–2210

Goodwin RG, Friend D, Ziegler SF, Jerzy R, Falk BA, Gimpel S, Cosman D, Dower SK, March CJ, Namen AE et al (1990) Cloning of the human and murine interleukin-7 receptors: demonstration of a soluble form and homology to a new receptor superfamily. Cell 60:941–951

Grabstein KH, Waldschmidt TJ, Finkelman FD, Hess BW, Alpert AR, Boiani NE, Namen AE, Morrissey PJ (1993) Inhibition of murine B and T lymphopoiesis in vivo by an anti-interleukin 7 monoclonal antibody. J Exp Med 178:257–264

Haas W, Pereira P, Tonegawa S (1993) Gamma/delta cells. Annu Rev Immunol 11:637–685

He YW, Malek TR (1996) Interleukin-7 receptor α is essential for the development of $\gamma\delta^+$ T cells, but not natural killer cells. J Exp Med 184:289–293

Hempel WM, Leduc I, Mathieu N, Tripathi RK, Ferrier P (1998) Accessibility control of V(D)J recombination: lessons from gene targeting. Adv Immunol 69:309–352

Huang J, Durum SK, Muegge K (2001) Histone acetylation and recombination at the TCRγ locus follows IL-7 induction. J Immunol 167:6073–6077

Ikuta K, Weissman IL (1991) The junctional modifications of a T cell receptor γ chain are determined at the level of thymic precursors. J Exp Med 174:1279–1282

Ikuta K, Kina T, MacNeil I, Uchida N, Peault B, Chien YH, Weissman IL (1990) A developmental switch in thymic lymphocyte maturation potential occurs at the level of hematopoietic stem cells. Cell 62:863–874

Ikuta K, Uchida N, Friedman J, Weissman IL (1992) Lymphocyte development from stem cells. Annu Rev Immunol 10:759–783

Inlay M, Alt FW, Baltimore D, Xu Y (2002) Essential roles of the κ light chain intronic enhancer and 3′ enhancer in κ rearrangement and demethylation. Nat Immunol 3:463–468

Isaksen DE, Baumann H, Trobridge PA, Farr AG, Levin SD, Ziegler SF (1999) Requirement for Stat5 in thymic stromal lymphopoietin-mediated signal transduction. J Immunol 163:5971–5977

Itohara S, Mombaerts P, Lafaille J, Iacomini J, Nelson A, Clarke AR, Hooper ML, Farr A, Tonegawa S (1993) T cell receptor δ gene mutant mice: Independent generation of αβ T cells and programmed rearrangements of γδ TCR genes. Cell 72:337–348

Kang J, Coles M, Raulet DH (1999) Defective development of γ/δ T cells in interleukin 7 receptor-deficient mice is due to impaired expression of T cell receptor γ genes. J Exp Med 190:973–982

Kappes DJ, Browne CP, Tonegawa S (1991) Identification of a T-cell-specific enhancer at the locus encoding T-cell antigen receptor γ chain. Proc Natl Acad Sci USA 88:2204–2208

Kawai K, Suzuki H, Tomiyama K, Minagawa M, Mak TW, Ohashi PS (1998) Requirement of the IL-2 receptor β chain for the development of Vγ3 dendritic epidermal T cells. J Invest Dermatol 110:961–965

Kondo M, Takeshita T, Higuchi M, Nakamura M, Sudo T, Nishikawa S, Sugamura K (1994) Functional participation of the IL-2 receptor γ chain in IL-7 receptor complexes. Science 263:1453–1454

Krangel MS (2003) Gene segment selection in V(D)J recombination: Accessibility and beyond. Nat Immunol 4:624–630

Kwon J, Morshead KB, Guyon JR, Kingston RE, Oettinger MA (2000) Histone acetylation and hSWI/SNF remodeling act in concert to stimulate V(D)J cleavage of nucleosomal DNA. Mol Cell 6:1037–1048

Lee HC, Ye SK, Honjo T, Ikuta K (2001) Induction of germline transcription in the human TCRγ locus by STAT5. J Immunol 167:320–326

Leonard WJ, O'Shea JJ (1998) Jaks and STATs: Biological implications. Annu Rev Immunol 16:293–322

Lodolce JP, Boone DL, Chai S, Swain RE, Dassopoulos T, Trettin S, Ma A (1998) IL-15 receptor maintains lymphoid homeostasis by supporting lymphocyte homing and proliferation. Immunity 9:669–676

Maki K, Ikuta K (2008) MEK1/2 induces STAT5-mediated germline transcription of the TCRγ locus in response to IL-7R signaling. J Immunol 181:494–502

Maki K, Sunaga S, Ikuta K (1996a) The V-J recombination of T cell receptor-γ genes is blocked in interleukin-7 receptor-deficient mice. J Exp Med 184:2423–2427

Maki K, Sunaga S, Komagata Y, Kodaira Y, Mabuchi A, Karasuyama H, Yokomuro K, Miyazaki JI, Ikuta K (1996b) Interleukin 7 receptor-deficient mice lack γδ T cells. Proc Natl Acad Sci USA 93:7172–7177

Maraskovsky E, O'Reilly LA, Teepe M, Corcoran LM, Peschon JJ, Strasser A (1997) Bcl-2 can rescue T lymphocyte development in interleukin-7 receptor-deficient mice but not in mutant *RAG-1*$^{-/-}$ mice. Cell 89:1011–1019

Masui N, Tani-ichi S, Maki K, Ikuta K (2008) Transcriptional activation of mouse TCR Jγ4 germline promoter by STAT5. Mol Immunol 45:849–855

Matsue H, Bergstresser PR, Takashima A (1993) Keratinocyte-derived IL-7 serves as a growth factor for dendritic epidermal T cells in mice. J Immunol. 151:6012–6019

Mazzucchelli R, Durum SK (2007) Interleukin-7 receptor expression: intelligent design. Nat Rev Immunol 7:144–154

McMurry MT, Krangel MS (2000) A role for histone acetylation in the developmental regulation of VDJ recombination. Science 287:495–498

Muegge K, Vila MP, Durum SK (1993) Interleukin-7: A cofactor for V(D)J rearrangement of the T cell receptor β gene. Science 261:93–95

Namen AE, Schmierer AE, March CJ, Overell RW, Park LS, Urdal DL, Mochizuki DY (1988) B cell precursor growth-promoting activity. Purification and characterization of a growth factor active on lymphocyte precursors. J Exp Med 167:988–1002

Noguchi M, Nakamura Y, Russell SM, Ziegler SF, Tsang M, Cao X, Leonard WJ (1993) Interleukin-2 receptor γ chain: A functional component of the interleukin-7 receptor. Science 262:1877–1880

Osipovich O, Cobb RM, Oestreich KJ, Pierce S, Ferrier P, Oltz EM (2007) Essential function for SWI-SNF chromatin-remodeling complexes in the promoter-directed assembly of *Tcrb* genes. Nat Immunol 8:809–816

Pallard C, Stegmann AP, van Kleffens T, Smart F, Venkitaraman A, Spits H (1999) Distinct roles of the phosphatidylinositol 3-kinase and STAT5 pathways in IL-7-mediated development of human thymocyte precursors. Immunity 10:525–535

Pandey A, Ozaki K, Baumann H, Levin SD, Puel A, Farr AG, Ziegler SF, Leonard WJ, Lodish HF (2000) Cloning of a receptor subunit required for signaling by thymic stromal lymphopoietin. Nat Immunol 1:59–64

Perlot T, Alt FW, Bassing CH, Suh H, Pinaud E (2005) Elucidation of IgH intronic enhancer functions via germ-line deletion. Proc Natl Acad Sci USA 102:14362–14367

Peschon JJ, Morrissey PJ, Grabstein KH, Ramsdell FJ, Maraskovsky E, Gliniak BC, Park LS, Ziegler SF, Williams DE, Ware CB, Meyer JD, Davison BL (1994) Early lymphocyte expansion is severely impaired in interleukin 7 receptor-deficient mice. J Exp Med 180:1955–1960

Plum J, De Smedt M, Leclercq G (1993) Exogenous IL-7 promotes the growth of $CD3^-CD4^-CD8^-$ $CD44^+CD25^{+/-}$ precursor cells and blocks the differentiation pathway of TCR-αβ cells in fetal thymus organ culture. J Immunol 150:2706–2716

Ray RJ, Furlonger C, Williams DE, Paige CJ (1996) Characterization of thymic stromal-derived lymphopoietin (TSLP) in murine B cell development in vitro. Eur J Immunol 26:10–16

Rodewald HR, Ogawa M, Haller C, Waskow C, DiSanto JP (1997) Pro-thymocyte expansion by c-kit and the common cytokine receptor γ chain is essential for repertoire formation. Immunity 6:265–272

Schlissel MS, Durum SD, Muegge K (2000) The interleukin 7 receptor is required for T cell receptor γ locus accessibility to the V(D)J recombinase. J Exp Med 191:1045–1050

Sleckman BP, Bardon CG, Ferrini R, Davidson L, Alt FW (1997) Function of the TCR α enhancer in αβ and γδ T cells. Immunity 7:505–515

Sleckman BP, Bassing CH, Bardon CG, Okada A, Khor B, Bories JC, Monroe R, Alt FW (1998) Accessibility control of variable region gene assembly during T-cell development. Immunol Rev 165:121–130

Spencer DM, Hsiang YH, Goldman JP, Raulet DH (1991) Identification of a T-cell-specific transcriptional enhancer located 3′ of Cγ1 in the murine T-cell receptor γ locus. Proc Natl Acad Sci USA 88:800–804

Suda T, Zlotnik A (1991) IL-7 maintains the T cell precursor potential of $CD3^-CD4^-CD8^-$ thymocytes. J Immunol 146:3068–3073

Sudo T, Ito M, Ogawa Y, Iizuka M, Kodama H, Kunisada T, Hayashi S, Ogawa M, Sakai K, Nishikawa S (1989) Interleukin 7 production and function in stromal cell-dependent B cell development. J Exp Med 170:333–338

Sudo T, Nishikawa S, Ohno N, Akiyama N, Tamakoshi M, Yoshida H, Nishikawa S (1993) Expression and function of the interleukin 7 receptor in murine lymphocytes. Proc Natl Acad Sci USA 90:9125–9129

Suzuki H, Duncan GS, Takimoto H, Mak TW (1997) Abnormal development of intestinal intraepithelial lymphocytes and peripheral natural killer cells in mice lacking the IL-2 receptor β chain. J Exp Med 185:499–505

Tani-ichi S, Satake M, Ikuta K (2009) Activation of the mouse T cell receptor γ enhancers by STAT5. Int Immunol 21:1079–1088

Tsuda S, Rieke S, Hashimoto Y, Nakauchi H, Takahama Y (1996) IL-7 supports D-J but not V-DJ rearrangement of TCR-β gene in fetal liver progenitor cells. J Immunol 156:3233–3242

Villey I, Caillol D, Selz F, Ferrier P, de Villartay JP (1996) Defect in rearrangement of the most 5' TCR-Jα following targeted deletion of T early α (TEA): Implications for TCR α locus accessibility. Immunity 5:331–342

von Freeden-Jeffry U, Vieira P, Lucian LA, McNeil T, Burdach SE, Murray R (1995) Lymphopenia in interleukin (IL)-7 gene-deleted mice identifies IL-7 as a nonredundant cytokine. J Exp Med 181:1519–1526

von Freeden-Jeffry U, Solvason N, Howard M, Murray R (1997) The earliest T lineage-committed cells depend on IL-7 for Bcl-2 expression and normal cell cycle progression. Immunity 7:147–154

Watanabe Y, Sudo T, Minato N, Ohnishi A, Katsura Y (1991) Interleukin 7 preferentially supports the growth of γδ T cell receptor-bearing T cells from fetal thymocytes in vitro. Int Immunol 3:1067–1075

Watanabe M, Ueno Y, Yajima T, Iwao Y, Tsuchiya M, Ishikawa H, Aiso S, Hibi T, Ishii H (1995) Interleukin 7 is produced by human intestinal epithelial cells and regulates the proliferation of intestinal mucosal lymphocytes. J Clin Invest 95:2945–2953

Weinstein Y, Morishita K, Cleveland JL, Ihle JN (1989) Interleukin 3 (IL-3) induces transcription from nonrearranged T cell receptor γ loci in IL-3-dependent cell lines. J Exp Med 169:2059–2071

Whitehurst CE, Chattopadhyay S, Chen J (1999) Control of V(D)J recombinational accessibility of the Dβ1 gene segment at the TCRβ locus by a germline promoter. Immunity 10:313–322

Xiong N, Raulet DH (2007) Development and selection of γδ T cells. Immunol Rev 215:15–31

Xiong N, Kang C, Raulet DH (2002) Redundant and unique roles of two enhancer elements in the TCRγ locus in gene regulation and γδ T cell development. Immunity 16:453–463

Ye SK, Maki K, Kitamura T, Sunaga S, Akashi K, Domen J, Weissman IL, Honjo T, Ikuta K (1999) Induction of germline transcription in the TCRγ locus by Stat5: Implications for accessibility control by the IL-7 receptor. Immunity 11:213–223

Ye SK, Agata Y, Lee HC, Kurooka H, Kitamura T, Shimizu A, Honjo T, Ikuta K (2001) The IL-7 receptor controls the accessibility of the TCRγ locus by Stat5 and histone acetylation. Immunity 15:813–823

Zhao H, Nguyen H, Kang J (2005) Interleukin 15 controls the generation of the restricted T cell receptor repertoire of γδ intestinal intraepithelial lymphocytes. Nat Immunol 6:1263–1271

Dendritic Cell Homeostasis: Physiology and Impact on Disease

Miriam Merad and Markus G. Manz

Abstract Dendritic cells (DCs) are a heterogeneous fraction of rare hematopoietic cells that coevolved with the formation of the adaptive immune system. DCs efficiently process and present antigen, move from sites of antigen uptake to sites of cellular interactions, and are critical in initiation of immune responses as well as in maintenance of self-tolerance. DCs are distributed throughout the body with the relatively highest concentrations in lymphoid organs and environmental contact sites. Steady-state DC half-lives account for days to up to a few weeks, and they need to be replaced via proliferating hematopoietic progenitors, monocytes, or locally dividing, tissue resident cells. In this review we integrate recent knowledge on DC progenitors, cytokines, and transcription factor usage to an emerging concept of in vivo DC homeostasis in steady-state and inflammatory conditions. We furthermore highlight how basic understanding of these differentiation mechanisms might help to improve our understanding of posttransplant immune reactions of DC malignancies and their respective therapies.

Introduction

Dendritic cells (DCs) are hematopoietic cells that belong to the antigen-presenting cell (APC) family, which also includes B cells and macrophages. Although Langerhans cells in the skin were described in 1868, their role as APCs was not appreciated until 1973 when Steinman and Cohn first identified DCs in mouse spleen as potent stimulators

M. Merad (✉)
Department of Gene and Cell medicine and the Department of Medicine, Mount Sinai School of Medicine, 1425 Madison Ave, New York, NY 10029
e-mail: miriam.merad@mssm.edu

M.G. Manz
Institute for Research in Biomedicine (IRB), Via Vincenzo Vela 6, 6500 Bellinzona, Switzerland
e-mail: manz@irb.unisi.ch

M. Kondo (ed.), *Hematopoietic Stem Cell Biology*, Stem Cell Biology and Regenerative Medicine, DOI 10.1007/978-1-60327-347-3_7,

of the primary immune response (Banchereau and Steinman 1998). Shortly thereafter, several groups reported the presence of DC in nonlymphoid tissues of rodents and humans and demonstrated early evidence that these cells contribute to heart and kidney transplant rejection (Hart 1997). However, the low number of DCs in vivo, the paucity of markers that distinguish them from monocytes/macrophages, and the problems involved in purifying these cells made for slow progress. In the 1990s, the development of methods to isolate and generate DCs from blood and bone marrow led to explosive growth of the DC field (Markowicz and Engleman 1990; Caux et al. 1992; Inaba et al. 1992a; Inaba et al. 1992b; Sallusto and Lanzavecchia 1994). Studies in the past decade have established the critical role of DCs in maintenance of immunological integrity and their importance in the development and potential treatment of human disease (Bashyam 2007), leading in 2007 to the attribution of the Albert Lasker Award for Basic Medical Research to Ralph Steinman (Rockefeller University, New York) in recognition for his discovery of DCs (Bashyam 2007).

Heterogeneity of Dendritic Cells

DCs form a heterogenous population of antigen presenting cells. DCs have been grouped based on phenotype, location, and biological function. Here we chose to describe the heterogeneity of DC populations present in lymphoid and nonlymphoid tissues, as understanding site specific DC functional diversity may help understand the mechanisms that regulate tissue immunity (Table 1).

DCs in Nonlymphoid Tissues

DCs are present in most nonlymphoid tissues. Tissue DCs have two main functions: the maintenance of peripheral tolerance (Steinman et al. 2003) to tissue antigens and

Table 1 Localization and frequency of dendritic cell (DC) populations in steady-state mice

		pDC	cDC		interstitial DC		LC
			CD8a–	CD8a+	CD103–	CD103+	
Hematopoieses sites	BM	1	–	–	0.02	0.08	–
	Thymus	0.2	0.05	0.25	–	–	–
Immune priming sites	Spleen	0.5	1	0.5	–	–	–
	Lymph node	0.2	0.2	0.1	0.25*	0.15*	0.1*
Environmental contact sites	Skin	↑	–	–	25	1	30–50**
	Lung	↑	–	–	1.5	0.5	–
	Intestine	↑	–	–	0.45	1.05	–
Filtering cites	Liver	10	–	–	+	+	–
	Kidney	↑	–	–	+	+	–

the induction of a specific immune response against pathogens that breach the tissues (Banchereau and Steinman 1998). DCs at environmental interfaces (i.e., skin, lung, gut, cornea, vagina) are more vulnerable to external stimuli, and several research groups are dedicated to understanding how these cells discriminate between pathogenic and nonpathogenic agents. Among interface DCs, epidermal DCs, also called Langerhans cells (LCs), are the most studied. LCs constitutively express major histocompatibility complex (MHC) II and high levels of the lectin langerin, which gives rise to the conspicuous birbeck granule, an intracytoplasmic organelle of which its origin remains a mystery since its discovery more than 50 years ago (Valladeau and Saeland 2005). Human but not mouse LCs express CD1a, a MHC class I homolog that is often used to identify LCs in humans (Valladeau and Saeland 2005). Recent data, however, have established that langerin expression is not specific for LCs. In mice, langerin is expressed at low levels on $CD8^+$ DCs in lymphoid organs. Langerin is also expressed on a population of DCs present in the lung and the dermis (Ginhoux et al. 2007; Poulin et al. 2007; Bursch et al. 2007). These langerin-positive interstitial DCs also coexpress $\alpha E\beta 7$ (also called CD103), a ligand of the cell adhesion molecule E-cadherin expressed by most epithelial cells (Cepek et al. 1994), suggesting a potential role for CD103 in the homing of DCs to epithelial sites. $CD103^+$ DCs play a nonredundant role in tissue immunity. Elimination of dermal $CD103^+$ DCs reduces cutaneous hypersensitivity response to dermal antigens (Bursch et al. 2007; Wang et al. 2008), while elimination of lung $CD103^+$ DCs upon influenza virus infection delays viral clearance and aggravates clinical symptoms (GeurtsvanKessel et al. 2008). $CD103^+$ DCs negative for langerin have also been identified in the gut. This population has been shown to be essential for the induction of regulatory T cells in the gut (Coombes et al. 2007; Sun et al. 2007).

Tissue DCs derive from a mobile pool of circulating precursors that seed the tissue in the steady state, with the exception of epidermal LCs that derive from hematopoietic precursors that reside in the skin. In the steady state tissue DCs capture cell-associated antigens obtained during normal tissue turnover and migrate at very low rate to the draining lymph node (LN), where they present antigens to T cells (Huang et al. 2000; Scheinecker et al. 2002). This process is thought to maintain peripheral tolerance to tissue antigens (Bonifaz et al. 2002; Hawiger et al. 2001; Kretschmer et al. 2005). DC traffic to the LN dramatically increases during inflammation, and the role of DCs in the induction of immunity against foreign antigens has now been clearly established (Jung et al. 2002; Norbury et al. 2002).

DCs in Lymphoid Tissues

Lymphoid tissue–resident DCs have been the most studied DC populations in mice and include thymic, spleen, and lymph node DCs. In contrast, human lymphoid-tissue DCs have been poorly characterized, and this remains one of the future challenges in the field.

Spleen

The spleen is populated via blood. Spleen DCs constitutively express MHC class II and CD11c and are further classified into two major subsets in mice that include $CD4^+CD8^-CD11b^+$ DCs that localize mostly in the marginal zone and $CD8^+CD4^-CD11b^-$DC that localize mostly in the T-cell zone (Shortman and Liu 2002). $CD4^-CD8^-CD11b^+$ DCs have also been identified and are called double negative DCs, although they are less well characterized. Studies have started to unravel functional specialization among the $CD8^+$ and $CD4^+$ DC subset. $CD8^+$ and $CD4^+$ DC subsets express distinct antigen receptors on the surface and have different antigen processing ability in vivo. $CD8^+$ DCs are specialized in MHC class I presentation and interact mostly with $CD8^+$ T cells, while $CD4^+$ DC subset is specialized in MHC II presentation. $CD8^+$ DCs have also been shown to cross-present cell-associated antigens, while $CD4^+$ DCs are unable to do so (den Haan et al. 2000).

Lymph Node (LN)

LN DCs are more heterogenous as they include blood-derived DCs that enter via high endothelial venules and tissue migratory DCs that enter the draining LN through the afferent lymphatics (Randolph et al. 2008). In the LN, DCs localize in the T zone but also near the B-cell follicles. DCs entering the LN via lymphatics present antigens that reflect the tissue environment from which they originate. For example, migratory epidermal Langerhans cells and dermal DCs transporting skin antigens are present in skin-draining LNs but are absent from mesenteric LNs. The phenotype of LN DCs is thus more diverse than in the spleen and includes the $CD8^+$, $CD4^+$, and double-negative spleen equivalent DCs in addition to tissue migratory DCs. It is important to note that human DCs do not express the CD8 marker, and whether a $CD8^+$ and $CD8^-$ mouse DC equivalent exists in humans remains to be determined.

Mucosa-Associated Lymphoid Tissues

Mucosa-associated lymphoid tissues include nasal-associated lymphoid tissue in the nasopharynx, Peyer's patches, and isolated lymphoid follicles in the small intestine, and isolated follicles and the appendix in the large intestine. These tissues are mostly populated by blood-derived cells, and the phenotype of DCs resembles those of spleen DCs (Iwasaki 2007).

Thymic DCs

Thymic DCs localize mostly in the thymic medulla. The majority of thymic DCs in mice are $CD8^+$ and are thought to be generated locally from thymic progenitors.

A minority of DCs is CD8⁻ and is best characterized by the expression of SIRP-α and is thought to derive independently of thymic progenitors (Wu and Shortman 2005). Thymic DCs play a critical role in negative selection of T cells, while they seem dispensable for the induction of positive T-cell selection (Brocker et al. 1997).

Plasmacytoid Dendritic Cells in Lymphoid and Nonlymphoid Tissues

Plasmacytoid dendritic cells (pDCs, also called natural interferon-producing cells) represent a subset of DCs recently identified in mice and humans (Liu 2005; Shortman and Liu 2002). pDCs are generated in the bone marrow (BM), circulate in blood, and are found in steady-state spleen, thymus, LN, and the liver. Human and mice pDCs enter the LN through the high endothelial venule (HEV) and accumulate in the paracortical T-cell–rich areas, a migration pattern that resembles lymphocytes but is clearly distinct from DCs (Colonna et al. 2004). Similar to DCs, pDCs express MHC class II molecules constitutively and lack most lineage markers. Human pDCs express very low to no level of CD11c, they express CD4 and CD45RA antigens, the c-type lectin receptor BDCA2, and the molecule BDCA4, a neuronal receptor often used to isolate pDCs in vivo. Human pDCs also express high levels of the interleukin 3 (IL-3) receptor (CD123). Murine pDCs express low levels of the integrin CD11c, low levels of the lineage markers CD45RA/B220^{+} and ly6C /GR-1^{+} molecules, and they lack CD11b but express pDCA1 and Siglec-H, a member of the sialic acid binding Ig-like lectin (Siglec) family, recently identified as a specific surface marker for mice pDCs (Shortman and Liu 2002; Colonna et al. 2004; Zhang et al. 2006). One important function of human and mice PDC is to produce a high amount of interferon α (IFN-α) in response to microbial pathogen. In addition to its antiviral properties, IFN-α can also promote Th1 differentiation and has been shown to contribute to autoimmune disease (Banchereau and Pascual 2006). The capacity of pDCs to capture and present antigens to T lymphocytes is still being examined in several settings. Freshly isolated human and mouse pDCs are very poor inducers of T-cell proliferation, however, upon activation pDCs can differentiate into mature DCs with a high level of MHC class II and costimulatory molecules and T-cell stimulatory activity (Colonna et al. 2004).

Turnover of Dendritic Cells

The exact lifespan of DCs in different tissues and the recruitment of new DCs remains unsettled mainly due to the lack of consensus on the best method to measure DC turnover in vivo. Several studies used the time for DC repopulation after lethal irradiation and hematopoietic cell transplantation as a measure of DC turnover, however, radiation-induced injuries may fasten DC repopulation in a way that does not necessarily reflect steady-state settings. Similar arguments apply to studies analyzing DC repopulation after in situ DC elimination (e.g., using toxins specifically targeted to CD11c$^+$). Other studies have used a DC labeling index after administration of thymidine or bromodeoxyuridine (BrdU) injection. However, because these agents incorporate irreversibly into the DNA, dividing it will label both proliferating cells and its progeny. Studies have also looked at the percentage of DCs in S/G2 phase using flow cytometry on freshly isolated DCs, but DC isolation procedure can theoretically modify DC proliferation status. Most conclusive results are likely obtained by the recent use of cellular transfer into nonconditioned animals and by studies on parabiosis mouse models in which congenic mice share their blood circulation for some period of time, providing a way to trace the rate of DC increase and decrease upon parabiosis and consecutive separation in steady state. The current results on DC turnover in different organs are summarized in Table 2.

Splenic DCs

In vivo BrdU labeling studies revealed that spleen DCs incorporate BrdU in less than 2 hours after injection and that 5% lymphoid organ DCs or their immediate progenitors are actively cycling at any given time (Kabashima et al. 2005; de Witte et al. 2007). Results from parabiotic animals showed that 3 weeks postparabiosis, 30% spleen DCs derive from blood-borne precursors of the donor parabionts. However, upon separation, donor-derived DCs fail to be sustained in the host parabiont and are entirely replaced by endogenous host-derived cells in 10–14 days. These results confirmed previous data showing that most spleen DCs are continuously replaced by blood-borne precursors and do not self-renew (Liu et al. 2007; Manz et al. 2001; Traver et al. 2000). Altogether these results suggest that DCs enter lymphoid tissues as progenitor cells with limited proliferation potential (Naik et al. 2006; Diao et al. 2006). This progenitor was subsequently defined as CD11c$^+$ MHC II$^-$ in the spleen and shown to proliferate for no more than three or four divisions prior to becoming DCs (Naik et al. 2006; Diao et al. 2006).

Thymic DCs

Thymic CD8$^+$ and CD8$^-$ DCs incorporate BrdU in a biphasic pattern with a very rapid uptake in the first 3 days followed by a lag period, giving rise to 80% labeled DCs in 10 days and suggesting some degree of heterogeneity among thymic DC populations (Kamath et al. 2002). Data from the parabiotic model have also identified two separate homeostatic patterns among thymic DCs (Donskoy and Goldschneider 2003). One DC population failed to equilibrate among parabionts, reaching, similar to thymocytes, less than 10% chimerism after 5 weeks of parabiosis, suggesting that this DC population derives from an intrathymic progenitor, as initially suggested by Wu and Shortman (2005). In contrast, a second thymic DC population rapidly exchanged among parabionts reached 50% chimerism in 5 weeks.

Table 2 Dendritic cell (DC) turnover in lymphoid and nonlymphoid tissues

DC subset	Specific markers	Sensitivity to γ-irradiation	Method	Turnover
Spleen PDC	MHC II^+ $CD11c^{int}$ $PDCA1^+$ $CD11b^-$ $B220^+$ $Gr\text{-}1^+$	Radiosensitive	**% Cells in cycleParabiosis** % Exchange time of repopulation upon separation	0.3% (Liu et al. 2007)16% exchange at 30 days 3 days (Liu et al. 2007)
Spleen DCs		Radiosensitive	**% Cells in cycleParabiosis** % Exchange Time of repopulation upon separation **Time of repopulation after toxin administration****	5% (Kabashima et al. 2005; Liu et al. 2007)29% exchange at 30 days 2 weeks (Liu et al. 2007) 6 days (Jung et al. 2002)
$CD8^+$ DCs	$Langerin^+$ $CD205^+$ $CD11b^-$	Radiosensitive	**BrdU**: 50% labeling	1.5 days (Kamath et al. 2000)
$CD4^+$ DCs	$Langerin^-$ $DCIR2^+$ $CD11b^+$	Radiosensitive	**BrdU**: 50% labeling	2.9 days (Kamath et al. 2000)
Thymic DCs	$CD8^+$ DCs	Radiosensitive	**BrdU**: 50% labeling**Parabiosis** % Exchange	Biphasic: one population turnover in 3 days one population turnover in 10 days (Kamath et al. 2000)Heterogenous exchange: One DC population exchange poorly (10% at 5 weeks) one population including $CD8^+$ and $CD8^-$ DCs exchange largely (50% at 8 weeks)
Peripheral LN	MHC II^+ $CD11c^+$ cells (includes resident DCs and migratory DCs)	Radiosensitive	**% Cells in cycleBrdU : 50% labeling** **Parabiosis** % Exchange Time of repopulation upon separation	5% (Liu et al. 2007)3 days (except for cutaneous LN DCs: 12.5% DCs labeled in 3 days due to the presence of long lived LCs) (Kamath et al. 2000) 29% exchange at 30 days 2 weeks (Liu et al. 2007)

(continued)

Table 2 (continued)

DC subset	Specific markers	Sensitivity to γ-irradiation	Method	Turnover
Epidermal LCs	Langerin⁺ MHC II⁺ CD11c^{lo} CD11b⁺ F4/80⁺	Radioresistant in mice (Merad et al. 2002a)	**% Cell in cycleParabiosis** % Exchange **BrdU** **Time of repopulation after elimination in situ (Lethal irradiation and BM transplantation)** **Human allogeneic transplantation**	2%–3% DC are cycling. Evidence for self-renewal (Merad et al. 2002a)No exchange at 6 months (Merad et al. 2002a) 30%–40% labeling in 3 weeks (Kamath et al. 2002; Merad et al. 2002b) Intravital imaging to measure influx or efflux of EGFP MHC II⁺ cells. LC half-life: 53–78 days (Vishwanath et al. 2006) 30%–50% Elimination in 7 days after x ray Local repopulation in 4 weeks independent of circulating cells. Human LCs persist for more than a year in a limb graft (Kanitakis et al. 2004) Human host LCs remain in the skin of patients more than a month after allogeneic hematopoietic cell transplant despite complete donor-derived chimerism in the blood (Collin et al. 2006)
Dermal DCs	Langerin– CD103⁻ CD11b⁺ (80% total dermal DCs)	Radiosensitive(20% radioresistant) (Bogunovic et al. 2006)	**% Cells in cycleParabiosis** % Exchange **BrdU** **Human allogeneic transplantation**	2%–3% DCs are cycling.20% Exchange at 6 months (Bogunovic et al. 2006) 60% labeling in 3 weeks (Bogunovic et al. 2006) Human host dermal DCs remain in the skin of patients 1 month after allogeneic hematopoietic cell transplant despite complete donor-derived chimerism in the blood (Bogunovic et al. 2006)

	Langerin$^+$ CD103$^+$ (20% total dermal DCs)	Radiosensitive (Ginhoux et al. 2007; Poulin et al. 2007; Bursch et al. 2007)	**% Cells in cycleTime of repopulation after toxin injection****	3–4% (Ginhoux et al. 2007)5 days (Ginhoux et al. 2007)
Airway epithelial DCs (Rat)	Langerin$^+$ CD103$^+$(Valladeau et al. 1999)	Radiosensitive(Holt et al. 1994)	**Time of repopulation Lethal irradiation and congenic BM transplantation)**	80% Diappear by day 3 (20% remain)7–10 days (Holt et al. 1994)
Intestinal DCs	CD103$^+$ and CD103$^-$ DCs	?	Thymidine injection in mesenteric lymphadenectomised rats	Labeled DCs appear in 3 days in the thoracic duct in (Pugh et al. 1983)
Vaginal DCs	langerin$^+$ (Iijima et al. 2007; Johansson-Lindbom et al. 2005; Sun et al. 2007); Coombes et al. 2007)	Radiosensitive (Iijima et al. 2007)	**BrdU: 50% labelingTime of repopulation after Lethal irradiation and congenic BM transplantation)** **Time of repopulation after toxin injection****	6 days (Iijima et al. 2007)13 days (Iijima et al. 2007) 10 days (Iijima et al. 2007)
Kidney DCs (Rat)	Langerin$^-$ CD103$^-$ CD11b$^+$ DCs and Langerin$^+$ CD103$^+$ DCs	Radiosensitive	**Time of repopulation after Lethal irradiation and BM transplantation**	7–20 days (Leszczynski et al. 1985a)
Heart DCs (Rat)		Radiosensitive	**Time of repopulation after Lethal irradiation and BM transplantation**	10–25 days (Leszczynski et al. 1985b; Hart and Fabre 1981)

However, the authors from this study were unable to establish phenotypical differences between these two DC populations, and more studies are needed to identify distinct phenotypical and functional attributes. A recent study, however, suggested that circulating DCs can enter the thymus to establish central tolerance against peripheral antigens (Bonasio et al. 2006). The turnover of these cells needs to be defined.

LN DCs

It is critical to take into account the origin of LN DCs in studies of DC turnover (see above). Blood-derived DCs or their local progenitors have similar turnover rates as spleen DCs, while the lifespan of tissue-derived DC varies, as discussed below.

Nonlymphoid Tissue DCs

Tissue DCs charged with tissue antigens migrate through afferent lymphatics to the T-cell areas of LN (Austyn et al. 1988; Randolph et al. 1999; Kissenpfennig et al. 2005). There is constant steady-state migration (Pugh et al. 1983; Hemmi et al. 2001) possibly to maintain peripheral tolerance to tissue antigens (Steinman et al. 2003; Huang et al. 2000; Scheinecker et al. 2002; Belz et al. 2002).

DC traffic in the afferent lymph to the draining LN increases manyfold in response to inflammatory signals, and antigen-presenting function is strongly enhanced by upregulation of MHC antigens, costimulatory molecules, and cytokines (reviewed in Caux et al. 2000). The efflux of DCs from the various organs is not known in detail, but it is clear that DCs are present in all lymphatic vessels examined (Kelly 1970; Bujdoso et al. 1989). The flow rate of DCs in lymphatics vessels varies with the size of organs and was estimated to be roughly 1×10^5/h in the rat mesenteric lymphatics vessels, 2×10^5/h in the limbs lymphatic vessels and 9×10^5/h in the liver lymphatics, which is much slower that the influx of blood-borne lymphocytes, which is around 30×10^6/h in the steady state (Fossum 1989).

In order to maintain a population of DCs in the periphery in light of a continuous migration to the local lymphoid tissue, DCs must constantly be replaced with new cells. The direct in vivo precursors of tissue DCs remain to be examined in detail, as discussed below. The mechanisms by which these precursors seed the peripheral tissues under steady-state conditions also remain largely undefined. Nonetheless, it is clear with the exception of LCs, tissue DCs derive from circulating precursors.

BM transplantation into congenic animals established that kidney and heart DCs are replaced in 2–4 weeks. In contrast DC repopulation in the vagina (Iijima et al. 2007), airway epithelia (Holt et al. 1994), and the gut is more rapid and occurs in 7–13 days (Bogunovic and Merad, unpublished). The labeling index of DCs isolated in the thoracic duct from mesenteric lymphadenectomized rats have been used to measure DC turnover in the gut. The first labeled DCs were detected in the thoracic duct at 24 hours and peaked at 3 days, suggesting that the turnover time of gut DCs is very rapid (Pugh et al. 1983). These results have yet to be reexamined in view of recent data showing that DCs or their immediate progenitors can also proliferate in situ in nonlymphoid tissue (Waskow et al. 2008) (see below).

Epidermal LC lifespan differs fundamentally from that of other DCs. In contrast to most DC populations, epidermal LCs resist lethal doses of irradiation and remain of host origin in BM chimeric animals more than 18 months after transplantation (Merad et al. 2002a). In addition, LCs do not mix between parabionts more than 6 months after parabiosis, suggesting that they do not derive from circulating precursors in the steady state (Merad et al. 2002a). Proliferating LCs can be observed scattered throughout the epidermis by electron microscopy (Czernielewski et al. 1985; Czernielewski and Demarchez 1987; Miyauchi and Hashimoto 1987; Giacometti and Montagna 1967) or by staining for Ki67 (a nuclear proliferation marker). We have recently estimated that approximately 2%–3% of freshly isolated langerin-positive cells are in S/G2/M phase in mice and humans (Collin, Bogunovic, and Merad, unpublished observations). Although self-renewal could be sufficient to maintain LC numbers most of the time, the existence of specialized local precursor cells has also been proposed. The bulge region of the hair follicle serves as a niche for keratinocytes, melanocytes, and mast-cell progenitors (Kumamoto et al. 2003; Blanpain and Fuchs 2006), and there is evidence to suggest that following skin injuries that specifically affect the epidermis but not the dermis, where the hair follicles are located, LCs can be repopulated from the follicles alone (Gilliam et al. 1998). It is therefore possible that both differentiated LCs, through self-renewal, and local hematopoietic precursor cells could contribute to LC homeostasis depending on physiological needs, as has been shown for skin stem cells (Kumamoto et al. 2003). Conditional depletion of LCs and careful monitoring of their repopulation should help to resolve these issues (Kissenpfennig et al. 2005; Bennett et al. 2007).

pDCs

In contrast to DCs, pDCs migrate mostly to lymphoid organs and do not efficiently migrate to peripheral tissue in the steady state with the exception of the liver. pDCs proliferation rate in lymphoid organs is very low and less than 0.3% pDCs are in the cell cycle. Upon continuous BrdU labeling over 14 days in vivo, 90% of pDCs were positive (O'Keeffe et al. 2002a), and upon separation of parabiotic mice, donor-derived pDC are replaced in 3 days (Liu et al. 2007). Thus, the lifespan of differentiated pDCs in the spleen and LN is very short.

Human DCs

Studies on DC homeostasis are more difficult in humans. Most information on DC turnover comes from data on patients who received sex-mismatched allogeneic hematopoietic cell transplants (allo-HCT). Although these results by no means reveal the steady-state homeostatic property of DCs, some of these results deserve specific attention. In patients that received allo-HCT, blood DCs and pDCs are replaced by donor-derived cells in less than a week posttransplant, while recipient LCs can be identified unequivocally in the epidermis more than 1 year after allo-HCT (Collin et al. 2006). Donor LCs have also been shown to persist for years in a recipient of a human limb graft (Kanitakis et al. 2004), suggesting that human LCs as their murine counterpart repopulate locally independent of circulating precursors.

Cytokines in Dendritic Cell Development

Granulocyte Macrophage Colony-Stimulating Factor (GM-CSF)

The discovery that GM-CSF induces the in vitro differentiation of mouse and human hematopoietic progenitors (Inaba et al. 1992a; Caux et al. 1992) and human monocytes (Sallusto and Lanzavecchia 1994) into DCs marks the acceleration of DC research. While hematopoietic progenitors divide and differentiate, monocytes differentiate but do not divide. However, for the first time large quantities of these in vivo rare cells could be generated, leading to extraordinary progress in understanding the physiology of DCs. Given the efficacy of GM-CSF in inducing DC differentiation in vitro, it came as a surprise that mice deficient in GM-CSF or its receptor do not lack cDCs, but only showed slightly reduced numbers, and injection of GM-CSF or transgenic expression did not lead to increased cDC numbers in lymphoid organs, demonstrating that this cytokine is neither essential for overall steady-state cDC maintenance, nor, at least under these conditions, seemed to greatly enhance DC development in vivo (Maraskovsky et al. 1996; Vremec et al. 1997) (Table 3).

GM-CSF is produced by several tissue stroma cells, and by activated T and natural killer (NK) cells (Hamilton 2008). In steady state, serum GM-CSF is not detectable by enzyme-linked immunosorbent assay (ELISA), while levels do increase in serum, and especially at sites of inflammation (e.g., in pulmonary infection) (Cheers et al. 1988; Johansson-Lindbom et al. 2005; Burgess et al. 1977). Upon injection of GM-CSF with engineered long half-life (pegylated GM-CSF) and adenovirus-driven massive overexpression in vivo, cDCs, preferentially $CD8a^-$ DCs, increased massively, and the generated DCs were inclined to production of the "inflammatory" cytokines tumor necrosis factor α (TNFa), IL-6, and inducible nitric oxide (iNOS) (Miller et al. 2002; Daro et al. 2000; O'Keeffe et al. 2002b; Serbina et al. 2003). Thus, although GM-CSF is dispensable for steady-state DC development, it can play a role in DC differentiation in inflammatory or emergency situations. Consistent with these data, clinical studies have been successfully using injection of GM-CSF to attract or generate DCs at disease sites (Simons et al. 1999; Dranoff et al. 1993). This in vivo inflammatory GM-CSF-driven DC differentiation pathway likely resembles the in vitro GM-CSF cultures and involves GM-CSFR (receptor) expressing monocytes.

Table 3 Cytokine effects on in vivo dendritic cell (DC) homeostasis

Cytokine		DC prog BM	S pDC	S cDC	Interst/dermal DC	LC	Citation
		(fold of WT)	(fold of WT)	(fold of WT)	(fold of WT)	(fold of WT)	
GM-CSF	GM-CSF$^{-/-}$	0.6–0.8 (abs)	1 (abs.+rel)	0.6–0.8 abs. + rel	0.8 (abs. + rel.)	0.6 (abs+ rel)	(Vremec et al. 1997; Kingston and Manz, unpublished)
	GM-CSFR$^{-/-}$	n.a.	n.a.	1	n.a.	n.a.	(Vremec et al. 1997)
	GM-CSF injection	n.a.	n.a.	1	n.a.	n.a.	(Maraskovsky et al. 1996)
	pegGM-CSF injection	n.a.	n.a.	7–12 (abs)	n.a.	n.a.	(O'Keeffe et al. 2002b; Daro et al. 2000)
	GM-CSF transgenic	n.a.	n.a.	1.5 abs., 1 rel	n.a.	n.a.	(Vremec et al. 1997)
	GM-CSF transient expr.	n.a.	n.a.	200 (abs)	n.a.	n.a.	(Miller et al. 2002)
Flt3L	Flt3$^{-/-}$	1	0.65	0.43	n.a.	n.a.	(Waskow et al. 2008; Kingston and Manz/Ginhoux and Merad, unpublished)
	Flt3L$^{-/-}$	0.3–0.7 (abs + re	0.1 (abs + rel)	0.1 (abs + rel)	0.5 (abs + rel.)	0.8 (abs + rel)	(McKenna et al. 2000) Kingston and Manz, unpublished
	Flt3L injection	2.5–10	28 (abs)	9–30 × abs+rel	n.a.	n.a.	(Maraskovsky et al. 1996; Pulendran et al. 1997; Waskow et al. 2008; Karsunky et al. 2003; Onai et al. 2007a; Daro et al. 2000; O'Keeffe et al. 2002b)
	Flt3L expr.	n.a.	>10	>10	>10	n.a.	(Miller et al. 2003; Manfra et al. 2003)

(continued)

Table 3 (continued)

Cytokine		DC prog BM	S pDC	S cDC	Interst/dermal DC	LC	Citation
		(fold of WT)	(fold of WT)	(fold of WT)	(fold of WT)	(fold of WT)	
	Flt3 sign. inhib.	n.a.	0.1 abs.+rel	0.1 (abs.+rel)	n.a.	n.a.	(Tussiwand et al. 2005)
Flt3L+GM-CSF	$Flt3^{-/-}$GM-CSFR-/-	n.a.	0.65 ("as $flt3^{-/-}$")	0.43 ("as $flt3^{-/-}$")	n.a.	n.a.	(Waskow et al. 2008)
	$Flt3L^{-/-}$GM-CSF-/-	0.15–0.25 (abs + rel)	0.1 (abs.+rel)	0.1 (abs.+rel)	0.25 (abs+rel.)	0.8 (abs+rel)	(McKenna et al. 2000; Kingston and Manz, unpublished)
M-CSF	$M\text{-}CSF^{-/-}$ (op/op)	n.a.	0.3 (abs)	0.6–1 (abs)	n.a.	0.8?-1 (abs)	(Witmer-Pack et al. 1993; Takahashi et al. 1993; Macdonald et al. 2005)
	$M\text{-}CSFR^{-/-}$	n.a.	1	1	n.a.	absent	(Ginhoux et al. 2006)
	M-CSF injection	n.a.	2.5 (abs.)	2.5 (abs.)	n.a.	n.a.	(Fancke et al. 2008)
TGFb	$TGFb1^{-/-}$	n.a.	n.a.	1	n.a.	absent	(Borkowski et al. 1996; Borkowski et al. 1997)

Flt3L

In sharp contrast to GM-CSF, genetic deletion of Flt3L in mice leads to about tenfold reduction of pDCs and cDCs in lymphoid organs (McKenna et al. 2000), as does in vivo treatment with tyrosine kinase inhibitors with high flt3 affinity (Tussiwand et al. 2005; Whartenby et al. 2005), while LCs seem only slightly or not reduced (Kingston and Manz, unpublished). Injection or conditional expression of the cytokine Flt3L leads to massive expansion of both pDCs and cDCs in all lymphoid and nonlymphoid organs, including the large and small intestines and the liver, with up to 30% of mouse spleens expressing CD11c (Maraskovsky et al. 1996; Pulendran et al. 1997; Karsunky et al. 2003; (Manfra et al. 2003; Miller et al. 2003; Bjorck 2001). Concurring with the in vivo Flt3L injection and deletion findings, it was shown that Flt3L as a single cytokine can drive the differentiation of mouse BM progenitors into all cDC subtypes (Brasel et al. 2000), and importantly, at the same time induce pDC differentiation, which was never observed in GM-CSF supplemented culture conditions (Gilliet et al. 2002). Similar to mice, Flt3L supports in vitro development of pDCs and cDCs from human hematopoietic progenitors (Blom et al. 2000; Spits et al. 2000; Chicha et al. 2004; Chen et al. 2004), and in vivo Flt3L injection in humans leads to massive expansion of blood pDCs and cDCs (Pulendran et al. 2000; Maraskovsky et al. 2000; Fong et al. 2001). Thus, in contrast to GM-CSF, Flt3L is both sufficient and essential for the development of the major lymphoid organ DC populations and serves as a key cytokine for steady-state DC maintenance.

The cognate receptor of Flt3L, the Fms-like tyrosine kinase 3 (flt3, also called fetal liver kinase 2, flk2) is a receptor tyrosine kinase with homology to c-Kit (the receptor for stem cell factor) and c-fms (the receptor for CSF-1, M-CSF) (Lyman and Jacobsen 1998; (Lyman et al. 1995). Flt3 is expressed on short-term HSCs in mice, in myeloid and lymphoid committed progenitors, and on mature lymphoid organ pDCs and cDCs, while flt3 expression is low on interstitial DCs and LCs and almost absent on monocytes (Karsunky et al. 2003; Whartenby et al. 2005; Kingston and Manz, unpublished). In vivo, pDC and cDC differentiation is confined to Flt3 expressing mouse bone marrow progenitors cells along both the lymphoid and myeloid pathways (D'Amico and Wu 2003; Karsunky et al. 2003). Spleen and lymph node DCs fail to develop from $Flt3^-$ hematopoietic progenitors (D'Amico and Wu 2003; Karsunky et al. 2003; Karsunky et al. 2005), and $Flt3^-$ progenitors fail to repopulate the LC pool in inflamed skin (Mende et al. 2006). However, enforced expression of Flt3 in $Flt3^-$ hematopoietic progenitors can rescue, and enforced expression of Flt3 in $Flt3^+$ progenitors enhances both pDC and cDC differentiation (Onai et al. 2006). Thus, Flt3L regulates DC development from $flt3^+$ progenitors to $flt3^+$ lymphoid organ DCs in vivo.

Flt3L exists in both soluble- and membrane-bound forms and is constitutively expressed by multiple tissue stroma cells and by activated T cells (Lyman and Jacobsen 1998; Lyman et al. 1995). Bioactive levels of Flt3L are measurable in serum in steady state and increase upon severe inflammation, for example, in some infections or autoimmune disease, and upon hematopoietic challenge, such as irradiation induced cytopenia. Interestingly, flt3 receptor deficient mice, mice treated with flt3 tyrosine kinase inhibitors, and sublethally irradiated mice have about tenfold increased levels of Flt3L (Tussiwand et al. 2005; Schmid and Manz, unpublished). Furthermore, diphtheria toxin–induced depletion of DCs in mice engineered to express the diphtheria toxin receptor under the promoter of CD11c leads to increased Flt3L levels in serum (S. Jung, personal communication). Together these data demonstrate that Flt3L is broadly available in steady state, fitting with the capacity of flt3L to continuously maintain all major lymphoid tissue steady-state DCs by guiding expansion and differentiation of early DC progenitors in bone marrow as well as immediate DC precursors in peripheral lymphoid tissues. The increase of Flt3L upon inflammation and DC depletion furthermore suggest a regulatory loop, tailored to ensure sufficient DC regeneration upon demand. It is important to note that supraphysiologic doses of Flt3L preferentially induce differentiation of pDCs and $CD8a^{+}$ DCs over $CD8a^{-}$ DCs (Maraskovsky et al. 1996; O'Keeffe et al. 2002b; Pulendran et al. 1997; Karsunky et al. 2003; Bjorck 2001; Onai et al. 2006). Furthermore, it should be noted that Flt3L is not an exclusive DC cytokine, but Flt3L also acts on early hematopoietic progenitors, myelomonocytic, and NK-cell differentiation (Lyman and Jacobsen 1998; Shaw et al. 1998).

CSF-1 (M-CSF)

CSF-1 is required for osteoclasts and macrophage development, and CSF-1 absence or absence of the CSF-1 receptor results in osteopetrosis (Yoshida et al. 1990; Dai et al. 2002). CSF-1 was first thought to be dispensable for DC development (Witmer-Pack et al. 1993; Takahashi et al. 1993), however, data from our laboratory established the critical, nonredundant role for CSF-1R in LC development (Ginhoux et al. 2006) and recent data suggest that CSF-1 is also critical for the development of DCs in the dermis, the gut, and the lungs (Ginhoux and Merad, unpublished). In addition, CSF-1 reporter mice revealed that CSF-1R is expressed in all DC subsets during differentiation, and absolute numbers of pDCs and cDCs are somewhat reduced in CSF-1 deficient mice (Macdonald et al. 2005). Also, CSF-1 can lead to increasing DC numbers in bone marrow progenitor cultures, and injection of CSF-1 increased both lymphoid organ pDCs and cDCs in mice (Onai et al. 2007a; Fancke et al. 2008).

CSF-1 is expressed by endothelia, stroma cells, osteoblasts, and macrophages, and, as Flt3L, is detectable in steady-state serum and increases upon inflammation (Hamilton 2008; Schmid and Manz, unpublished). Consistently, we found that monocytes repopulate LCs in inflamed skin in a CSF-1R-dependent manner (Ginhoux et al. 2006), but the exact role of Csf-1 in LC and DC development remains to be examined. In vivo data in mice have shown that bcl-2 overexpression in BM progenitor that carry the spontaneous null Csf-1op/op mutation can rescue osteoclast differentiation, suggesting that csf-1 plays a role in macrophages survival in tissues. However, recent data in our laboratory found no role for bcl2 in csf-1-induced LC development and enforced bcl-2 expression in csf-1$r^{-/-}$ BM progenitors does not rescue LC repopulation in inflamed s kin (Merad, unpublished).

Transforming Growth Factor-β

Transforming growth factor-β (TGF-β) has multiple roles in regulation of homeostatic and inflammatory hematopoiesis (Fortunel et al. 2000). In addition to CSF-1, TGF-β is a nonredundant cytokine for LC development in vivo in mice (Borkowski et al. 1996; Borkowski et al. 1997) and similarly supports LC development in vitro from human hematopoietic progenitors (Strobl et al. 1996; Caux et al. 1999). Interestingly, TGF-β acts upstream of transcription factors Id2 and Runx3, and deletion of those also leads to LC deficiency in vivo (Fainaru et al. 2004; Hacker et al. 2003) (discussed below).

In the skin, keratinocytes are a large source of TGF-β, and it was been assumed that exogenous TGF-β was critical for LC development (Borkowski et al. 1997). Recent data, however, have challenged this view, as mice in which an absence of TGF-β secretion is restricted to langerin-positive cells cannot develop epidermal LCs (Kaplan et al. 2007).

IL-4, TNFα, LTβ, and G-CSF

Although Flt3L, GM-CSF, CSF-1, and TGF-β are major known cytokines for in vivo DC development in steady state and upon demand, other cytokines are described that execute more subtle effects within the complex setting of steady-state and inflammatory DC homeostasis, and we only mention them briefly here.

GM-CSF alone, as discussed above, is sufficient for generation of functional DCs from mouse whole bone marrow progenitors and blood cells, and this became a standard protocol for mouse in vitro DC generation (Inaba et al. 1992a; Inaba et al. 1992b). However, the initial hallmark study on in vitro differentiation of human monocytes to DCs reported that addition of IL-4 to GM-CSF in cultures enhanced cellular recovery as well as a DC characteristic phenotype ($CD1^{+}CD14^{-}$) and function (mixed lymphocyte reactions, presentation of soluble antigen to specific T cells) and likely suppresses macrophage differentiation compared to DC generation from monocytes with GM-CSF alone or with GM-CSF and TNFα (Sallusto and Lanzavecchia 1994). Thus, for human monocyte to DC differentiation GM-CSF plus IL-4 became the standard protocol maintained today. While IL-4 is not essential for in vivo mouse DC differentiation, follow-up studies on mouse GM-CSF cultured whole bone marrow cells revealed that IL-4 added to DC maturation and increased T-cell immunogenicity (Lutz et al. 2002). The exact role of IL-4 in vivo and specifically the question of whether IL-4 has a role in steady-state DC maintenance or only inflammatory DC activation still needs to be addressed (Lutz 2004).

In the seminal study by Caux et al., CD34-expressing progenitor cells isolated either from human bone marrow or umbilical cord blood were shown to give rise to both monocytes and DC in vitro in the presence of GM-CSF and TNFα. These results contrast with data in the mouse in which GM-CSF alone is sufficient for the proliferation and differentiation of DCs from bone marrow progenitors. The exact role of TNFα in DC differentiation from hematopoietic progenitors is still unclear, but it has been suggested that at least part of the mechanism of action of TNFα can be explained by the inhibition of $CD34^+$ cell differentiation into granulocytes (Santiago-Schwarz et al. 1993) or other myeloid lineages (Canque et al. 2000). In mice, TNFα has been shown to play a critical role in DC migration from tissue to LN during inflammatory injuries (Cumberbatch and Kimber 1992; Cumberbatch et al. 1997). Mice deficient in TNFα can form DCs but have a reduced cutaneous hypersensitivity response after skin immunization (Pasparakis et al. 1996).

Other cytokines have been shown to play a role in DC development or maintenance. For example, lymphotoxine β (LTβ), a member of the TNF family, has been shown to play a role in DC maintenance in lymphoid organs and to contribute to $CD8^-$ cDC homeostasis in the spleen (Kabashima et al. 2005). Another one is G-CSF, shown to increase DC numbers in blood and lymphoid tissues, most likely due to mobilization but not differentiation effects (O'Keeffe et al. 2002b; Pulendran et al. 2000).

The complex interaction of cytokines required for optimal DC differentiation should be addressed by investigating combined cytokine effects in vivo but also by live in situ visualization via genetic labeling of cytokine producing cells and cytokine receptor expressing DC progenitor cells in vivo.

Transcription Factors in Dendritic Cell Development

Environmental stimuli, such as cytokines, need to be translated on the subcellular level. In hematopoiesis, few transcription factors that are both essential and specific for the development of one selective lineage or sublineage diversification, yet are still compatible with viability, have been described. A prominent example for the former is Pax5, with its deletion leading to blockage of B-cell development (Nutt et al. 1999). An example for the latter is FoxP3, with its deletion leading to complete blockage of regulatory T-cell development (Fontenot et al. 2003; Hori et al. 2003). Most transcription factors, however, are not only active in specification of one cell type, but in concerted and graded action shape the development and diversification of different mature lineages (e.g., reviewed in Orkin and Zon 2008; Laiosa et al. 2006). Deletion of transcription factors active in development and diversification of multiple lineages consequently lead to multiple hematopoietic alterations, and resulting homeostatic imbalances often complicate dissection of primary from secondary effects.

Thus far, no transcription factor exclusively active on and essential to all DC development or to DC diversification has been described. However, by studying respective transcription factor knockout mice, a substantial amount of knowledge has been gained on the relevance of several multiple lineage active transcription factors in both global DC development and, specifically, DC diversification. A summary is shown in Table 4. Of note, since little or no information is available on transcription factor dependency of nonlymphoid tissue interstitial DCs, these are not discussed here.

Table 4 Transcription factor effects on in vivo dendritic dell (DC) homeostasis

Transcription	DC prog	S pDC	S cDC	LC	Citation
Factor	**BM**	CD8a–/CD8a+			
$STAT3^{-/-}$	n.a.	n.a.	0.1/0.1	n.a.	(Laouar et al. 2003)
$Gfi\text{-}1^{-/-}$	0.2	0.5	0.5/0.5	2	(Rathinam et al. 2005)
$STAT5^{-/-}$	n.a.	0.4	0.4/0.3	n.a.	(Esashi et al. 2008)
$Ikaros^{-/-}$	n.a.	n.a.	absent/0.15	n.a.	(Wu et al. 1997)
Ikaros $DN^{-/-}$	n.a.	n.a.	absent	n.a.	(Wu et al. 1997)
Ikaros $Ik^{L/L}$	n.a.	<0.1	0.8/1.1	n.a.	(Allman et al. 2006)
$XBP\text{-}1^{-/-}$	n.a.	0.3	0.6	n.a.	(Iwakoshi et al. 2007)
$RelB^{-/-}$	n.a.	n.a.	absent/1	1	(Wu et al. 1998)
$PU.1^{-/-}$	n.a.	n.a.	<0.1/1	n.a.	(Anderson et al. 2000; Guerriero et al. 2000)
$IRF\text{-}2^{-/-}$	n.a.	n.a.	0.25/1	0.7	(Ichikawa et al. 2004)
$IRF\text{-}4^{-/-}$	n.a.	0.5–1	0.3/1	n.a.	(Suzuki et al. 2004; Tamura et al. 2005)
$IRF\text{-}8^{-/-}$	1(?)	0.1	1/0.1	0.5	(Aliberti et al. 2003; Schiavoni et al. 2002; Schiavoni et al. 2004; Tsujimura et al. 2003)
$Id2^{-/-}$	n.a.	1.5	1(?)/<01	absent	(Hacker et al. 2003)
$Runx3^{-/-}$	n.a.	n.a.	0.8(?)/2(?)	absent	(Fainaru et al. 2004)

Values are given as fold of wild type, n.a. = not available

Transcription Factors Affecting "Global" DC Development

Consistent with the critical role of FLt3L in DC development, hematopoietic deletion of signal transducer and activator of transcription 3 (STAT3), a transcription factor in downstream flt3 signaling, led to ablation of Flt3L-dependent DC development, resembling the phenotype of Flt3L deficient mice (Laouar et al. 2003). Conclusively, overexpression and activation of STAT3 in flt3 negative hematopoietic progenitors rescues both their cDC and pDC differentiation potential (Onai et al. 2006). Furthermore, deletion of the transcriptional repressor Gfi-1, which is involved in regulating STAT3 activation, leads to reduction of all lymphoid tissue DCs, while LC numbers were increased (Rathinam et al. 2005). Notably, Gfi-1-deficient mice display multiple hematopoietic disturbances, and Gfi-1 is also involved in DC versus macrophage diversification (Hock and Orkin 2006; Rathinam et al. 2005).

The most recent data shed some light on how Flt3L-dependent steady-state versus GM-CSF dependent inflammatory DC developmental signals are translated on the transcriptional level in DC progenitors. Although STAT3 deficiency blocks Flt3L mediated pDC and cDC development, it does not block GM-CSF mediated cDC development (Laouar et al. 2003). Specifically, GM-CSF downstream activated STAT5 supports cDC development, while it suppresses the development of pDCs via direct suppression of IRF-8 (also called ICSBP, described below) (Esashi et al. 2008). However, besides activation of STAT5, GM-CSF also leads to activation of STAT3 and to IRF-4 expression, a transcription factor important in cDC development, described below (Esashi et al. 2008). Thus, in Flt3 signaling–mediated steady-state pDC and cDC development, STAT3 activation is a nonredundant requirement, while STAT5 is nonessential; in GM-CSF signaling–mediated cDC development, STAT3 is dispensable and activated STAT5 suppresses pDC development. However, in the absence of STAT5, pDCs develop in GM-CSF stimulated cultures, possibly via GM-CSF mediated STAT3 activation. Also, GM-CSF stimulated STAT5-deficient progenitors seem to produce fewer cDCs, and hematopoietic STAT5 deficient animals display somewhat reduced lymphoid organ cDC numbers, paralleling the findings in GM-CSF deficient animals (Esashi et al. 2008; Onai and Manz 2008; Vremec et al. 1997; Kingston and Manz, unpublished). Thus it is likely that beyond suppression of pDC development, STAT5 is also directly involved in GM-CSF-mediated cDC development, an issue that needs further clarification. Notably, GM-CSF downstream STAT5 activation and consecutive pDC suppression seem not to impact steady-state lymphoid organ pDC maintenance, as GM-CSF deficient mice do not show increased pDC numbers (Kingston and Manz, unpublished).

Ikaros

Ikaros is a zinc finger DNA binding transcriptional regulator, expressed and important in maintenance of hematopoietic stem cells and in the early development of multiple lymphoid lineages (Georgopoulos 2002). Several Ikaros mutant mice have been generated. An Ikaros null mutation led to the absence of CD8α^+ cDC and reduction of CD8α^- cDC, a dominant-negative mutation to complete cDC ablation, and a truncated, low-level–expressed protein led to massive reduction of pDCs, while cDCs were only slightly affected (Wu et al. 1997; Allman et al. 2006). Although observed differences in the mutants still need to be elucidated, these findings show that the requirement of Ikaros at early stages of hematopoietic development affects DC development as well.

XBP-1

More recently the transcription factor XBP-1, involved in the signaling pathway of endoplasmic reticulum stress, was shown to affect pDCs and cDCs. XBP-1 deficient mice developed normal numbers of flt3$^+$ progenitor cells, however, differentiating cells displayed an increased sensitivity to apoptotic cell death, leading to reduced mature cell populations (Iwakoshi et al. 2007).

Transcription Factors Affecting DC Subtype Diversification

RelB

The NF-κB/Rel family member RelB is expressed in lymphoid organs. $RelB^{-/-}$mice have an altered thymic organ structure and develop myeloid hyperplasia and multiorgan inflammation (Burkly et al. 1995; Weih et al. 1995). RelB is expressed in cDCs, with relatively higher amounts in $CD8a^-$ cDCs, and RelB-deficient mice show massively reduced numbers of this cDC population (Wu et al. 1998). In line with these findings, similar alterations of the lymphoid organ cDC compartment are observed in mice deficient in TRAF6, a TNF receptor-associated family protein, acting upstream of the NF-κB cascade (Kobayashi et al. 2003).

PU.1

The ets-family of DNA binding proteins member PU.1 is a transcription factor expressed in hematopoietic cells (Klemsz et al. 1990), and its deletion causes embryonic or neonatal death (Scott et al. 1994; McKercher et al. 1996). PU.1-deficient mice, those constituted with PU.1 deficient hematopoietic cells, in addition to other hematopoietic defects, lack or have severely reduced $CD8a^-$, or $CD8a^-$ and $CD8a^+$ cDCs, and GM-CSF is not sufficient to induce DC differentiation from PU-1 deficient progenitors in vitro (Anderson et al. 2000; Guerriero et al. 2000). Downstream flt3 signaling leads to PU-1 expression (Mizuki et al. 2003), and PU.1 cooperatively with C/EBPa activates myeloid development-associated cytokine receptor genes, including G-CSFR, M-CSFR, and GM-CSFR (Friedman 2002). Interestingly, enforced PU.1 expression in $flt3^-$megakaryocyte-erythrocyte lineage committed hematopoietic progenitors was sufficient to permit the development of both pDCs and cDCs as well as other myeloid lineages (Onai et al. 2006). In human $CD34^+$ progenitor cells, knockdown of PU.1 inhibited development pDCs, as did the knockdown of another ETS transcription factor, Spi-B (Schotte et al. 2004). Although the effects of PU.1 deletion on pDC development in mice has not been reported, the above described research raises the expectation that PU.1 likely also has a role in mouse pDC development in vivo.

The interferon regulatory factors (IRF) 2, 4, and 8 (formerly also called PU.1 interaction partner, Pip, or interferon consensus sequence binding protein, ICSBP) were recently described as key transcription factors involved in DC subset diversification. IRF-2 deficient mice display reduced $CD8a^-$ cDCs and slightly reduced LCs (Ichikawa et al. 2004). IRF-4 deficient mice have reduced $CD8a^-$ cDCs and slightly reduced pDCs (Suzuki et al. 2004; Tamura et al. 2005), and IRF-8 knockout mice have reduced $CD8a^+$ DCs, pDCs, and LCs, while numbers of early DC progenitors in BM seem to be conserved (Aliberti et al. 2003; Schiavoni et al. 2002; Schiavoni et al. 2004; Tsujimura et al. 2003). Interestingly, while both IRF-4 and -8 show some overlapping activity, in vitro DC differentiation upon GM-CSF stimulation depends preferentially on IRF-4, while DC differentiation upon flt3L preferentially depends on IRF-8 (Tamura et al. 2005), fitting to some extent the DC phenotypes observed in respective cytokine and IRF deficient mice.

Id2

Helix-loop-helix (HLH) proteins that act as transcriptional activators are inhibited by counter-acting HLH proteins, the Id (inhibitors of DNA binding) proteins. Id2 is induced during GM-CSF (and likely flt3L) driven DC development in vitro, and Id2 deficient mice have moderately increased pDCs, severely reduced $CD8\alpha^+$ DCs, and lack LCs (Hacker et al. 2003). Interestingly, LC deficiency in TGF-β knockout mice can be directly linked to the LC deficiency observed in Id2 deficient mice, as TGF-β delivers the upstream signal for Id2 induction (Hacker et al. 2003). In human $CD34^+$ hematopoietic progenitor cells, ectopic expression of Id2 and Id3 led to inhibition of pDC but not cDC development in vitro (Spits et al. 2000), while ectopic expression of the transcriptional activating HLH protein E2A, which is inhibited by Id2, stimulated pDC development (similar as Spi-B, discussed above) (Schotte et al. 2004). These observations in human cells fit the finding that Id2 deficient mice have elevated pDC numbers (Hacker et al. 2003).

Runx3

Another transcription factor mediating TGF-β responses is Runx3, a member of the runt domain family of transcription factors. Lack of Runx3 in mature DCs results in loss of TGF-β-mediated inhibition of maturation, therefore leading to DC activation and inflammation (Fainaru et al. 2004). Importantly, lack of appropriate TGF-β induced Runx3 signaling results, as with Id2 deficiency, in the absence of LCs (Fainaru et al. 2004).

Differentiation of Dendritic Cell from Hematopoietic Stem and Progenitor Cells

Early Hematopoietic DC Development

Hematopoiesis is a hierarchically structured organ system with a very high cellular turnover that must be tightly regulated according to demand. In adult mammals, a small fraction of self-renewing hematopoietic stem cells (HSCs) in bone marrow gives rise to multipotent progenitors (MPPs) that then generate progenitors with gradually restricted developmental options that finally differentiate to mature cells (Kondo et al. 2003; Orkin and Zon 2008). The differentiation of HSC into mature hematopoietic cells involves early loss of self-renewal ability from HSCs to MPPs, accompanied by an increase of proliferative capacity per unit of time, which is again gradually lost during the functional maturation of lineage progeny. However, some rare defined lineage cells, such as memory T and B cells, do regain self-renewal and proliferation, but not multilineage differentiation potential, a process accompanied by reactivation of a HSC self-renewal transcriptional profile (Luckey et al. 2006).

Based on cell surface antigen expression and response to environmental stimuli, several early developmental intermediates with restricted differentiation options have been defined in mice and men. Prominent examples are clonal common lymphoid progenitors (CLPs) that harbor all lymphoid, but not myeloid, and clonal common myeloid progenitors (CMPs) that harbor all myeloid, but not lymphoid, developmental options (Akashi et al. 2000; Kondo et al. 1997; Karsunky et al. 2008; Galy et al. 1995; Manz et al. 2002). These progenitors have been an invaluable tool in understanding the control of physiologic and pathologic early developmental steps. However, in contrast to frequent assumptions, proof of existence of such progenitors does not disprove alternatives. Indeed, more recent studies are helping to refine our current understanding of the hematopoietic developmental tree, revealing overlapping and graded stages of development in very early lineage commitment. Although this is important and adds a note of caution to any evaluation of lineage commitment, the controversies discussed in very early lymphomyeloid lineage choices might not directly impact downstream DC development, as discussed below, and we thus refer to specific recent reviews in the field (e.g., in Iwasaki and Akashi 2007; Buza-Vidas et al. 2007).

Lymphoid and myeloid committed progenitors as CLPs and CMPs maintain developmental options for all DC subtypes in both mice and humans (Traver et al. 2000; Manz et al. 2001; Wu et al. 2001; Shigematsu et al. 2004; Chicha et al. 2004; Ishikawa et al. 2007; Karsunky et al. 2005). Specifically, it was demonstrated that that all lymphoid organ pDCs and cDCs are generated upon recipient irradiation and transfer of mouse CMPs, granulocyte-macrophage progenitors (GMPs), CLPs, and thymic pro-T1 cells, while pDC and cDC differentiation potential is lost once definitive megakaryocyte-erythrocyte and B-cell commitment occurs (Traver et al. 2000; Manz et al. 2001; Wu et al. 2001; Shigematsu et al. 2004; Chicha et al. 2004; Ishikawa et al. 2007; Karsunky et al. 2005). The 2–4 week transient generation of DCs from all these progenitors revealed a strong bias toward $CD8a^+$ DC generation not only in thymus but also in spleen, while $CD8a^+$ DCs in steady-state spleen represent the minor cDC fraction (discussed above). Thus, the interpretations of initial hallmark progenitor transfer studies that used early thymocyte progenitors and observed $CD8a^+$ biased read out in irradiated mice, leading to claims of "lymphoid-lineage derived $CD8a^+$ DCs," needed to be revised, as preferential $CD8a^+$ DC read out in spleen is induced by environmental factors postirradiation (Ardavin et al. 1993; Wu et al. 1996; Traver et al. 2000; Manz et al. 2001; Wu et al. 2001). DC differentiation was transient and disappeared by 2–4 weeks, suggesting that lymphoid organ DCs do not have self-renewal potential and are constantly maintained by new blood precursor input, as mentioned above.

Subsequent studies revealed that the maintenance of DC developmental options is intimately linked to the capacity of progenitors to respond to the above discussed nonredundant DC development relevant cytokine, Flt3L. It was shown that during early lineage commitment, DC developmental options segregate with flt3 expression, present on all the above mentioned progenitors with DC developmental potential, while absent on those without (Karsunky et al. 2003; D'Amico and Wu 2003; Mende et al. 2006). Importantly, enforced expression and signaling of human flt3 in naturally flt3$^-$ progenitors induced in vitro and in vivo pDC and cDC differentiation and enforced expression in naturally flt3$^+$ progenitors enhanced their pDC and cDC differentiation, thus demonstrating that flt3 indeed delivers an instructive signal to activate DC differentiation programs from early progenitors to mature steady-state DCs (Onai et al. 2006; Onai et al. 2007b). Because alternative lympho-myeloid developmental programs of flt3 transduced progenitors were not shut down, it is critical to stress that flt3 signaling is not immediately deterministic but primarily opens access to a DC differentiation program that might be used if no alternative, competing signals occur, leading to a hypothetic gradual "flt3-licence" model in DC differentiation (Onai et al. 2007b).

Gradual Downstream Restriction to the DC Lineages

Given the fact that DC developmental options segregate with flt3$^+$ early lympho- and myeloid differentiation, the subsequent question was whether downstream progenitors with further DC-restricted differentiation potential or exclusive DC potential and stepwise reduced proliferation potential existed. Several of these have been recently described, and we here divide them into three groups (Table 5).

The first group comprises cells in an early hematopoietic hierarchy with a high proliferative capacity, expressing some level of surface CD117 (c-kit, the receptor for stem cell factor) but not a mature lineage marker including CD11c. Given their burst size, clonal assays with robust read out in different cellular subpopulations in vitro were feasible, and, upon transfer in irradiation conditioned hosts, these cells produce several fold their input numbers' progeny at the peak of read out in spleen.

Lin$^-$c-kit$^+$CX$_3$CR1$^+$ cells, accounting for 0.5% of bone marrow in *Cx$_3$cr1-gfp* knockin mice, differentiated exclusively into monocytes, macrophages, and cDCs but not into pDCs on a clonal level in vitro, and as a population in vivo in irradiated and nonirradiated mice, and were thus named "macrophage and DC progenitors" (MDPs) (Fogg et al. 2006). Maximum MDP offspring cells in spleen were generated in 7 days after sublethal irradiation and transplantation, with an approximate 1:10 MDP input to spleen cDC output ratio.

Based on relevant cytokine receptor expression for DC development, we identified that a cell population identified as lin$^-$CD117^{int}CD135$^+$CD115$^+$CD127$^-$ accounted for 0.1% of adult mouse bone marrow and was able to differentiate exclusively into pDCs and cDCs clonally in vitro and as a population in vivo (Onai

Table 5 Immediate in vivo dendritic cell (DC) progenitors and precursors

Group	Precursor population name	Site of isolation	Ex vivo sorting immunophenotype	Frequency	Differentiation potential in vitro	clonal potential in vitro	Differentiation potential in vivo	Conditions	Peak day of progeny and fold expansion of input in spleen at peak	Citation
I	MDPs (Mac. and DC Prog.)	BM	$Lin^-CD117^+CX_3CR1^+ CD11b^-$	0.5%	cDC, Mac	cDC, Mac	S, Perit: Mo, cDC, Mac	U, RT, IF Perit.	day 7, 10 × (RT)	(Fogg et al. 2006)
I	CDPs (Common DC Progenitors)	BM	$Lin^-CD117^{int}CD135^+ CD115^+$ CD127	0.1%	pDC, cDC	pDC, cDC	S, LN: pDC, cDC	U, RT	day 10, 7–8 × (RT)	(Onai et al. 2007a)
I	Pro-DCs	BM	$Lin^-CD117^{int}CD135^+ CD16/32^{lo}$	n.a.	pDC, cDC	pDC, cDC	S: pDC, cDC	U	day 8?, 0.05×[a]	(Naik et al. 2007)
II	DC progenitors B220+	BM	$CD11c^+MHCclassII^- B220^+$	0.3%	pDC?,cDC	n.a.	S: pDC, cDC	RT	day 7?, 0.01 x (RT)	(Diao et al. 2004)
II	DC progenitors B220–	BM	$CD11c^+MHCclassII^- B220^-$	0.2%	cDC	n.a.	S: cDC	RT	day 7?, 0.03 × (RT)	(Diao et al. 2004; Diao et al. 2006)
II	Preimmunocyte	BM	$CD11c^+CD31^+Ly6C^+$	0.5–1%	pDC, cDC, Mac	n.a.	n.a.	n.a.	n.a.	(Bruno et al. 2001)
II	DC precursors	Blood	$CD11c^+MHCclassII^-$	5%	n.a.	n.a.	S: pDC, cDC	RT	day 14, 1×	(del Hoyo et al. 2002)[b]
II	Pre-cDCs	Spleen	$CD11c^{int}CD45RA^{lo} CD43^{int}SIRP\text{-}\alpha^{int}$	0.05%	cDC	n.a.	S: cDC	U	day 5, 0.02×*	(Naik et al. 2006)
III	Monocytes	BM	$Ly6C^{hi}$ ($Gr\text{-}1^{hi}$)	-	n.a.	n.a.	DC	IF speen	day 2	(Naik et al. 2006)
III	Monocytes	Blood	$Gr\text{-}1^{hi}$	-	n.a.	n.a.	LC	IF skin	-	(Ginhoux et al. 2006)
III	Monocytes	BM	$Gr\text{-}1^{hi}$ inflammatory	-	n.a.	n.a.	DC	IF intestine+ lung		(Varol et al. 2007)

(continued)

Table 5 (continued)

Group	Precursor population name	Site of isolation	Ex vivo sorting immunophenotype	Frequency	Differentiation potential in vitro	clonal potential in vitro	Differentiation potential in vivo	Conditions	Peak day of progeny and fold expansion of input in spleen at peak	Citation
III	Monocytes	BM	$Gr\text{-}1^{hi}$ and $Gr\text{-}1^{-}$	-	n.a.	n.a.	DC	IF lung		(Landsman et al. 2007)
III	Monocytes	Blood	$Gr\text{-}1^{hi}$	-	n.a.	n.a.	DC	IF vagina		(Iijima et al. 2007)

Group 1: Lin-$CD117^{+/lo}$ progenitors with high proliferation potential; Group 2: $CD11c^{+/lo}$ progenitors with low proliferation potential; Group 3: Immediate DC precursors without proliferation potential

Data acquried in cellular transfer models: U = unconditioned animals, i.e., "steady-state" differentiation upon transfer in nonirradiated and noninflamed animals

RT = Irradiation before transfer

IF= Induction of inflammation pre or post transfer

[a]Expansion in nonirradiated animals cannot be compared with expansion in irradiated animals

[b]Population contained $DX5^{+}$ NK cells, purification of restricted blood DC precursors remains to be established (corrigendum in Nature, 13 May 2004)

et al. 2007a). Referring on their capacity to generate pDCs and cDCs clonally but no other cells in vitro in all assays applied, and in analogy to the previously described CLPs and CMPs progenitor, we named these cells common DC progenitors (CDPs). Maximum CDP offspring cells in spleen were reached at about 10 days after irradiation and transplantation, with a 1:7–8 CDP input to spleen pDC and cDC output ratio, and cells went through more than six divisions as determined by in vivo CFSE dilution (Onai et al. 2007a). Importantly, after in vivo transfer, CDPs efficiently generated pDCs and cDCs in bone marrow, spleen, and lymph nodes even in unconditioned mice, a setting closely mimicking the steady-state situation. Also, both CDPs and their offspring cell generation were enhanced by in vivo Flt3L injection, confirming and extending the previous finding that Flt3L drives both, DC progenitor development as well as full differentiation of DCs (Karsunky et al. 2003; Onai et al. 2007a). A likely identical cell population termed pro-DCs, defined first in vitro in Flt3L cultures and then in vivo in the bone marrow by a $lin^-CD117^{int}CD135^+CD115^+CD16/32^{lo}$ surface phenotype, gives similar rise to both pDCs and cDCs being cloned in vitro and on a population basis in vivo in unconditioned recipients (Naik et al. 2007).

The second group consists of further differentiated populations that are CD117 negative and express CD11c but not MHC class II surface molecules, with a low proliferative capacity. In this group, clonal in vitro assays for different DC subtypes were not feasible with current methods, and, upon transfer in irradiation conditioned hosts, the recovery of progeny cells at peak offspring development in spleen is several fold lower that input numbers. Furthermore, some populations of this group were described not only in bone marrow, but also in blood and spleen.

Bone marrow $CD11c^+MHC\text{-}II^-$ cells can be further divided into $B220^+$ and $B220^-$ cells, accounting for 0.3% and 0.2% of bone marrow nucleated cells, respectively. Upon transfer into irradiated mice, $CD11c^+$MHC class II^- $B220^+$ cells generated exclusively pDCs and cDCs, while $CD11c^+MHC\text{-}II^-$ $B220^-$ cells generated exclusively cDCs, with at 7 days posttransfer 0.01- and 0.03-fold expansion in spleen, respectively (Diao et al. 2006; Diao et al. 2004). A similar population, $CD11c^+CD31^+Ly6C^+MHC\text{-}II^-$ B220— cells, accounts for 0.5%–1% of bone marrow and 1%–2% of blood cells and gives rise to pDCs, cDCs, and macrophages without proliferation in vitro, which were called preimmunocytes (Bruno et al. 2001). No in vivo transfer experiments were reported.

Peripheral blood $CD11c^+MHC\text{-}II^-$ DC-precursors (a population later found to be highly contaminated by DX5+ NK cells) account for about 5% of nucleated cells, divide approximately three times upon in vivo transfer into sublethally irradiated mice, and generate spleen pDCs, and cDCs but no other cells in vivo, with an approximate 1:1 input cell to spleen cell read out ratio at 2 weeks posttransfer (del Hoyo et al. 2002).

Finally, $CD11c^{int}CD45RA^{lo}CD43^{int}SIRP\text{-}\alpha^{int}MHC\text{-}II^-$ pre-cDCs in spleen generated all spleen cDCs, but no pDCs or other cells on a population basis in vitro and upon transfer into nonirradiated recipients in vivo (Naik et al. 2006).

The third group consists of nonproliferating Gr-1^{hi} monocytes with immediate DC precursor potential upon in vivo transfer. DC differentiation in this group is not, however, a default pathway, as it was only described in context of inflammatory stimuli at sites of tissue involvement (Naik et al. 2006; Randolph et al. 1999; Ginhoux et al. 2006; Iijima et al. 2007; Landsman et al. 2007; Varol et al. 2007). Thus, monocytes do not likely contribute to the majority of DCs in lymphoid organs at steady state.

Conclusions and Controversies Regarding Groups One and Two DC Progenitor Populations

As DC progenitor populations in groups one and two were identified at similar times by different groups using slightly different isolation methods, subsequent characterization, and in vitro and in vivo read out systems, some definitive conclusions on early DC development can be drawn, but some controversies also arose.

As definitive conclusion, the progenitor populations of groups one and two show that, as for other hematopoietic lineages, DC lineage development follows a path of commitment that leads first to lineage and then subpopulation restriction and is accompanied by decreasing immediate proliferative bust size. Most of this path is likely achieved within the bone marrow. Also, group one progenitor cells definitively proof the existence of cells with either clonal cDC and MDPs or pDC and cDC (CDP, pro-DCs) potential, which was equally confirmed on the respective population basis in vitro and in vivo.

The controversies revolve around the following questions: Are defined populations within groups one and two each homogenous? Are they strictly separated and coexisting or are they instead one and the same, or, especially if they are not homogenous, are they partially overlapping? If the same or partially overlapping, are the observed biological differences possibly due to minor technical differences in isolation and read out methods used?

By phenotype, all group one progenitors, MDPs, CDPs, and pro-DCs, reside within the lin^- bone marrow fraction with MDPs expressing likely somewhat higher levels of CD117 and CDPs and pro-DCs expressing CD135 at the surface, while CD135 expression on MDP was only reported at the mRNA level. All three do not generate megacaryocyte-erythrocyte, granulocyte, and lymphoid read out and possess similar proliferation capacity in vitro and upon in vivo transfer, making all of them candidates to contribute majorly to steady-state DC generation. However, while CDPs and pro-DCs respond to the DC cytokine flt3L, respective responses of MDPs are less well characterized, although a recent study that used a progenitor population containing MDPs suggested an Flt3L-driven proliferative response (Waskow et al. 2008). A key biological difference lies in the finding that MDPs clonally and on a population basis efficiently gave rise to macrophages and cDCs and lacked pDC differentiation potential (Fogg et al. 2006), while CDPs and pro-DCs exclusively generated pDCs and cDCs, but no or very low macrophage or monocyte read out (Onai et al. 2007a; Naik et al. 2007). Lack of macrophage read out from CDPs, but not pro-DCs, could in theory be attributed to the blocking

M-CSFR antibody (AFS-98) used for prospective isolation. However, M-CSF induced enhancement of DC development (Onai et al. 2007a) and all subsequent experiments showed, that this antibody does not substantially affect read out as long as it is not added continuously to cultures (Onai and Manz, unpublished). This parallels the observation that the blocking IL-7Rα antibody A7R34 used for CLP isolation does not inhibit B-cell read out from these IL-7 dependent cells (Kondo et al. 1997; Manz et al. 2001). If these findings hold true by all means, MDPs on the one side and CDPs and pro-DCs on the other side must at least differ on the fraction of those cells that give clonal read out to cDCs and macrophages, and pDCs and cDCs, respectively. Of note, recent further analysis of MDPs revealed some pDC read out, at least on the population level (Auffray et al. 2008).

Group two progenitors, due to their relatively low proliferative burst size, have not been evaluated as rigorously as group one progenitors and thus within populations there might be high heterogeneity. In terms of immunophenotype, they are all $CD11c^{+}$ but still $MCH\text{-}II^{-}$ and are heterogeneous by several additional surface markers analyzed. Of interest, bone marrow DC progenitors B220+ (Diao et al. 2004) and pre-immunocytes (Bruno et al. 2001), as well as blood DC precursors (del Hoyo et al. 2002) share on a population basis many surface markers and the capacity to give rise to pDCs and cDCs. Thus, these populations might have substantial overlap. The of the group two progenitors best characterized population are the pre-cDCs isolated from spleen (Naik et al. 2006). As a population, these progenitors have lost monocyte/macrophage and pDC potential but still go through two or three divisions and give rise to all splenic cDC populations, presenting evidence that (1) monocytes are not the major progenitors of spleen cDC in steady state, and (2) some and possibly most cDCs enter the spleen in an immature state, ready to go through further expansion and terminal differentiation. Whether bipotent pDC and cDC progenitors (i.e., CDP, pro-DCs or their respective offspring) can also directly travel through blood and enter lymphatic organs, or if pDCs only enter in a cDC separated fashion, still needs to be determined. Of interest, a minute spleen population with CDP phenotype and in vitro pDC and cDC differentiation capacity was already demonstrated (Onai et al. 2007a).

Future direct comparison of the progenitors in groups one and two, as well as improved technologies for in vivo tracking, possibly with the demission of transfer studies, will be critical to solve these issues.

The LC Exception

It is important to note that this DC differentiation model does not apply to LCs. We have shown that in contrast to most DC populations, LCs are maintained locally and independently of circulating precursors in the steady state (see also the chapter on developmental biology of mammalian T-cell progenitors). In contrast, in inflamed settings, LCs are replaced by circulating Gr-1hi monocytes in an M-CSFR-dependent manner.

How LC homeostasis is maintained during the steady state is unclear. We and others showed that LCs do proliferate in the skin. Although self-renewal could be sufficient to maintain LC numbers most of the time, the existence of specialized local precursor cells has also been proposed. The bulge region of the hair follicle serves as a niche for keratinocytes, melanocytes, and mast-cell progenitors (Kumamoto et al. 2003; Blanpain and Fuchs 2006), and there is evidence to suggest that following skin injuries that specifically affect the epidermis, but not the dermis where the hair follicles are located, LCs can be repopulated from the follicles alone (Gilliam et al. 1998) (Fig. 1). It is therefore possible that both differentiated LCs, through self-renewal, and local hematopoietic precursor cells could contribute to LC homeostasis depending on physiological needs, as has been shown for skin stem cells (Kumamoto et al. 2003). Conditional depletion of LCs and careful monitoring of their repopulation should help to resolve these issues (Kissenpfennig et al. 2005; Bennett et al. 2007).

A DC Homeostasis Integrated View

An integrated view on DC homeostasis is starting to emerge from the studies described above (Fig. 1). All current data indicate that homeostasis of lymphoid and nonlymphoid tissue DCs, with the exception of the LCs and thymic DCs, relies on continuous input from blood-born cells (Manz et al. 2001; Wu et al. 2001; Merad et al. 2002a; Liu et al. 2007). Nonlymphoid tissue DCs migrate to the draining LN to present tissue-derived antigens to T cells in order to prime or modulate T-cell specific response. In contrast to lymphocytes, most DCs do not recirculate in the blood and have a limited half-life in lymphoid organs. In lymphoid tissues two major cDC subsets exist, the $CD8^+$ and $CD8^-$ DC subsets with distinct roles in T-cell immunity (Shortman and Liu 2002). In nonlymphoid tissues, two DC subsets, $CD103^+$ and $CD103^-$,have been described and the exact role of these subsets is being unraveled. Several studies already underline the critical and nonredundant function of $CD103^+$ DCs in the induction of lung antiviral responses, lung allergic responses, and gut homing T cells, but also in the induction of gut regulatory T cells (GeurtsvanKessel et al. 2008; Coombes et al. 2007; Sun et al. 2007; Sung et al. 2006; Hintzen et al. 2006). Furthermore, pDCs or natural interferon-producing cells are located in lymphoid tissues, bone marrow, and liver but rarely found in other tissues in the steady state. pDCs are the largest producers of IFN-α and can also differentiate into antigen-presenting cells able to prime naïve T cells, providing a remarkable link between innate and adaptive immune immunity (Liu 2005).

In the thymus, the majority of cDCs are derived locally from early thymocyte progenitors, most likely driven by flt3L (Ardavin et al. 1993; Manz et al. 2001; Wu et al. 2001). Circulating DCs can also enter the thymus, and one recent study suggests that tissue migratory DCs can recirculate through the blood and migrate back to the thymus to mediate central tolerance (Bonasio et al. 2006).

In the steady state, DC-restricted progenitors (Naik et al. 2006) (type 2 progenitors, Table 4), and most likely earlier progenitors (type 1 progenitors, Table 4) (Onai et al. 2007a), seed the spleen prior to differentiating into end-stage differentiated cDCs and pDCs. Alternatively, the pDC and cDC developmental pathways part earlier in bone marrow or blood, and pDCs are recruited separately. Similar DC differentiation processes may also occur in the LN, although this remains to be determined. Steady-state DC homeostasis in lymphoid organs is primarily driven by Flt3L and correlates with flt3 expression on proliferative DC progenitors (McKenna et al. 2000; Laouar et al. 2003; Karsunky et al. 2003; D'Amico and Wu 2003; Onai et al. 2007a; Fogg et al. 2006; Naik et al. 2007). Interestingly, inhibition of flt3 signaling or depletion of CD11c$^+$ DCs leads to an increase of serum flt3L (Tussiwand et al. 2005; S. Jung, personal communication). If this is simply due to DC depletion and a consecutive lack of a scavenging flt3 receptor, or if a direct feedback loop exists to signal increased demand, needs to be determined.

Homeostasis of nonlymphoid tissue DC in steady state is far less studied. Although it is clear that in order to maintain a population of DC in the periphery, in the face of a continuous migration to the local lymphoid tissue, DC must be constantly replaced with new cells. The direct in vivo precursors of tissue DC remain to be examined in detail. It has been suggested that monocytes, DC committed progenitors, and HSC contribute to the DC pool in the steady state and that flt3L participates in the process. However, the exact contribution of each potential progenitor remains unclear. Monocytes infiltrate nonlymphoid tissues in steady state, although the mechanisms that regulate this process are unclear. Monocytes express little or no flt3 (Karsunky et al. 2003), but do express GM-CSFR and M-CSFR, and it is possible that by analogy to in vitro findings that they require low amounts of tissue GM-CSF or M-CSF to differentiate into DCs (Fig. 1). In contrast to most DCs, LCs repopulate locally in the steady state either through self-renewal or through a local hematopoietic precursor that takes residence in the skin.

The development of DCs in inflamed nonlymphoid tissues has been more thoroughly studied. Monocytes differentiate into DCs in inflamed nonlymphoid tissues and in the spleen (Table 4, type 3 progenitors; Naik et al. 2006; Randolph et al. 1999; Iijima et al. 2007; Ginhoux et al. 2006; Landsman et al. 2007; Varol et al. 2007). This DC differentiation pathway is likely driven by an inflammatory cytokine network including GM-CSF, TNFα, and IL-4. In addition, flt3L increases monocyte numbers (Karsunky et al. 2003), and Flt3L levels increase upon systemic infection and is produced by activated T cells. This might add to the enhancement of the flt3L-driven pathway. HSCs, known to circulate continuously in small numbers in steady-state blood (Wright et al. 2001), were also identified in nonlymphoid tissues and in circulating lymph (Massberg et al. 2007). In response to inflammatory stimuli, HSCs gave rise to DCs directly at sites of inflammation (Massberg et al. 2007). These results reveal a potential mechanism to develop emergency hematopoiesis at an inflamed site, which is consistent with previous findings showing that early myeloid and lymphoid progenitors express toll-like receptors (TLRs) and can differentiate into DCs in response to TLR agonists, short-cutting usual differentiation steps (Nagai et al. 2006).

It is clear that mouse and human DCs share several differentiation mechanisms, including their dependence on Flt3L and GM-CSF to develop. Thus, while it is likely that the above described homeostatic and inflammatory DC maintenance pathways also apply to humans, it remains a challenge to be directly assessed in the future.

Dendritic Cells in Hematologic Disease

DC Neoplasms

DC neoplasms are rare and remain poorly defined. They are also referred to as histiocytosis syndrome, a family of disease involving the proliferation of histiocytes (Favara et al. 1997). The term histiocyte was originally used to designate a large cell in lymphoid organs that had no specific morphology but had voluminous cytoplasm and irregular nuclei histiocytosis. This term was subsequently used to refer to the monocyte/macrophage lineage and then extended to include LCs and DCs but also follicular DCs, a population of stromal cells that are present in germinal centers (Cline 1994). Although nearly a century has passed since histiocytic disorders were recognized (Coppes-Zantinga and Egeler 2002), their pathophysiology remains an enigma, and treatment is nonspecific. These problems underscore the need for an improved understanding of the etiology and pathogenesis of histiocytosis.

The three main DC neoplasms reported in the literature include Langerhans cell histiocytosis (LCH), pDC leukemia/hematodermic neoplasm, interdigitating DC sarcoma, and follicular DC sarcoma. Follicular DCs are not of hematopoietic origin and will not be described here. The low prevalence of DC tumors has been attributed to DC capacity to initiate immune response and used as an argument for tumor immunosurveillance. This concept has not been clearly addressed, however, the very low incidence rate of DC malignancy is intriguing. An alternative hypothesis is that the short DC half-life reduces the risk for acquiring genetic mutations. Consistent with this hypothesis, the most frequent DC neoplasm is the Langerhans cell histiocytosis that derives from long-term locally repopulating LCs. This hypothesis is also consistent with recent understanding of the origin of cancer-initiating cells, also called cancer stem cells (Clarke et al. 2006).

In recent years it was shown that about one third of acute myeloid leukemias (AML) carry flt3 mutations that include internal tandem duplications (ITDs, 20% of AML) and kinase domain (KD, approximately 10% of AML) mutations (Small 2006). Both mutations lead to constitutive activation of flt3 signaling. Given the relevance of flt3/flt3L for DC homeostasis, it is intriguing that these mutations are not associated with DC neoplasia. The reasons for this are unclear, but several hypotheses can be proposed. First, flt3 is expressed on ST-HSCs, multipotent, and early myeloid and lymphoid hematopoietic progenitors, and thus alternations in this limited self-renewing population might read out into other hematopoietic cell lineages. Second, mouse models with enforced expression of Flt3-ITD in hematopoiesis develop solely myeloproliferative disease, and full transformation to leukemia is only achieved in combination with other leukemia-associated mutations (Small 2006). Thus, in multistep leukemia development, flt3 mutations represent only one factor contributing to the evolvement of leukemic diseases, and additional factors might determine the hematopoietic lineage type. Third, downstream signaling of mutated, constitutively active flt3 might not be the same as signaling from wild-type, cognate ligand stimulated flt3, thus causing different effects on the cellular signaling pathways. Furthermore, mutated flt3 might not signal as wild-type flt3 from the cell membrane but could involve different cellular compartments. Indeed, Flt3-ITD represses PU.1 and c/EBPα and activates STAT5 in a SOCS-resistant fashion (Choudhary et al. 2007).

Langerhans Cell Histiocytosis

Previously called histiocytosis X, LCH is the most common of the rare group of disorders that constitute histiocytosis (Coppes-Zantinga and Egeler 2002), with an approximate incidence of three to seven cases per million, predominantly but not exclusively in children (Bechan et al. 2006). It affects all age groups, but the features are much better defined in children than in adults (Arico 2004). The clinical spectrum of disease is wide but fatal in up to 25% of cases. LCH patients can present with three major phenotypes. Two third of LCH patients present as a single-system disease commonly in bone (Cline 1994). Single bone LCH lesions have also been called syndrome of Hand-Schuller-Christian or eosinophilic granuloma (Coppes-Zantinga and Egeler 2002). Many of these patients require minimal treatment or the lesion resolves spontaneously. At the other extreme, patients develop a leukemia-like disorder also called Abt-Letterer-Siwe disease, in which multiple organs are affected, leading to organ failure, and patients often require chemotherapy (Cline 1994; Coppes-Zantinga and Egeler 2002; Bechan et al. 2006). Between these extremes are patients with multisystem disease without organ failure in whom the disease runs a fluctuating course and may eventually burn out often leaving serious residual disabilities. The origin of LCH remains unclear and the treatment is often difficult to establish. The hypothesis suggesting that LCH derives from an accumulation of LCs is based on phenotypical and ultrastructural observations showing that LCH lesions are infiltrated by CD1a$^+$ (Cepek et al. 1994), langerin-positive (Chikwava and Jaffe 2004) cells that may contain Birbeck granules (Nezelof et al. 1973). It has been difficult to understand how LCs, which are normally restricted to stratified epithelia, could give rise to such a multifocal disorder. The recent discovery by our and other laboratories that langerin-positive DCs are present in most of the tissue types that are affected by LCH raises the possibility that these cells are also involved in LCH.

A central question about LCH is whether it is a true neoplastic or a reactive disorder. Data showing that LCH cells have an activated phenotype that includes upregulation of costimulatory molecules such as CD40, CD80, CD86 (Geissmann et al. 2001; Emile et al. 1995), the frequent presence of granulomas that include eosinophils, macrophages, and T cells, together with the absence of cellular atypia and genetic abnormality in LCH cells have been used to suggest their reactive origin (Nezelof and Basset 2004). Arguments for a malignant transformation are based on studies showing that LCH cells may be clonal and proliferate in situ (Yu et al. 1994; Willman et al. 1994). However, DC clonality may result from clustering of DC emerging from a common local progenitor, as shown in mice. Some proliferation of LCs in LCH lesions may not be fundamentally pathological as there is evidence that local proliferation is a normal homeostatic mechanism of LC (discussed above) (Merad et al. 2002a; Bogunovic et al. 2006). Since Flt3 ligand might control the proliferation of immediate DC progenitors in peripheral tissue (Karsunky et al. 2003; Waskow et al. 2008), it is intriguing that a study on 24 patients with LCH revealed significantly elevated serum levels of Flt3 ligand and CSF-1 (Rolland et al. 2005). Higher levels of these cytokines correlated with patients having more extensive disease (Rolland et al. 2005). Serum levels of Flt3 ligand and CSF-1 were the highest in high-risk patients with extensive skin and/or multisystem involvement, while patients with bone lesions had relatively higher levels of CSF-1 and stem cell factor (Rolland et al. 2005). These studies, together with our findings showing that CSF-1 receptor is required for LC development (Ginhoux et al. 2006), argue for the potential benefit of drugs modulating Flt3 ligand (Tussiwand et al. 2005) and CSF-1 effects for the treatment of LCH. However, it remains to be determined if these cytokines are involved in initiation of disease. Of interest, recent data revealed elevated IL-17A serum levels in LCH patients and demonstrated the role of IL-17 in the development of multiple giant cells that are often present in LCH lesions (Coury et al. 2008).

Other factors may play a role in the development of LCH. The antiapoptotic gene Bcl2 is expressed in many lesional cells in LCH, while Bcl2 is not expressed in normal LCs, suggesting that the cellular mechanisms that detect DNA damage and apoptosis are activated in LCH cells (Schouten et al. 2002). Southern blot hybridization analysis revealed no evidence for Bcl2 gene rearrangements, suggesting that LCH cells may respond to signals that regulate normal Bcl2 expression (Savell et al. 1998). Signals provided by activated T cells, such as CD40 ligand, upregulates Bcl2 in human DCs, a phenomenon that correlates with a resistance to Fas-mediated apoptosis (Bjorck et al. 1997). Another TNF-related molecule expressed exclusively on T cells (TRANCE) upregulates the antiapoptotic molecule Bcl-xL in DCs, which enhances their viability in vitro (Wong et al. 1997). T lymphocytes represent 20% of the lesional infiltrate in LCH, and about 80% of these T cells are memory cells (Bechan et al. 2006). Thus it is possible that upregulation of antiapoptotic molecules on normal langerin-positive DCs induced by TNF-related molecules expressed by activated T cells play a role in LCH. An intriguing study has recently shown that transgenic mice expressing the simian virus 40 (SV40)–derived oncogenes under the CD11c promoter (Steiner et al. 2008) developed a dramatic increase of $CD8^+$ DCs in spleen and liver at around 2 months of age. $CD8^+$ DCs expressed langerin and were highly proliferative. Granuloma, which are almost always present in LCH, were absent in these mice. Although this disease is evocative of LCH, the high proliferation rate of $CD8^+$ cells and the absence of granuloma can also suggest a DC sarcoma (Pileri et al. 2002).

Early pDC Leukemia/Lymphoma (pDCL, CD4+CD56+ Hematodermic Neoplasm)

This disease was initially described in the World Health Organization (WHO) classification as blastic CD56+ NK-cell lymphoma. Recent advances in characterization of this malignancy, however, revealed a close relationship to pDCs, and it was subsequently classified in the 2005 WHO/EORTC (European Organisation for Research and Treatment of Cancer) classification of cutaneous lymphomas as a separate entity termed "early pDC leukemia/lymphoma" (Willemze et al. 2005). pDCL represent less than 1% of acute leukemias (Jacob et al. 2003) and 0.7% of cutaneous lymphomas (Ng et al. 2006). pDCL cells coexpress CD4, CD56, high levels of CD123 (IL-3 receptor), BDCA-2, BDCA-4, HLA-DR, GM-CSFR, and transcription factors such as IRF-8 and PU.1, but few or no B, T, NK, lymphoid-, or myeloid-specific markers, thus resembling a pDC phenotype (Marafioti et al. 2008; Dijkman et al. 2007; Chaperot et al. 2001; Garnache-Ottou et al. 2007; Herling and Jones 2007). Furthermore, similar to their healthy counterpart, pDCL cells can mature into DCs that efficiently prime T cells in the presence of IL-3 and CD40L and can also produce IFN-α upon viral stimulation (Chaperot et al. 2001). Interestingly, although 13q12 deletions that led to the loss of flt3 were detected, expression of flt3 was found to be elevated in pDCL (Dijkman et al. 2007), and activating flt3 mutations have not been reported so far. pDCLs usually manifest as isolated cutaneous lesions that rapidly evolve to multiple sites and proliferate in the blood, BM, LN, and other organs such as the spleen, liver, central nervous system, tonsils, lungs, kidneys, and muscles (Herling and Jones 2007). Skin involvement is typically diffusely dermal, and epidermal tropism is almost always absent (Ng et al. 2006). This is consistent with the distribution of pDCs, which are found in inflamed dermis but always absent from the epidermis. The overall prognosis is dismal with a median overall survival of 9–12 months irrespective of initial pattern of disease (Herling and Jones 2007). Long-lasting remissions have been rarely reported and usually occur in younger patients treated with allogeneic stem cell transplantation (Reimer et al. 2003).

Dendritic Cell Sarcoma

Dendritic cell sarcoma is a very rare entity with less than 50 cases described so far (Gaertner et al. 2001). DC sarcoma affects mostly the LN, but cases of DCs affecting extranodal sites including the skin, BM, lungs, liver, kidney, and spleen have been described. Histologically, mitotic figures, sometimes multipolar, are frequent and provide a crucial element for distinguishing this condition from a reactive process. Accumulation of inflammatory cells is frequent and, among them, activated macrophages and lymphocytes are usually present and tend to confuse the histopathologic picture. DC sarcoma expresses CD68 and S100 and no maker of T, B, or NK lineage (Favara et al. 1997; Gogusev and Nezelof 1998). CD1a and langerin can be expressed, and it is then called LC sarcoma. Similar to LCH, multinucleated cells are common, and the presence of lymphocyte and plasma cells are also common, but in contrast to LCH, LC sarcoma displays cytological atypia (Gaertner et al. 2001; Andriko et al. 1998; Weiss et al. 1986; Rousselet et al. 1994). The course is often progressive and can be rapidly fatal (Gaertner et al. 2001). There is no consensus on treatment and various regimens have been tried, including chemotherapy, radiotherapy, surgery, or combination therapy. The prognosis is difficult to estimate, given the relative paucity of cases, but it appears poor.

DCs in Allogeneic Hematopoietic Cell Transplantation (allo-HCT)

Allo-HCT has been a therapeutic modality in humans for almost half a century (Baron et al. 2003; Ford et al. 1956). Engraftment of allogeneic stem cells is facilitated by myelosuppressive and immunosuppressive conditioning given just prior to the infusion of stem cells. It is now well established that the immunological activity of donor T cells in allo-HCT grafts is a critical factor in eradicating residual recipient hematopoiesis and malignancy, also called graft-versus-leukemia response (GVL) (Weiden et al. 1979; Weiden et al. 1981; Horowitz et al. 1990). The potential of donor T cells to secure remission is most dramatically demonstrated by the use of donor lymphocyte infusion (DLI) to treat posttransplant relapses (Kolb et al. 1990; Kolb et al. 1995; Kolb et al. 2004). However, in addition to their favorable attack of the diseased host hematopoietic system, allo-reactive T cells also lead to a major posttransplant immune complication called graft-versus-host disease (GVHD). There is ample evidence in animal models of transplantation that both GVHD and GVL are dependent on host APC, of which DCs are the most potent. Functional recipient, but not donor, APCs are required for GVL mediated by $CD4^+$ and $CD8^+$ cells (Matte et al. 2004; Reddy et al. 2005). For CD8 T cell–mediated GVL, donor alloantigens must be present on both recipient APCs and tumor; tumor lines expressing costimulatory molecules are also unable to substitute for professional APCs (Reddy et al. 2005). Of interest, stably engrafted donor APCs, while not required for GVL, are able to mediate some CD8 T-cell–dependent GVL activity at lower tumor burden while sparing the effect of GVHD (Matte et al. 2004; Reddy et al. 2005; Chakraverty et al. 2006).

GVL appears to occur at a lower threshold than GVHD in studies of escalated DLI in mice and humans, suggesting that access to a narrow therapeutic window between GVL and GVHD is possible (Johnson and Truitt 1995; Mackinnon et al. 1995). Changing the density or location of host DC in the recipient may be able influence the therapeutic advantage in favor of GVL. In murine models, recipient DCs present in the spleen or the LN are in contact with donor T cells and are sufficient to prime robust GVL responses (Reddy et al. 2005; Chakraverty et al. 2006; Mapara et al. 2002). In contrast, recipient peripheral tissues are not infiltrated by allo-reactive donor T cells in the absence of inflammation (Chakraverty et al. 2006). The interaction of donor T cells with lymphoid organ DCs, but not tissue DC populations, may achieve a selective benefit in promoting GVL without GVHD (Reddy et al. 2005; Chakraverty et al. 2006; Mapara et al. 2002). This argument underpins the logic of delayed T cell add-back strategies (Barrett et al. 1998) and preemptive DLI (Peggs et al. 2004), which allow the inflammatory insult of conditioning to subside prior to the infusion of donor T cells.

Finally, it has been proposed that in certain malignancies, APCs may arise directly from the malignant clone, serving to prime donor T cells directly to tumor-associated antigens (Kolb et al. 2004). In chronic myelogenous leukemia, bcr-abl$^+$ DC can be derived in vitro from monocytes (Terme et al. 2005), but the spontaneous generation of these cells in vivo has never been examined. A similar argument applies to B cells in follicular lymphoma (Vyth-Dreese et al. 1995). It is likely that the susceptibility of these diseases to GVL is not only due to efficient priming of donor T cells, but also to the slow kinetics of disease and relatively normal phenotype of the leukaemic clone (Matte et al. 2004). In other diseases where GVL is less effective, relapse can easily occur during GVHD, suggesting that the target cells escape recognition or overwhelm the effectors rather than a fault with T-cell priming.

The importance of recipient DCs in initiating GVHD has also been clearly established. In a mouse model of allo-HCT, recipient splenic DCs were shown to persist for only 5 days after irradiation, but allo-reactive T cells express activation antigens after 6 hours and begin to proliferate within 24 hours (Zhang et al. 2002). Chimeric animals with defective APCs are generated by a preparatory transplant, and the effect on GVHD was tested with a second experimental transplant. These maneuvers show attenuation of acute CD8-mediated GVHD if MHC class I or costimulatory molecules are deleted from recipient DCs or if recipient DCs are first replaced by donor cells (Zhang et al. 2002; Shlomchik et al. 1999; Teshima et al. 2002). Donor DCs play a supplementary role and are able to augment acute GVHD through cross-presentation (Matte et al. 2004). Further work confirms the importance of DCs compared with other APCs and demonstrates that recipient DC "add-back" is sufficient to induce GVHD (Duffner et al. 2004; Xia et al. 2006). It is also possible that the potential of some host DC subsets, in particular LCs, to resist the transplantation regimen could play some part in the predilection of GVHD for certain target organs. In mice transplanted with T-cell–depleted BM, persistent recipient cutaneous DCs are sufficient to generate cutaneous GVHD upon DLI (Merad et al. 2004). In contrast to skin, other GVHD target organs have not yet been closely scrutinized. The gut and associated Peyer's patches may harbor radio-resistant DC precursors, even though rapid turnover of mesenteric LN DC is apparent (Merad et al. 2002a). In the liver, the bulk of parenchymal DCs rapidly equilibrate with the blood (Merad et al. 2004), but this does not exclude a niche population of cycling DCs in focal GVHD targets, such as the portal triad of the liver.

The role of cutaneous DCs in cutaneous GVHD is starting to be unraveled in human patients that receive allo-HCT. In contrast to blood DCs and monocytes, human LCs and dermal DCs survive conditioning therapy in significant numbers (Collin et al. 2006; Fagnoni et al. 2004) and may persist after transplantation (Perreault et al. 1985; Bogunovic et al. 2006). A recent study on a limited number of patients treated with reduced intensity and nonmyeloablative regimen suggests that recipient cutaneous DC survival is linked to the intensity of conditioning and to GVHD (Collin et al. 2006; Bogunovic et al. 2006). As cutaneous DCs are themselves a target of GVHD, it is very difficult to prove a causal link between recipient DC and GVHD leveling in humans. To directly examine the role of host DCs in post-transplant immunity, we have initiated a large prospective study to examine correlations between cutaneous DC chimerism and GVHD kinetics in cohorts of patients transplanted with different regimens (Mielacrek, Collin, Merad et al., in preparation). We will also examine in this context the correlation between DC chimerism and antitumor relapse, as it is possible that the disappearance of most remaining DCs may also correlate with a relapse risk factor.

Future Directions in Research and Possible Clinical Implementation

Thirty-five years after their discovery, DCs have now emerged as key modulators of several immune processes, including antimicrobial immunity, tumor immunity, autoimmunity, atherosclerosis, and posttransplantation immune responses. These immune functions are maintained by a heterogeneous population of DCs with specialized functions. It is critical to understand the mechanisms that regulate the homeostasis of DC subsets to better utilize their therapeutic potential. This will require better understanding the mechanisms that regulate DC development, trafficking, and localization in specialized niches in vivo. These studies will be possible through the currently evolving labeling and in vivo imaging technology. Furthermore, selective targeting of DCs in vivo or ex vivo (e.g., via specific antibody or small molecules) may lead to new immunomodulation strategies that may be beneficial for the treatment of several human disease, as discussed above.

A main challenge for the future is to translate what we have learned from the mouse into humans and better explore the diversity of human DC populations, in particular, the diversity of lymphoid organs DCs. Another challenge will be to develop more relevant mouse models for the study of the human immune system (Aspord et al. 2007; Manz 2007).

References

Akashi K, Traver D, Miyamoto T, Weissman IL (2000) A clonogenic common myeloid progenitor that gives rise to all myeloid lineages. Nature 404:193–197

Aliberti J, Schulz O, Pennington DJ, Tsujimura H, Reis e Sousa C, Ozato K, Sher A (2003) Essential role for ICSBP in the in vivo development of murine CD8alpha + dendritic cells. Blood 101:305–310

Allman D, Dalod M, Asselin-Paturel C, Delale T, Robbins SH, Trinchieri G, Biron CA, Kastner P, Chan S (2006) Ikaros is required for plasmacytoid dendritic cell differentiation. Blood 108:4025–4034

Anderson KL, Perkin H, Surh CD, Venturini S, Maki RA, Torbett BE (2000) Transcription factor PU.1 is necessary for development of thymic and myeloid progenitor-derived dendritic cells. J Immunol 164:1855–1861

Andriko JW, Kaldjian EP, Tsokos M, Abbondanzo SL, Jaffe ES (1998) Reticulum cell neoplasms of lymph nodes: A clinicopathologic study of 11 cases with recognition of a new subtype derived from fibroblastic reticular cells. Am J Surg Pathol 22:1048–1058

Ardavin C, Wu L, Li CL, Shortman K (1993) Thymic dendritic cells and T cells develop simultaneously in the thymus from a common precursor population. Nature 362:761–763

Arico M (2004) Langerhans cell histiocytosis in adults: more questions than answers? Eur J Cancer 40:1467–1473

Aspord C, Pedroza-Gonzalez A, Gallegos M, Tindle S, Burton EC, Su D, Marches F, Banchereau J, Palucka AK (2007) Breast cancer instructs dendritic cells to prime interleukin 13-secreting CD4+ T cells that facilitate tumor development. J Exp Med 204:1037–1047

Auffray C, Emre Y, Geissmann F (2008) Homeostasis of dendritic cell pool in lymphoid organs. Nat Immunol 9:584–586

Austyn JM, Kupiec-Weglinski JW, Hankins DF, Morris PJ (1988) Migration patterns of dendritic cells in the mouse. Homing to T cell-dependent areas of spleen, and binding within marginal zone. J Exp Med 167:646–651

Banchereau J, Pascual V (2006) Type I interferon in systemic lupus erythematosus and other autoimmune diseases. Immunity 25:383–392

Banchereau J, Steinman RM (1998) Dendritic cells and the control of immunity. Nature 392:245–252

Baron F, Storb R, Little MT (2003) Hematopoietic cell transplantation: five decades of progress. Arch Med Res 34:528–544

Barrett AJ, Mavroudis D, Tisdale J, Molldrem J, Clave E, Dunbar C, Cottler-Fox M, Phang S, Carter C, Okunnieff P, Young NS, Read EJ (1998) T cell-depleted bone marrow transplantation and delayed T cell add-back to control acute GVHD and conserve a graft-versus-leukemia effect. Bone Marrow Transplant 21:543–551

Bashyam H (2007) Ralph Steinman: Dendritic cells bring home the Lasker. J Exp Med 204:2245–2248

Bechan GI, Egeler RM, Arceci RJ (2006) Biology of Langerhans cells and Langerhans cell histiocytosis. Int Rev Cytol 254:1–43

Belz GT, Behrens GM, Smith CM, Miller JF, Jones C, Lejon K, Fathman CG, Mueller SN, Shortman K, Carbone FR, Heath WR (2002) The CD8alpha(+) dendritic cell is responsible for inducing peripheral self-tolerance to tissue-associated antigens. J Exp Med 196:1099–1104

Bennett CL, Noordegraaf M, Martina CA, Clausen BE (2007) Langerhans cells are required for efficient presentation of topically applied hapten to T cells. J Immunol 179:6830–6835

Bjorck P (2001) Isolation and characterization of plasmacytoid dendritic cells from Flt3 ligand and granulocyte-macrophage colony-stimulating factor-treated mice. Blood 98:3520–3526

Bjorck P, Banchereau J, Flores-Romo L (1997) CD40 ligation counteracts Fas-induced apoptosis of human dendritic cells. Int Immunol 9:365–372

Blanpain C, Fuchs E (2006) Epidermal stem cells of the skin. Annu Rev Cell Dev Biol 22:339–373

Blom B, Ho S, Antonenko S, Liu YJ (2000) Generation of interferon alpha-producing predendritic cell (Pre-DC)2 from human CD34(+) hematopoietic stem cells. J Exp Med 192:1785–1796

Bogunovic M, Ginhoux F, Wagers A, Loubeau M, Isola LM, Lubrano L, Najfeld V, Phelps RG, Grosskreutz C, Scigliano E, Frenette PS, Merad M (2006) Identification of a radio-resistant and cycling dermal dendritic cell population in mice and men. J Exp Med 203:2627–2638

Bonasio R, Scimone ML, Schaerli P, Grabie N, Lichtman AH, von Andrian UH (2006) Clonal deletion of thymocytes by circulating dendritic cells homing to the thymus. Nat Immunol 7:1092–1100

Bonifaz L, Bonnyay D, Mahnke K, Rivera M, Nussenzweig MC, Steinman RM (2002) Efficient targeting of protein antigen to the dendritic cell receptor DEC-205 in the steady state leads to antigen presentation on major histocompatibility complex class I products and peripheral CD8+ T cell tolerance. J Exp Med 196:1627–1638

Borkowski TA, Letterio JJ, Farr AG, Udey MC (1996) A role for endogenous transforming growth factor beta 1 in Langerhans cell biology: The skin of transforming growth factor beta 1 null mice is devoid of epidermal Langerhans cells. J Exp Med 184:2417–2422

Borkowski TA, Letterio JJ, Mackall CL, Saitoh A, Wang XJ, Roop DR, Gress RE, Udey MC (1997) A role for TGFbeta1 in Langerhans cell biology. Further characterization of the epidermal Langerhans cell defect in TGFbeta1 null mice. J Clin Invest 100:575–581

Brasel K, De Smedt T, Smith JL, Maliszewski CR (2000) Generation of murine dendritic cells from flt3-ligand-supplemented bone marrow cultures. Blood 96:3029–3039

Brocker T, Riedinger M, Karjalainen K (1997) Targeted expression of major histocompatibility complex (MHC) class II molecules demonstrates that dendritic cells can induce negative but not positive selection of thymocytes in vivo. J Exp Med 185:541–550

Bruno L, Seidl T, Lanzavecchia A (2001) Mouse pre-immunocytes as non-proliferating multipotent precursors of macrophages, interferon-producing cells, CD8alpha(+) and CD8alpha(–) dendritic cells. Eur J Immunol 31:3403–3412

Bujdoso R, Hopkins J, Dutia BM, Young P, McConnell I (1989) Characterization of sheep afferent lymph dendritic cells and their role in antigen carriage. J Exp Med 170:1285–1301

Burgess AW, Camakaris J, Metcalf D (1977) Purification and properties of colony-stimulating factor from mouse lung-conditioned medium. J Biol Chem 252:1998–2003

Burkly L, Hession C, Ogata L, Reilly C, Marconi LA, Olson D, Tizard R, Cate R, Lo D (1995) Expression of relB is required for the development of thymic medulla and dendritic cells. Nature 373:531–536

Bursch LS, Wang L, Igyarto B, Kissenpfennig A, Malissen B, Kaplan DH, Hogquist KA (2007) Identification of a novel population of langerin+ dendritic cells. J Exp Med 204:3147–3156

Buza-Vidas N, Luc S, Jacobsen SE (2007) Delineation of the earliest lineage commitment steps of haematopoietic stem cells: New developments, controversies and major challenges. Curr Opin Hematol 14:315–321

Canque B, Camus S, Dalloul A, Kahn E, Yagello M, Dezutter-Dambuyant C, Schmitt D, Schmitt C, Gluckman JC (2000) Characterization of dendritic cell differentiation pathways from cord blood CD34(+)CD7(+)CD45RA(+) hematopoietic progenitor cells. Blood 96:3748–3756

Caux C, Dezutter-Dambuyant C, Schmitt D, Banchereau J (1992) GM-CSF and TNF-alpha cooperate in the generation of dendritic Langerhans cells. Nature 360:258–261

Caux C, Massacrier C, Dubois B, Valladeau J, Dezutter-Dambuyant C, Durand I, Schmitt D, Saeland S (1999) Respective involvement of TGF-beta and IL-4 in the development of Langerhans cells and non-Langerhans dendritic cells from CD34+ progenitors. J Leukoc Biol 66:781–791

Caux C, Ait-Yahia S, Chemin K, de Bouteiller O, Dieu-Nosjean MC, Homey B, Massacrier C, Vanbervliet B, Zlotnik A, Vicari A (2000) Dendritic cell biology and regulation of dendritic cell trafficking by chemokines. Springer Semin Immunopathol 22:345–369

Cepek KL, Shaw SK, Parker CM, Russell GJ, Morrow JS, Rimm DL, Brenner MB (1994) Adhesion between epithelial cells and T lymphocytes mediated by E-cadherin and the alpha E beta 7 integrin. Nature 372:190–193

Chakraverty R, Cote D, Buchli J, Cotter P, Hsu R, Zhao G, Sachs T, Pitsillides CM, Bronson R, Means T, Lin C, Sykes M (2006) An inflammatory checkpoint regulates recruitment of graft-versus-host reactive T cells to peripheral tissues. J Exp Med 203:2021–2031

Chaperot L, Bendriss N, Manches O, Gressin R, Maynadie M, Trimoreau F, Orfeuvre H, Corront B, Feuillard J, Sotto JJ, Bensa JC, Briere F, Plumas J, Jacob MC (2001) Identification of a leukemic counterpart of the plasmacytoid dendritic cells. Blood 97:3210–3217

Cheers C, Haigh AM, Kelso A, Metcalf D, Stanley ER, Young AM (1988) Production of colony-stimulating factors (CSFs) during infection: Separate determinations of macrophage-, granulocyte-, granulocyte-macrophage-, and multi-CSFs. Infect Immun 56:247–251

Chen W, Antonenko S, Sederstrom JM, Liang X, Chan AS, Kanzler H, Blom B, Blazar BR, Liu YJ (2004) Thrombopoietin cooperates with FLT3-ligand in the generation of plasmacytoid dendritic cell precursors from human hematopoietic progenitors. Blood 103:2547–2553

Chicha L, Jarrossay D, Manz MG (2004) Clonal type I interferon-producing and dendritic cell precursors are contained in both human lymphoid and myeloid progenitor populations. J Exp Med 200:1519–1524

Chikwava K, Jaffe R (2004) Langerin (CD207) staining in normal pediatric tissues, reactive lymph nodes, and childhood histiocytic disorders. Pediatr Dev Pathol 7:607–614

Choudhary C, Brandts C, Schwable J, Tickenbrock L, Sargin B, Ueker A, Bohmer FD, Berdel WE, Muller-Tidow C, Serve H (2007) Activation mechanisms of STAT5 by oncogenic Flt3-ITD. Blood 110:370–374

Clarke MF, Dick JE, Dirks PB, Eaves CJ, Jamieson CH, Jones DL, Visvader J, Weissman IL, Wahl GM (2006) Cancer stem cells–perspectives on current status and future directions: AACR Workshop on cancer stem cells. Cancer Res 66:9339–9344

Cline MJ (1994) Histiocytes and histiocytosis. Blood 84:2840–2853

Collin MP, Hart DN, Jackson GH, Cook G, Cavet J, Mackinnon S, Middleton PG, Dickinson AM (2006) The fate of human Langerhans cells in hematopoietic stem cell transplantation. J Exp Med 203:27–33

Colonna M, Trinchieri G, Liu YJ (2004) Plasmacytoid dendritic cells in immunity. Nat Immunol 5:1219–1226

Coombes JL, Siddiqui KR, Arancibia-Carcamo CV, Hall J, Sun CM, Belkaid Y, Powrie F (2007) A functionally specialized population of mucosal CD103+ DCs induces Foxp3+ regulatory T cells via a TGF-beta and retinoic acid-dependent mechanism. J Exp Med 204:1757–1764

Coppes-Zantinga A, Egeler RM (2002) The Langerhans cell histiocytosis X files revealed. Br J Haematol 116:3–9

Coury F, Annels N, Rivollier A, Olsson S, Santoro A, Speziani C, Azocar O, Flacher M, Djebali S, Tebib J, Brytting M, Egeler RM, Rabourdin-Combe C, Henter JI, Arico M, Delprat C (2008) Langerhans cell histiocytosis reveals a new IL-17A-dependent pathway of dendritic cell fusion. Nat Med 14:81–87

Cumberbatch M, Kimber I (1992) Dermal tumour necrosis factor-alpha induces dendritic cell migration to draining lymph nodes, and possibly provides one stimulus for Langerhans' cell migration. Immunology 75:257–263

Cumberbatch M, Dearman RJ, Kimber I (1997) Langerhans cells require signals from both tumour necrosis factor-alpha and interleukin-1 beta for migration. Immunology 92:388–395

Czernielewski JM, Demarchez M (1987) Further evidence for the self-reproducing capacity of Langerhans cells in human skin. J Invest Dermatol 88:17–20

Czernielewski J, Vaigot P, Prunieras M (1985) Epidermal Langerhans cells–a cycling cell population. J Invest Dermatol 84:424–426

Dai XM, Ryan GR, Hapel AJ, Dominguez MG, Russell RG, Kapp S, Sylvestre V, Stanley ER (2002) Targeted disruption of the mouse colony-stimulating factor 1 receptor gene results in osteopetrosis, mononuclear phagocyte deficiency, increased primitive progenitor cell frequencies, and reproductive defects. Blood 99:111–120

D'Amico A, Wu L (2003) The early progenitors of mouse dendritic cells and plasmacytoid predendritic cells are within the bone marrow hemopoietic precursors expressing Flt3. J Exp Med 198:293–303

Daro E, Pulendran B, Brasel K, Teepe M, Pettit D, Lynch DH, Vremec D, Robb L, Shortman K, McKenna HJ, Maliszewski CR, Maraskovsky E (2000) Polyethylene glycol-modified GM-CSF expands CD11b(high)CD11c(high) but notCD11b(low)CD11c(high) murine dendritic cells in vivo: A comparative analysis with Flt3 ligand. J Immunol 165:49–58

de Witte L, Nabatov A, Pion M, Fluitsma D, de Jong MA, de Gruijl T, Piguet V, van Kooyk Y, Geijtenbeek TB (2007) Langerin is a natural barrier to HIV-1 transmission by Langerhans cells. Nat Med 13:367–371

del Hoyo GM, Martin P, Vargas HH, Ruiz S, Arias CF, Ardavin C (2002) Characterization of a common precursor population for dendritic cells. Nature 415:1043–1047/Eratum in Nature (2004 May 13) 429(6988):205

den Haan JM, Lehar SM, Bevan MJ (2000) CD8(+) but not CD8(-) dendritic cells cross-prime cytotoxic T cells in vivo. J Exp Med 192:1685–1696

Diao J, Winter E, Chen W, Cantin C, Cattral MS (2004) Characterization of distinct conventional and plasmacytoid dendritic cell-committed precursors in murine bone marrow. J Immunol 173:1826–1833

Diao J, Winter E, Cantin C, Chen W, Xu L, Kelvin D, Phillips J, Cattral MS (2006) In situ replication of immediate dendritic cell (DC) precursors contributes to conventional DC homeostasis in lymphoid tissue. J Immunol 176:7196–7206

Dijkman R, van Doorn R, Szuhai K, Willemze R, Vermeer MH, Tensen CP (2007) Gene-expression profiling and array-based CGH classify CD4+CD56+ hematodermic neoplasm and cutaneous myelomonocytic leukemia as distinct disease entities. Blood 109:1720–1727

Donskoy E, Goldschneider I (2003) Two developmentally distinct populations of dendritic cells inhabit the adult mouse thymus: Demonstration by differential importation of hematogenous precursors under steady state conditions. J Immunol 170:3514–3521

Dranoff G, Jaffee E, Lazenby A, Golumbek P, Levitsky H, Brose K, Jackson V, Hamada H, Pardoll D, Mulligan RC (1993) Vaccination with irradiated tumor cells engineered to secrete murine granulocyte-macrophage colony-stimulating factor stimulates potent, specific, and long-lasting anti-tumor immunity. Proc Natl Acad Sci USA 90:3539–3543

Duffner UA, Maeda Y, Cooke KR, Reddy P, Ordemann R, Liu C, Ferrara JL, Teshima T (2004) Host dendritic cells alone are sufficient to initiate acute graft-versus-host disease. J Immunol 172:7393–7398

Emile JF, Fraitag S, Leborgne M, de Prost Y, Brousse N (1995) In situ expression of activation markers by Langerhans' cells containing GM-CSF. Adv Exp Med Biol 378:101–103

Esashi E, Wang YH, Perng O, Qin XF, Liu YJ, Watowich SS (2008) The signal transducer STAT5 inhibits plasmacytoid dendritic cell development by suppressing transcription factor IRF8. Immunity 28:509–520

Fagnoni FF, Oliviero B, Giorgiani G, De Stefano P, Deho A, Zibera C, Gibelli N, Maccario R, Da Prada G, Zecca M, Locatelli F (2004) Reconstitution dynamics of plasmacytoid and myeloid dendritic cell precursors after allogeneic myeloablative hematopoietic stem cell transplantation. Blood 104:281–289

Fainaru O, Woolf E, Lotem J, Yarmus M, Brenner O, Goldenberg D, Negreanu V, Bernstein Y, Levanon D, Jung S, Groner Y (2004) Runx3 regulates mouse TGF-beta-mediated dendritic cell function and its absence results in airway inflammation. Embo J 23:969–979

Fancke B, Suter M, Hochrein H, O'Keeffe M (2008) M-CSF: A novel plasmacytoid and conventional dendritic cell poietin. Blood 111:150–159

Favara BE, Feller AC, Pauli M, Jaffe ES, Weiss LM, Arico M, Bucsky P, Egeler RM, Elinder G, Gadner H, Gresik M, Henter JI, Imashuku S, Janka-Schaub G, Jaffe R, Ladisch S, Nezelof C, Pritchard J (1997) Contemporary classification of histiocytic disorders. The WHO Committee On Histiocytic/Reticulum Cell Proliferations. Reclassification Working Group of the Histiocyte Society. Med Pediatr Oncol 29:157–166

Fogg DK, Sibon C, Miled C, Jung S, Aucouturier P, Littman DR, Cumano A, Geissmann F (2006) A clonogenic bone marrow progenitor specific for macrophages and dendritic cells. Science 311:83–87

Fong L, Hou Y, Rivas A, Benike C, Yuen A, Fisher GA, Davis MM, Engleman EG (2001) Altered peptide ligand vaccination with Flt3 ligand expanded dendritic cells for tumor immunotherapy. Proc Natl Acad Sci USA 26:26

Fontenot JD, Gavin MA, Rudensky AY (2003) Foxp3 programs the development and function of CD4+CD25+ regulatory T cells. Nat Immunol 4:330–336

Ford CE, Hamerton JL, Barnes DW, Loutit JF (1956) Cytological identification of radiation-chimaeras. Nature 177:452–454

Fortunel NO, Hatzfeld A, Hatzfeld JA (2000) Transforming growth factor-beta: Pleiotropic role in the regulation of hematopoiesis. Blood 96:2022–2036

Fossum S (1989) The life history of dendritic leukocytes (DL). Curr Top Pathol 79:101–124

Friedman AD (2002) Transcriptional regulation of granulocyte and monocyte development. Oncogene 21:3377–3390

Gaertner EM, Tsokos M, Derringer GA, Neuhauser TS, Arciero C, Andriko JA (2001) Interdigitating dendritic cell sarcoma. A report of four cases and review of the literature. Am J Clin Pathol 115:589–597

Galy A, Travis M, Cen D, Chen B (1995) Human T, B, natural killer, and dendritic cells arise from a common bone marrow progenitor cell subset. Immunity 3:459–473

Garnache-Ottou F, Feuillard J, Saas P (2007) Plasmacytoid dendritic cell leukaemia/lymphoma: Towards a well defined entity? Br J Haematol 136:539–548

Geissmann F, Lepelletier Y, Fraitag S, Valladeau J, Bodemer C, Debre M, Leborgne M, Saeland S, Brousse N (2001) Differentiation of Langerhans cells in Langerhans cell histiocytosis. Blood 97:1241–1248

Georgopoulos K (2002) Haematopoietic cell-fate decisions, chromatin regulation and Ikaros. Nat Rev Immunol 2:162–174

GeurtsvanKessel CH, Willart MA, van Rijt LS, Muskens F, Kool M, Baas C, Thielemans K, Bennett C, Clausen BE, Hoogsteden HC, Osterhaus AD, Rimmelzwaan GF, Lambrecht BN (2008) Clearance of influenza virus from the lung depends on migratory langerin+CD11b- but not plasmacytoid dendritic cells. J Exp Med 205:1621–1634

Giacometti L, Montagna W (1967) Langerhans cells: Uptake of tritiated thymidine. Science 157:439–440

Gilliam AC, Kremer IB, Yoshida Y, Stevens SR, Tootell E, Teunissen MB, Hammerberg C, Cooper KD (1998) The human hair follicle: A reservoir of CD40+ B7-deficient Langerhans cells that repopulate epidermis after UVB exposure. J Invest Dermatol 110:422–427

Gilliet M, Boonstra A, Paturel C, Antonenko S, Xu XL, Trinchieri G, O'Garra A, Liu YJ (2002) The development of murine plasmacytoid dendritic cell precursors is differentially regulated by FLT3-ligand and granulocyte/macrophage colony-stimulating factor. J Exp Med 195:953–958

Ginhoux F, Tacke F, Angeli V, Bogunovic M, Loubeau M, Dai XM, Stanley ER, Randolph GJ, Merad M (2006) Langerhans cells arise from monocytes in vivo. Nat Immunol 7:265–273

Ginhoux F, Collin MP, Bogunovic M, Abel M, Leboeuf M, Helft J, Ochando J, Kissenpfennig A, Malissen B, Grisotto M, Snoeck H, Randolph G, Merad M (2007) Blood-derived dermal langerin+ dendritic cells survey the skin in the steady state. J Exp Med 204:3133–3146

Gogusev J, Nezelof C (1998) Malignant histiocytosis. Histologic, cytochemical, chromosomal, and molecular data with a nosologic discussion. Hematol Oncol Clin North Am 12:445–463

Guerriero A, Langmuir PB, Spain LM, Scott EW (2000) PU.1 is required for myeloid-derived but not lymphoid-derived dendritic cells. Blood 95:879–885

Hacker C, Kirsch RD, Ju XS, Hieronymus T, Gust TC, Kuhl C, Jorgas T, Kurz SM, Rose-John S, Yokota Y, Zenke M (2003) Transcriptional profiling identifies Id2 function in dendritic cell development. Nat Immunol 4:380–386

Hamilton JA (2008) Colony-stimulating factors in inflammation and autoimmunity. Nat Rev Immunol 8:533–544

Hart DN (1997) Dendritic cells: Unique leukocyte populations which control the primary immune response. Blood 90:3245–3287

Hart DN, Fabre JW (1981) Demonstration and characterization of Ia-positive dendritic cells in the interstitial connective tissues of rat heart and other tissues, but not brain. J Exp Med 154:347–361

Hawiger D, Inaba K, Dorsett Y, Guo M, Mahnke K, Rivera M, Ravetch JV, Steinman RM, Nussenzweig MC (2001) Dendritic cells induce peripheral T cell unresponsiveness under steady state conditions in vivo. J Exp Med 194:769–779

Hemmi H, Yoshino M, Yamazaki H, Naito M, Iyoda T, Omatsu Y, Shimoyama S, Letterio JJ, Nakabayashi T, Tagaya H, Yamane T, Ogawa M, Nishikawa S, Ryoke K, Inaba K, Hayashi S, Kunisada T (2001) Skin antigens in the steady state are trafficked to regional lymph nodes by transforming growth factor-beta1-dependent cells. Int Immunol 13:695–704

Herling M, Jones D (2007) CD4+/CD56+ hematodermic tumor: The features of an evolving entity and its relationship to dendritic cells. Am J Clin Pathol 127:687–700

Hintzen G, Ohl L, del Rio ML, Rodriguez-Barbosa JI, Pabst O, Kocks JR, Krege J, Hardtke S, Forster R (2006) Induction of tolerance to innocuous inhaled antigen relies on a CCR7-dependent dendritic cell-mediated antigen transport to the bronchial lymph node. J Immunol 177:7346–7354

Hock H, Orkin SH (2006) Zinc-finger transcription factor Gfi-1: Versatile regulator of lymphocytes, neutrophils and hematopoietic stem cells. Curr Opin Hematol 13:1–6

Holt PG, Haining S, Nelson DJ, Sedgwick JD (1994) Origin and steady-state turnover of class II MHC-bearing dendritic cells in the epithelium of the conducting airways. J Immunol 153:256–261

Hori S, Nomura T, Sakaguchi S (2003) Control of regulatory T cell development by the transcription factor Foxp3. Science 299:1057–1061

Horowitz MM, Gale RP, Sondel PM, Goldman JM, Kersey J, Kolb HJ, Rimm AA, Ringden O, Rozman C, Speck B et al (1990) Graft-versus-leukemia reactions after bone marrow transplantation. Blood 75:555–562

Huang FP, Platt N, Wykes M, Major JR, Powell TJ, Jenkins CD, MacPherson GG (2000) A discrete subpopulation of dendritic cells transports apoptotic intestinal epithelial cells to T cell areas of mesenteric lymph nodes. J Exp Med 191:435–444

Ichikawa E, Hida S, Omatsu Y, Shimoyama S, Takahara K, Miyagawa S, Inaba K, Taki S (2004) Defective development of splenic and epidermal CD4+ dendritic cells in mice deficient for IFN regulatory factor-2. Proc Natl Acad Sci USA 101:3909–3914

Iijima N, Linehan MM, Saeland S, Iwasaki A (2007) Vaginal epithelial dendritic cells renew from bone marrow precursors. Proc Natl Acad Sci USA 104:19061–19066

Inaba K, Inaba M, Romani N, Aya H, Deguchi M, Ikehara S, Muramatsu S, Steinman RM (1992a) Generation of large numbers of dendritic cells from mouse bone marrow cultures supplemented with granulocyte/macrophage colony-stimulating factor. J Exp Med 176:1693–1702

Inaba K, Steinman RM, Pack MW, Aya H, Inaba M, Sudo T, Wolpe S, Schuler G (1992b) Identification of proliferating dendritic cell precursors in mouse blood. J Exp Med 175:1157–1167

Ishikawa F, Niiro H, Iino T, Yoshida S, Saito N, Onohara S, Miyamoto T, Minagawa H, Fujii S, Shultz LD, Harada M, Akashi K (2007) The developmental program of human dendritic cells is operated independently of conventional myeloid and lymphoid pathways. Blood 110:3591–3660

Iwakoshi NN, Pypaert M, Glimcher LH (2007) The transcription factor XBP-1 is essential for the development and survival of dendritic cells. J Exp Med 204:2267–2275

Iwasaki A (2007) Mucosal dendritic cells. Annu Rev Immunol 25:381–418

Iwasaki H, Akashi K (2007) Myeloid lineage commitment from the hematopoietic stem cell. Immunity 26:726–740

Jacob MC, Chaperot L, Mossuz P, Feuillard J, Valensi F, Leroux D, Bene MC, Bensa JC, Briere F, Plumas J (2003) CD4+ CD56+ lineage negative malignancies: A new entity developed from malignant early plasmacytoid dendritic cells. Haematologica 88:941–955

Johansson-Lindbom B, Svensson M, Pabst O, Palmqvist C, Marquez G, Forster R, Agace WW (2005) Functional specialization of gut CD103+ dendritic cells in the regulation of tissue-selective T cell homing. J Exp Med 202:1063–1073

Johnson BD, Truitt RL (1995) Delayed infusion of immunocompetent donor cells after bone marrow transplantation breaks graft-host tolerance allows for persistent antileukemic reactivity without severe graft-versus-host disease. Blood 85:3302–3312

Jung S, Unutmaz D, Wong P, Sano G, De los Santos K, Sparwasser T, Wu S, Vuthoori S, Ko K, Zavala F, Pamer EG, Littman DR, Lang RA (2002) In vivo depletion of CD11c(+) dendritic cells abrogates priming of CD8(+) T cells by exogenous cell-associated antigens. Immunity 17:211–220

Kabashima K, Banks TA, Ansel KM, Lu TT, Ware CF, Cyster JG (2005) Intrinsic lymphotoxin-beta receptor requirement for homeostasis of lymphoid tissue dendritic cells. Immunity 22:439–450

Kamath AT, Pooley J, O'Keeffe MA, Vremec D, Zhan Y, Lew AM, D'Amico A, Wu L, Tough DF, Shortman K (2000) The development, maturation, and turnover rate of mouse spleen dendritic cell populations. J Immunol 165:6762–6770

Kamath AT, Henri S, Battye F, Tough DF, Shortman K (2002) Developmental kinetics and lifespan of dendritic cells in mouse lymphoid organs. Blood 100:1734–1741

Kanitakis J, Petruzzo P, Dubernard JM (2004) Turnover of epidermal Langerhans' cells. N Engl J Med 351:2661–2662

Kaplan DH, Li MO, Jenison MC, Shlomchik WD, Flavell RA, Shlomchik MJ (2007) Autocrine/paracrine TGFbeta1 is required for the development of epidermal Langerhans cells. J Exp Med 204:2545–2552

Karsunky H, Merad M, Cozzio A, Weissman IL, Manz MG (2003) Flt3 ligand regulates dendritic cell development from Flt3+ lymphoid and myeloid-committed progenitors to Flt3+ dendritic cells in vivo. J Exp Med 198:305–313

Karsunky H, Merad M, Mende I, Manz MG, Engleman EG, Weissman IL (2005) Developmental origin of interferon-alpha-producing dendritic cells from hematopoietic precursors. Exp Hematol 33:173–181

Karsunky H, Inlay MA, Serwold T, Bhattacharya D, Weissman IL (2008) Flk2+ common lymphoid progenitors possess equivalent differentiation potential for the B and T lineages. Blood 111:5562–5570

Kelly RH (1970) Localization of afferent lymph cells within the draining node during a primary immune response. Nature 227:510–513

Kissenpfennig A, Henri S, Dubois B, Laplace-Builhe C, Perrin P, Romani N, Tripp CH, Douillard P, Leserman L, Kaiserlian D, Saeland S, Davoust J, Malissen B (2005) Dynamics and function of Langerhans cells in vivo: Dermal dendritic cells colonize lymph node areas distinct from slower migrating Langerhans cells. Immunity 22:643–654

Klemsz MJ, McKercher SR, Celada A, Van Beveren C, Maki RA (1990) The macrophage and B cell-specific transcription factor PU.1 is related to the ets oncogene [see comments]. Cell 61:113–124

Kobayashi T, Walsh PT, Walsh MC, Speirs KM, Chiffoleau E, King CG, Hancock WW, Caamano JH, Hunter CA, Scott P, Turka LA, Choi Y (2003) TRAF6 is a critical factor for dendritic cell maturation and development. Immunity 19:353–363

Kolb HJ, Mittermuller J, Clemm C, Holler E, Ledderose G, Brehm G, Heim M, Wilmanns W (1990) Donor leukocyte transfusions for treatment of recurrent chronic myelogenous leukemia in marrow transplant patients. Blood 76:2462–2465

Kolb HJ, Schattenberg A, Goldman JM, Hertenstein B, Jacobsen N, Arcese W, Ljungman P, Ferrant A, Verdonck L, Niederwieser D, van Rhee F, Mittermueller J, de Witte T, Holler E, Ansari H (1995) Graft-versus-leukemia effect of donor lymphocyte transfusions in marrow grafted patients. Blood 86:2041–2050

Kolb HJ, Schmid C, Barrett AJ, Schendel DJ (2004) Graft-versus-leukemia reactions in allogeneic chimeras. Blood 103:767–776

Kondo M, Weissman IL, Akashi K (1997) Identification of clonogenic common lymphoid progenitors in mouse bone marrow. Cell 91:661–672

Kondo M, Wagers AJ, Manz MG, Prohaska SS, Scherer DC, Beilhack GF, Shizuru JA, Weissman IL (2003) Biology of hematopoietic stem cells and progenitors: Implications for clinical application. Annu Rev Immunol 21:759–806

Kretschmer K, Apostolou I, Hawiger D, Khazaie K, Nussenzweig MC, von Boehmer H (2005) Inducing and expanding regulatory T cell populations by foreign antigen. Nat Immunol 6:1219–1227

Kumamoto T, Shalhevet D, Matsue H, Mummert ME, Ward BR, Jester JV, Takashima A (2003) Hair follicles serve as local reservoirs of skin mast cell precursors. Blood 102:1654–1660

Laiosa CV, Stadtfeld M, Graf T (2006) Determinants of lymphoid-myeloid lineage diversification. Annu Rev Immunol 24:705–738

Landsman L, Varol C, Jung S (2007) Distinct differentiation potential of blood monocyte subsets in the lung. J Immunol 178:2000–2007

Laouar Y, Welte T, Fu XY, Flavell RA (2003) STAT3 is required for Flt3L-dependent dendritic cell differentiation. Immunity 19:903–912

Leszczynski D, Renkonen R, Hayry P (1985a) Localization and turnover rate of rat renal 'dendritic' cells. Scand J Immunol 21:355–360

Leszczynski D, Renkonen R, Hayry P (1985b) Turnover of dendritic cells in rat heart. Scand J Immunol 22:351–355

Liu YJ (2005) IPC: Professional type 1 interferon-producing cells and plasmacytoid dendritic cell precursors. Annu Rev Immunol 23:275–306

Liu K, Waskow C, Liu X, Yao K, Hoh J, Nussenzweig M (2007) Origin of dendritic cells in peripheral lymphoid organs of mice. Nat Immunol 8:578–583

Luckey CJ, Bhattacharya D, Goldrath AW, Weissman IL, Benoist C, Mathis D (2006) Memory T and memory B cells share a transcriptional program of self-renewal with long-term hematopoietic stem cells. Proc Natl Acad Sci USA 103:3304–3309

Lutz MB (2004) IL-3 in dendritic cell development and function: A comparison with GM-CSF and IL-4. Immunobiology 209:79–87

Lutz MB, Schnare M, Menges M, Rossner S, Rollinghoff M, Schuler G, Gessner A (2002) Differential functions of IL-4 receptor types I and II for dendritic cell maturation and IL-12 production and their dependency on GM-CSF. J Immunol 169:3574–3580

Lyman SD, Jacobsen SE (1998) c-kit ligand and Flt3 ligand: stem/progenitor cell factors with overlapping yet distinct activities. Blood 91:1101–1134

Lyman SD, James L, Escobar S, Downey H, de Vries P, Brasel K, Stocking K, Beckmann MP, Copeland NG, Cleveland LS et al (1995) Identification of soluble and membrane-bound isoforms of the murine flt3 ligand generated by alternative splicing of mRNAs. Oncogene 10:149–157

Macdonald KP, Rowe V, Bofinger HM, Thomas R, Sasmono T, Hume DA, Hill GR (2005) The colony-stimulating factor 1 receptor is expressed on dendritic cells during differentiation and regulates their expansion. J Immunol 175:1399–13405

Mackinnon S, Papadopoulos EB, Carabasi MH, Reich L, Collins NH, Boulad F, Castro-Malaspina H, Childs BH, Gillio AP, Kernan NA et al (1995) Adoptive immunotherapy evaluating escalating doses of donor leukocytes for relapse of chronic myeloid leukemia after bone marrow transplantation: Separation of graft-versus-leukemia responses from graft-versus-host disease. Blood 86:1261–1268

Manfra DJ, Chen SC, Jensen KK, Fine JS, Wiekowski MT, Lira SA (2003) Conditional expression of murine Flt3 ligand leads to expansion of multiple dendritic cell subsets in peripheral blood and tissues of transgenic mice. J Immunol 170:2843–2852

Manz MG (2007) Human-hemato-lymphoid-system mice: Opportunities and challenges. Immunity 26:537–541

Manz MG, Traver D, Miyamoto T, Weissman IL, Akashi K (2001) Dendritic cell potentials of early lymphoid and myeloid progenitors. Blood 97:3333–3341

Manz MG, Miyamoto T, Akashi K, Weissman IL (2002) Prospective isolation of human clonogenic common myeloid progenitors. Proc Natl Acad Sci USA 99:11872–11877

Mapara MY, Kim YM, Wang SP, Bronson R, Sachs DH, Sykes M (2002) Donor lymphocyte infusions mediate superior graft-versus-leukemia effects in mixed compared to fully allogeneic chimeras: A critical role for host antigen-presenting cells. Blood 100:1903–1909

Marafioti T, Paterson JC, Ballabio E, Reichard KK, Tedoldi S, Hollowood K, Dictor M, Hansmann ML, Pileri SA, Dyer MJ, Sozzani S, Dikic I, Shaw AS, Petrella T, Stein H, Isaacson PG, Facchetti F, Mason DY (2008) Novel markers of normal and neoplastic human plasmacytoid dendritic cells. Blood 111:3778–3792

Maraskovsky E, Brasel K, Teepe M, Roux ER, Lyman SD, Shortman K, McKenna HJ (1996) Dramatic increase in the numbers of functionally mature dendritic cells in Flt3 ligand-treated mice: Multiple dendritic cell subpopulations identified. J Exp Med 184:1953–1962

Maraskovsky E, Daro E, Roux E, Teepe M, Maliszewski CR, Hoek J, Caron D, Lebsack ME, McKenna HJ (2000) In vivo generation of human dendritic cell subsets by Flt3 ligand. Blood 96:878–884

Markowicz S, Engleman EG (1990) Granulocyte-macrophage colony-stimulating factor promotes differentiation and survival of human peripheral blood dendritic cells in vitro. J Clin Invest 85:955–961

Massberg S, Schaerli P, Knezevic-Maramica I, Kollnberger M, Tubo N, Moseman EA, Huff IV, Junt T, Wagers AJ, Mazo IB, von Andrian UH (2007) Immunosurveillance by hematopoietic progenitor cells trafficking through blood, lymph, and peripheral tissues. Cell 131:994–1008

Matte CC, Liu J, Cormier J, Anderson BE, Athanasiadis I, Jain D, McNiff J, Shlomchik WD (2004) Donor APCs are required for maximal GVHD but not for GVL. Nat Med 10:987–992

McKenna HJ, Stocking KL, Miller RE, Brasel K, De Smedt T, Maraskovsky E, Maliszewski CR, Lynch DH, Smith J, Pulendran B, Roux ER, Teepe M, Lyman SD, Peschon JJ (2000) Mice lacking flt3 ligand have deficient hematopoiesis affecting hematopoietic progenitor cells, dendritic cells, and natural killer cells. Blood 95:3489–3497

McKercher SR, Torbett BE, Anderson KL, Henkel GW, Vestal DJ, Baribault H, Klemsz M, Feeney AJ, Wu GE, Paige CJ, Maki RA (1996) Targeted disruption of the PU.1 gene results in multiple hematopoietic abnormalities. Embo J 15:5647–5658

Mende I, Karsunky H, Weissman IL, Engleman EG, Merad M (2006) Flk2+ myeloid progenitors are the main source of Langerhans cells. Blood 107:1383–1390

Merad M, Manz MG, Karsunky H, Wagers A, Peters W, Charo I, Weissman IL, Cyster JG, Engleman EG (2002a) Langerhans cells renew in the skin throughout life under steady-state conditions. Nat Immunol 3:1135–1141

Merad M, Sugie T, Engleman EG, Fong L (2002b) In vivo manipulation of dendritic cells to induce therapeutic immunity. Blood 99:1676–1682

Merad M, Hoffmann P, Ranheim E, Slaymaker S, Manz MG, Lira SA, Charo I, Cook DN, Weissman IL, Strober S, Engleman EG (2004) Depletion of host Langerhans cells before transplantation of donor alloreactive T cells prevents skin graft-versus-host disease. Nat Med 10:510–517

Miller G, Pillarisetty VG, Shah AB, Lahrs S, Xing Z, DeMatteo RP (2002) Endogenous granulocyte-macrophage colony-stimulating factor overexpression in vivo results in the long-term recruitment of a distinct dendritic cell population with enhanced immunostimulatory function. J Immunol 169:2875–2885

Miller G, Pillarisetty VG, Shah AB, Lahrs S, DeMatteo RP (2003) Murine Flt3 ligand expands distinct dendritic cells with both tolerogenic and immunogenic properties. J Immunol 170:3554–3564

Miyauchi S, Hashimoto K (1987) Epidermal Langerhans cells undergo mitosis during the early recovery phase after ultraviolet-B irradiation. J Invest Dermatol 88:703–708

Mizuki M, Schwable J, Steur C, Choudhary C, Agrawal S, Sargin B, Steffen B, Matsumura I, Kanakura Y, Bohmer FD, Muller-Tidow C, Berdel WE, Serve H (2003) Suppression of myeloid transcription factors and induction of STAT response genes by AML-specific Flt3 mutations. Blood 101:3164–3173

Nagai Y, Garrett KP, Ohta S, Bahrun U, Kouro T, Akira S, Takatsu K, Kincade PW (2006) Toll-like receptors on hematopoietic progenitor cells stimulate innate immune system replenishment. Immunity 24:801–812

Naik SH, Metcalf D, van Nieuwenhuijze A, Wicks I, Wu L, O'Keeffe M, Shortman K (2006) Intrasplenic steady-state dendritic cell precursors that are distinct from monocytes. Nat Immunol 7: 663-671

Naik SH, Sathe P, Park HY, Metcalf D, Proietto AI, Dakic A, Carotta S, O'Keeffe M, Bahlo M, Papenfuss A, Kwak JY, Wu L, Shortman K (2007) Development of plasmacytoid and conventional dendritic cell subtypes from single precursor cells derived in vitro and in vivo. Nat Immunol 8:1217–1226

Nezelof C, Basset F (2004) An hypothesis Langerhans cell histiocytosis: the failure of the immune system to switch from an innate to an adaptive mode. Pediatr Blood Cancer 42:398–400

Nezelof C, Basset F, Rousseau MF (1973) Histiocytosis X histogenetic arguments for a Langerhans cell origin. Biomedicine 18:365–371

Ng AP, Lade S, Rutherford T, McCormack C, Prince HM, Westerman DA (2006) Primary cutaneous CD4+/CD56+ hematodermic neoplasm (blastic NK-cell lymphoma): A report of five cases. Haematologica 91:143–144

Norbury CC, Malide D, Gibbs JS, Bennink JR, Yewdell JW (2002) Visualizing priming of virus-specific CD8+ T cells by infected dendritic cells in vivo. Nat Immunol 3:265–271

Nutt SL, Heavey B, Rolink AG, Busslinger M (1999) Commitment to the B-lymphoid lineage depends on the transcription factor Pax5 [comments]. Nature 401:556–562

O'Keeffe M, Hochrein H, Vremec D, Caminschi I, Miller JL, Anders EM, Wu L, Lahoud MH, Henri S, Scott B, Hertzog P, Tatarczuch L, Shortman K (2002a) Mouse plasmacytoid cells: Long-lived cells, heterogeneous in surface phenotype and function, that differentiate into CD8(+) dendritic cells only after microbial stimulus. J Exp Med 196:1307–1319

O'Keeffe M, Hochrein H, Vremec D, Pooley J, Evans R, Woulfe S, Shortman K (2002b) Effects of administration of progenipoietin 1, Flt-3 ligand, granulocyte colony-stimulating factor, and pegylated granulocyte-macrophage colony-stimulating factor on dendritic cell subsets in mice. Blood 99:2122–2130

Onai N, Manz MG (2008) The STATs on dendritic cell development. Immunity 28:490–492

Onai N, Obata-Onai A, Tussiwand R, Lanzavecchia A, Manz MG (2006) Activation of the Flt3 signal transduction cascade rescues and enhances type I interferon-producing and dendritic cell development. J Exp Med 203:227–238

Onai N, Obata-Onai A, Schmid MA, Ohteki T, Jarrossay D, Manz MG (2007a) Identification of clonogenic common Flt3+M-CSFR+ plasmacytoid and conventional dendritic cell progenitors in mouse bone marrow. Nat Immunol 8:1207–1216

Onai N, Obata-Onai A, Schmid MA, Manz MG (2007) Flt3 in regulation of type-I interferon producing and dendritic cell development. Ann N Y Acad Sci 1106: 253-261

Orkin SH, Zon LI (2008) Hematopoiesis: An evolving paradigm for stem cell biology. Cell 132:631–644

Pasparakis M, Alexopoulou L, Episkopou V, Kollias G (1996) Immune and inflammatory responses in TNF alpha-deficient mice: a critical requirement for TNF alpha in the formation of primary B cell follicles, follicular dendritic cell networks and germinal centers, and in the maturation of the humoral immune response. J Exp Med 184:1397–1411

Peggs KS, Thomson K, Hart DP, Geary J, Morris EC, Yong K, Goldstone AH, Linch DC, Mackinnon S (2004) Dose-escalated donor lymphocyte infusions following reduced intensity transplantation: Toxicity, chimerism, and disease responses. Blood 103:1548–1556

Perreault C, Pelletier M, Belanger R, Boileau J, Bonny Y, David M, Gyger M, Landry D, Montplaisir S (1985) Persistence of host Langerhans cells following allogeneic bone marrow transplantation: Possible relationship with acute graft-versus-host disease. Br J Haematol 60:253–260

Pileri SA, Grogan TM, Harris NL, Banks P, Campo E, Chan JK, Favera RD, Delsol G, De Wolf-Peeters C, Falini B, Gascoyne RD, Gaulard P, Gatter KC, Isaacson PG, Jaffe ES, Kluin P, Knowles DM, Mason DY, Mori S, Muller-Hermelink HK, Piris MA, Ralfkiaer E, Stein H, Su IJ, Warnke RA, Weiss LM (2002) Tumours of histiocytes and accessory dendritic cells: An immunohistochemical approach to classification from the International Lymphoma Study Group based on 61 cases. Histopathology 41:1–29

Poulin LF, Henri S, de Bovis B, Devilard E, Kissenpfennig A, Malissen B (2007) The dermis contains langerin+ dendritic cells that develop and function independently of epidermal Langerhans cells. J Exp Med 204:3119–3131

Pugh CW, MacPherson GG, Steer HW (1983) Characterization of nonlymphoid cells derived from rat peripheral lymph. J Exp Med 157:1758–1779

Pulendran B, Lingappa J, Kennedy MK, Smith J, Teepe M, Rudensky A, Maliszewski CR, Maraskovsky E (1997) Developmental pathways of dendritic cells in vivo: Distinct function, phenotype, and localization of dendritic cell subsets in FLT3 ligand-treated mice. J Immunol 159:2222–2231

Pulendran B, Banchereau J, Burkeholder S, Kraus E, Guinet E, Chalouni C, Caron D, Maliszewski C, Davoust J, Fay J, Palucka K (2000) Flt3-ligand and granulocyte colony-stimulating factor mobilize distinct human dendritic cell subsets in vivo. J Immunol 165:566–572

Randolph GJ, Inaba K, Robbiani DF, Steinman RM, Muller WA (1999) Differentiation of phagocytic monocytes into lymph node dendritic cells in vivo. Immunity 11:753–761

Randolph GJ, Ochando J, Partida-Sanchez S (2008) Migration of dendritic cell subsets and their precursors. Annu Rev Immunol 26:293–316

Rathinam C, Geffers R, Yucel R, Buer J, Welte K, Moroy T, Klein C (2005) The transcriptional repressor Gfi1 controls STAT3-dependent dendritic cell development and function. Immunity 22:717–728

Reddy P, Maeda Y, Liu C, Krijanovski OI, Korngold R, Ferrara JL (2005) A crucial role for antigen-presenting cells and alloantigen expression in graft-versus-leukemia responses. Nat Med 11:1244–1249

Reimer P, Rudiger T, Kraemer D, Kunzmann V, Weissinger F, Zettl A, Konrad Muller-Hermelink H, Wilhelm M (2003) What is CD4+CD56+ malignancy and how should it be treated? Bone Marrow Transplant 32:637–646

Rolland A, Guyon L, Gill M, Cai YH, Banchereau J, McClain K, Palucka AK (2005) Increased blood myeloid dendritic cells and dendritic cell-poietins in Langerhans cell histiocytosis. J Immunol 174:3067–3071

Rousselet MC, Francois S, Croue A, Maigre M, Saint-Andre JP, Ifrah N (1994) A lymph node interdigitating reticulum cell sarcoma. Arch Pathol Lab Med 118:183–188

Sallusto F, Lanzavecchia A (1994) Efficient presentation of soluble antigen by cultured human dendritic cells is maintained by granulocyte/macrophage colony-stimulating factor plus interleukin 4 and downregulated by tumor necrosis factor alpha. J Exp Med 179:1109–1118

Santiago-Schwarz F, Divaris N, Kay C, Carsons SE (1993) Mechanisms of tumor necrosis factor-granulocyte-macrophage colony-stimulating factor-induced dendritic cell development. Blood 82:3019–3028

Savell VH Jr, Sherman T, Scheuermann RH, Siddiqui AM, Margraf LR (1998) Bcl-2 expression in Langerhans' cell histiocytosis. Pediatr Dev Pathol 1:210–215

Scheinecker C, McHugh R, Shevach EM, Germain RN (2002) Constitutive presentation of a natural tissue autoantigen exclusively by dendritic cells in the draining lymph node. J Exp Med 196:1079–1090

Schiavoni G, Mattei F, Sestili P, Borghi P, Venditti M, Morse HC 3rd, Belardelli F, Gabriele L (2002) ICSBP is essential for the development of mouse type I interferon-producing cells and for the generation and activation of CD8alpha(+) dendritic cells. J Exp Med 196:1415–1425

Schiavoni G, Mattei F, Borghi P, Sestili P, Venditti M, Morse HC 3rd, Belardelli F, Gabriele L (2004) ICSBP is critically involved in the normal development and trafficking of Langerhans cells and dermal dendritic cells. Blood 103:2221–2228

Schotte R, Nagasawa M, Weijer K, Spits H, Blom B (2004) The ETS transcription factor Spi-B is required for human plasmacytoid dendritic cell development. J Exp Med 200:1503–1509

Schouten B, Egeler RM, Leenen PJ, Taminiau AH, van den Broek LJ, Hogendoorn PC (2002) Expression of cell cycle-related gene products in Langerhans cell histiocytosis. J Pediatr Hematol Oncol 24:727–732

Scott EW, Simon MC, Anastasi J, Singh H (1994) Requirement of transcription factor PU.1 in the development of multiple hematopoietic lineages. Science 265:1573–1577

Serbina NV, Salazar-Mather TP, Biron CA, Kuziel WA, Pamer EG (2003) TNF/iNOS-producing dendritic cells mediate innate immune defense against bacterial infection. Immunity 19:59–70

Shaw SG, Maung AA, Steptoe RJ, Thomson AW, Vujanovic NL (1998) Expansion of functional NK cells in multiple tissue compartments of mice treated with Flt3-ligand: Implications for anti-cancer and anti-viral therapy. J Immunol 161:2817–2824

Shigematsu H, Reizis B, Iwasaki H, Mizuno S, Hu D, Traver D, Leder P, Sakaguchi N, Akashi K (2004) Plasmacytoid dendritic cells activate lymphoid-specific genetic programs irrespective of their cellular origin. Immunity 21:43–53

Shlomchik WD, Couzens MS, Tang CB, McNiff J, Robert ME, Liu J, Shlomchik MJ, Emerson SG (1999) Prevention of graft versus host disease by inactivation of host antigen-presenting cells. Science 285:412–415

Shortman K, Liu YJ (2002) Mouse and human dendritic cell subtypes. Nat Rev Immunol 2002:151–161

Simons JW, Mikhak B, Chang JF, DeMarzo AM, Carducci MA, Lim M, Weber CE, Baccala AA, Goemann MA, Clift SM, Ando DG, Levitsky HI, Cohen LK, Sanda MG, Mulligan RC, Partin AW, Carter HB, Piantadosi S, Marshall FF, Nelson WG (1999) Induction of immunity to prostate cancer antigens: Results of a clinical trial of vaccination with irradiated autologous prostate tumor cells engineered to secrete granulocyte-macrophage colony-stimulating factor using ex vivo gene transfer. Cancer Res 59:5160–5168

Small D (2006) FLT3 mutations: Biology and treatment. Hematol Am Soc Hematol Educ Prog 1: 178–184

Spits H, Couwenberg F, Bakker AQ, Weijer K, Uittenbogaart CH (2000) Id2 and Id3 inhibit development of CD34(+) stem cells into predendritic cell (pre-DC)2 but not into pre-DC1. Evidence for a lymphoid origin of pre-DC2. J Exp Med 192:1775–1784

Steiner QG, Otten LA, Hicks MJ, Kaya G, Grosjean F, Saeuberli E, Lavanchy C, Beermann F, McClain KL, Acha-Orbea H (2008) In vivo transformation of mouse conventional CD8alpha+ dendritic cells leads to progressive multisystem histiocytosis. Blood 111:2073–2082

Steinman RM, Hawiger D, Nussenzweig MC (2003) Tolerogenic dendritic cells. Annu Rev Immunol 21:685–711

Strobl H, Riedl E, Scheinecker C, Bello-Fernandez C, Pickl WF, Rappersberger K, Majdic O, Knapp W (1996) TGF-beta 1 promotes in vitro development of dendritic cells from CD34+ hemopoietic progenitors. J Immunol 157:1499–1507

Sun CM, Hall JA, Blank RB, Bouladoux N, Oukka M, Mora JR, Belkaid Y (2007) Small intestine lamina propria dendritic cells promote de novo generation of Foxp3 T reg cells via retinoic acid. J Exp Med 204:1775–1785

Sung SS, Fu SM, Rose CE Jr, Gaskin F, Ju ST, Beaty SR (2006) A major lung CD103 (alphaE)-beta7 integrin-positive epithelial dendritic cell population expressing Langerin and tight junction proteins. J Immunol 176:2161–2172

Suzuki S, Honma K, Matsuyama T, Suzuki K, Toriyama K, Akitoyo I, Yamamoto K, Suematsu T, Nakamura M, Yui K, Kumatori A (2004) Critical roles of interferon regulatory factor 4 in CD11bhighCD8alpha- dendritic cell development. Proc Natl Acad Sci USA 101:8981–8986

Takahashi K, Naito M, Shultz LD, Hayashi S, Nishikawa S (1993) Differentiation of dendritic cell populations in macrophage colony-stimulating factor-deficient mice homozygous for the osteopetrosis (op) mutation. J Leukoc Biol 53:19–28

Tamura T, Tailor P, Yamaoka K, Kong HJ, Tsujimura H, O'Shea JJ, Singh H, Ozato K (2005) IFN regulatory factor-4 and -8 govern dendritic cell subset development and their functional diversity. J Immunol 174:2573–2581

Terme M, Borg C, Guilhot F, Masurier C, Flament C, Wagner EF, Caillat-Zucman S, Bernheim A, Turhan AG, Caignard A, Zitvogel L (2005) BCR/ABL promotes dendritic cell-mediated natural killer cell activation. Cancer Res 65:6409–6417

Teshima T, Ordemann R, Reddy P, Gagin S, Liu C, Cooke KR, Ferrara J (2002) Acute graft-versus-host disease docs not require alloantigen expression on host epithelium. Nat Med 6:575–581

Traver D, Akashi K, Manz M, Merad M, Miyamoto T, Engleman EG, Weissman IL (2000) Development of CD8alpha-positive dendritic cells from a common myeloid progenitor. Science 290:2152–2154

Tsujimura H, Tamura T, Ozato K (2003) Cutting edge: IFN consensus sequence binding protein/IFN regulatory factor 8 drives the development of type I IFN-producing plasmacytoid dendritic cells. J Immunol 170:1131–1135

Tussiwand R, Onai N, Mazzucchelli L, Manz MG (2005) Inhibition of natural type I IFN-producing and dendritic cell development by a small molecule receptor tyrosine kinase inhibitor with Flt3 affinity. J Immunol 175:3674–3680

Valladeau J, Saeland S (2005) Cutaneous dendritic cells. Semin Immunol 17:273–283

Valladeau J, Duvert-Frances V, Pin JJ, Dezutter-Dambuyant C, Vincent C, Massacrier C, Vincent J, Yoneda K, Banchereau J, Caux C, Davoust J, Saeland S (1999) The monoclonal antibody DCGM4 recognizes langerin, a protein specific of Langerhans cells, and is rapidly internalized from the cell surface. Eur J Immunol 29:2695–2704

Varol C, Landsman L, Fogg DK, Greenshtein L, Gildor B, Margalit R, Kalchenko V, Geissmann F, Jung S (2007) Monocytes give rise to mucosal, but not splenic, conventional dendritic cells. J Exp Med 204:171–280

Vishwanath M, Nishibu A, Saeland S, Ward BR, Mizumoto N, Ploegh HL, Boes M, Takashima A (2006) Development of intravital intermittent confocal imaging system for studying Langerhans cell turnover. J Invest Dermatol 126:2452–2457

Vremec D, Lieschke GJ, Dunn AR, Robb L, Metcalf D, Shortman K (1997) The influence of granulocyte/macrophage colony-stimulating factor on dendritic cell levels in mouse lymphoid organs. Eur J Immunol 27:40–44

Vyth-Dreese FA, Dellemijn TA, van Oostveen JW, Feltkamp CA, Hekman A (1995) Functional expression of adhesion receptors and costimulatory molecules by fresh and immortalized B-cell non-Hodgkin's lymphoma cells. Blood 85:2802–2812
Wang L, Bursch LS, Kissenpfennig A, Malissen B, Jameson SC, Hogquist KA (2008) Langerin expressing cells promote skin immune responses under defined conditions. J Immunol 180:4722–4727
Waskow C, Liu K, Darrasse-Jeze G, Guermonprez P, Ginhoux F, Merad M, Shengelia T, Yao K, Nussenzweig M (2008) The receptor tyrosine kinase Flt3 is required for dendritic cell development in peripheral lymphoid tissues. Nat Immunol 9:676–683
Weiden PL, Flournoy N, Thomas ED, Prentice R, Fefer A, Buckner CD, Storb R (1979) Antileukemic effect of graft-versus-host disease in human recipients of allogeneic-marrow grafts. N Engl J Med 300:1068–1073
Weiden PL, Sullivan KM, Flournoy N, Storb R, Thomas ED (1981) Antileukemic effect of chronic graft-versus-host disease: Contribution to improved survival after allogeneic marrow transplantation. N Engl J Med 304:1529–1533
Weih F, Carrasco D, Durham SK, Barton DS, Rizzo CA, Ryseck RP, Lira SA, Bravo R (1995) Multiorgan inflammation and hematopoietic abnormalities in mice with a targeted disruption of RelB, a member of the NF-kappa B/Rel family. Cell 80:331–340
Weiss LM, Azzi R, Dorfman RF, Warnke RA (1986) Sinusoidal hematolymphoid malignancy ("malignant histiocytosis") presenting as atypical sinusoidal proliferation. A study of nine cases. Cancer 58:1681–1688
Whartenby KA, Calabresi PA, McCadden E, Nguyen B, Kardian D, Wang T, Mosse C, Pardoll DM, Small D (2005) Inhibition of FLT3 signaling targets DCs to ameliorate autoimmune disease. Proc Natl Acad Sci USA 102:16741–16746
Willemze R, Jaffe ES, Burg G, Cerroni L, Berti E, Swerdlow SH, Ralfkiaer E, Chimenti S, Diaz-Perez JL, Duncan LM, Grange F, Harris NL, Kempf W, Kerl H, Kurrer M, Knobler R, Pimpinelli N, Sander C, Santucci M, Sterry W, Vermeer MH, Wechsler J, Whittaker S, Meijer CJ (2005) WHO-EORTC classification for cutaneous lymphomas. Blood 105:3768–3785
Willman CL, Busque L, Griffith BB, Favara BE, McClain KL, Duncan MH, Gilliland DG (1994) Langerhans'-cell histiocytosis (histiocytosis X)—a clonal proliferative disease. N Engl J Med 331:154–160
Witmer-Pack MD, Hughes DA, Schuler G, Lawson L, McWilliam A, Inaba K, Steinman RM, Gordon S (1993) Identification of macrophages and dendritic cells in the osteopetrotic (op/op) mouse. J Cell Sci 104(Pt 4):1021–1029
Wong BR, Josien R, Lee SY, Sauter B, Li HL, Steinman RM, Choi Y (1997) TRANCE (tumor necrosis factor [TNF]-related activation-induced cytokine), a new TNF family member predominantly expressed in T cells, is a dendritic cell-specific survival factor. J Exp Med 186:2075–2080
Wright DE, Wagers AJ, Gulati AP, Johnson FL, Weissman IL (2001) Physiological migration of hematopoietic stem and progenitor cells. Science 294:1933–1936
Wu L, Shortman K (2005) Heterogeneity of thymic dendritic cells. Semin Immunol 17:304–312
Wu L, Li CL, Shortman K (1996) Thymic dendritic cell precursors: Relationship to the T lymphocyte lineage and phenotype of the dendritic cell progeny. J Exp Med 184:903–911
Wu L, Nichogiannopoulou A, Shortman K, Georgopoulos K (1997) Cell-autonomous defects in dendritic cell populations of Ikaros mutant mice point to a developmental relationship with the lymphoid lineage. Immunity 7:483–492
Wu L, D'Amico A, Winkel KD, Suter M, Lo D, Shortman K (1998) RelB is essential for the development of myeloid-related CD8alpha– dendritic cells but not of lymphoid-related CD8alpha+ dendritic cells. Immunity 9:839–847
Wu L, D'Amico A, Hochrein H, O'Keeffe M, Shortman K, Lucas K (2001) Development of thymic and splenic dendritic cell populations from different hemopoietic precursors. Blood 98:3376–3382
Xia G, Truitt RL, Johnson BD (2006) Graft-versus-leukemia and graft-versus-host reactions after donor lymphocyte infusion are initiated by host-type antigen-presenting cells and regulated by regulatory T cells in early and long-term chimeras. Biol Blood Marrow Transplant 12:397–407

Yoshida H, Hayashi S, Kunisada T, Ogawa M, Nishikawa S, Okamura H, Sudo T, Shultz LD (1990) The murine mutation osteopetrosis is in the coding region of the macrophage colony stimulating factor gene. Nature 345:442–444

Yu RC, Chu C, Buluwela L, Chu AC (1994) Clonal proliferation of Langerhans cells in Langerhans cell histiocytosis. Lancet 343:767–768

Zhang Y, Louboutin JP, Zhu J, Rivera AJ, Emerson SG (2002) Preterminal host dendritic cells in irradiated mice prime CD8+ T cell-mediated acute graft-versus-host disease. JCI 109:1335–1344

Zhang J, Raper A, Sugita N, Hingorani R, Salio M, Palmowski MJ, Cerundolo V, Crocker PR (2006) Characterization of Siglec-H as a novel endocytic receptor expressed on murine plasmacytoid dendritic cell precursors. Blood 107:3600–3608

Wnt in Hematopoietic and Leukemic Stem Cells

Kim-Hien T. Dao and H.M. Jamieson Catriona

Abstract When first discovered, it was clear that Wnts and their downstream targets played a pivotal role in the evolution of cancer in mice. However, recent research has underscored the importance of the Wnt pathway in essential regenerative and repair processes, particularly in the maintenance of self-renewing, blood-forming stem cells. Seminal studies involving a multiplicity of human cancers revealed epigenetic and genetic changes that activated the Wnt pathway, primarily through deregulation of the destruction complex, rendering it a vital potential therapeutic target in cancer. Wnt pathway deregulation is particularly prominent in leukemia, which has provided an important paradigm for dissecting the molecular mechanisms involved in Wnt pathway deregulation in a cell type and context-specific manner. Perhaps most importantly, the discovery of leukemia stem cells provided the impetus for investigating the role of the Wnt pathway in fueling their characteristically unbridled self-renewal capacity. In this review, we examine the contribution of the Wnt pathway to normal hematopoietic stem cell (HSC) versus leukemic stem cell (LSC) propagation in order to establish whether a therapeutic index is likely to exist that will warrant the development of Wnt inhibitor clinical trials in cancer.

Wnt Pathway

The Wnt family of proteins is comprised of lipid-modified proteins that are excreted in the microenvironment to act in an autocrine and paracrine manner to activate signaling pathways involved in multiple developmental events during embryogenesis (Nusse et al. 1991; Reya et al. 2003; Staal and Clevers 2005; Willert et al. 2003; Hope et al. 2004; Jamieson et al. 2004). The wingless (Wg) gene was first identified in wingless *Drosophila* mutants that displayed defective segment polarity and limb

K.-H.T. Dao (✉) and H.M.J. Catriona
University of California at San Diego, School of Medicine, Moores Cancer Center, 3855, Health Sciences Road, Mail Code 0820, La Jolla, CA 92093, USA

M. Kondo (ed.), *Hematopoietic Stem Cell Biology,* Stem Cell Biology and Regenerative Medicine, DOI 10.1007/978-1-60327-347-3_8,

formation (Nusslein-Volhard and Wieschaus 1980; Wu and Cohen 2002). In a completely different area of research, the integration, or INT proto-oncogenes were identified as genes near multiple integration sites of mouse mammary tumor virus (MMTV) that result in their overexpression and oncogenic potential (Staal and Clevers 2005; Willert et al. 2003). Subsequently, it was discovered that the wingless gene is identical to the INT-1 gene (Rijsewijk et al. 1987). As a result, the family of proteins was named the Wnt family of proteins (Nusse et al. 1991), now expanded to include a number of proteins across invertebrate and vertebrate species based on conserved amino acid sequences. The readers are referred to an excellent and comprehensive review of the role of the Wnt pathway in development and disease (Logan and Nusse 2004). So far, 19 Wnt genes have been described in the mouse and human genome (Miller 2002). Classical Wnts are Wnt-1, -3a, -8, and -8b and nonclassical are Wnt-4, -5a, and -11. The former group is termed "classical" because they act through the canonical pathway, resulting in β-catenin nuclear accumulation and transcriptional activation of target genes. The latter group activates Wnt/noncanonical pathways, including Wnt/calcium and Wnt/planar cell polarity.

The Wnt proteins contain a highly conserved cysteine-rich motif that is required for palmitoylation. Removal of palmitoylation, by either enzymatic methods or by site-directed mutagenesis of a conserved cysteine, result in insoluble and biologically inactive forms of purified Wnt proteins (Nusslein-Volhard and Wieschaus 1980). The precise function of palmitoylation is unknown, but this posttranslational modification is integral to Wnt protein secretion and activity. The hydrophobicity of Wnt proteins may localize the proteins to membranes and limit its solubility and paracrine action in the microenvironment. Much has been learned about the concentration or gradient-dependent action of the Wnts in the microenvironment (Cadigan et al. 1998; Zecca et al. 1996), but additional long-range effects mediated by carrier molecules or vesicle transport are also emerging (Greco et al. 2001). The activity of Wnt proteins are further modulated by secreted frizzled-related proteins (sFRPs), Wnt inhibitory factors (WIFs), and heparin-sulfated forms of proteoglycans (HSPGs) (Hoang et al. 1996; Rattner et al. 1997; Hsieh et al. 1999; Lin and Perrimon 2000; Wang et al. 1997). The former two harbor striking similarities to the extracellular domains of Wnt receptors but do not have the transmembrane or intracellular domains to transduce the signal intracellularly. Other roles for these proteins remain to be determined, including the transport of Wnt proteins or protection of Wnt proteins from degradation.

The focus of the remainder of this review will be on the Wnt/canonical pathway, which ultimately leads to β-catenin accumulation. The primary receptors for the Wnt proteins are the Frizzled (Fzd) family of proteins, which are seven-pass transmembrane receptors with an N-terminal region called the cysteine-rich domain (CRD), which is essential for Wnt binding and ligand-dependent activation of Wnt signaling (Bhanot et al. 1996). So far, in humans, ten Fzd receptors have been identified, and in mice, there are nine of these receptors. Activation of the canonical pathway involves Fzd-1, -7, and -8 receptors, and the noncanonical pathway via Fzd-2, -3, -4, and -6 receptors. Complicating the biology of the Fzd receptors is the potential for oligomerization (Kaykas et al. 2004), which has characterized. In addition, there are nonconventional Fzd ligand Hendrickx and Leyns 2008). Heterotrimeric G-protein coupled sig the Fzd receptors has been demonstrated in the canonical and non ways, but its role under physiologic conditions and cell-specific being elucidated (Katanaev et al. 2005; Sheldahl et al. 1999; Slusars

A coreceptor in the low-density lipoprotein-related protein (L required for efficient binding of Wnt to Fzd receptors (Tamai et al. 2000; Wehrli et al. 2000). In vertebrates, this role is fulfilled by LRP-5 or -6 receptor proteins. Another level of modulation of Wnt protein activity is the presence of Dickkopf (Dkk) (Wang et al. 1997; Bhanot et al. 1996) and Wise proteins (Itasaki et al. 2003), which bind to LRPs and modulate the availability of LRP co-receptors by inducing their internalization or by competitive binding with Wnt proteins, respectively. Another family of Wnt receptor is Derailed, a receptor tyrosine kinase of the RYK subfamily (Hendrickx and Leyns 2008). The cellular context and tissue-specific function of Wnt activation involving this receptor remains to be determined, but early work in *Drosophila* indicates a role in axonal guidance in the central nervous system (Yoshikawa et al. 2003).

Wnt-activated signal transduction is dependent on disheveled (Dsh) protein, which is a cytoplasmic protein that binds to a C-terminal Lys-Thr-X-X-X-Trp motif, highly conserved on Fz receptors (Katanaev et al. 2005). Wnt activation results in phosphorylation of Dsh, and this is thought to modulate the affinity of Dsh to the Fz receptors and perhaps to other signaling proteins (Yanagawa et al. 1995). Dsh is necessary for recruitment of Axin to the membrane (Cliffe et al. 2003). The LRP co-receptor also transduces ligand-mediated activation of the Wnt pathway by direct binding with Axin. The Pro-Pro-Pro-Ser/Trp-Pro motif on the cytoplasmic tail of LRP is phosphorylated by Wnt activation, thereby facilitating Axin binding (Tamai et al. 2004). Both Dsh and Axin contain a DIX domain, which permits homodimer and heterodimer formation (Hedgepeth et al. 1999; Hsu et al. 1999; Sakanaka and Williams 1999; Itoh et al. 2000). The exact function of dimerization is unclear, but it likely serves to recruit and modulate other signaling protein complexes into a subcellular region of Wnt activation.

In the absence of Wnt activation, β-catenin resides in a destruction complex with scaffolding proteins APC and Axin (Hart et al. 1998; Kishida et al. 1998), which facilitates interaction of β-catenin with casein kinase Iα (CKIα) and glycogen synthase kinase 3 (GSK3) (Amit et al. 2002; Liu et al. 2002; Yanagawa et al. 2002; Yost et al. 1996). Both are serine/threonine kinases. This "destruction complex" allows for priming phosphorylation of β-catenin on serine 45 by CKIα (Liu et al. 2002; Gao et al. 2002), followed by more efficient phosphorylation of β-catenin on threonine 41, serine 37, and serine 33 by GSK3 (Lagna et al. 1999; Winston et al., 1999). Phosphorylation of β-catenin on these serine residues is recognized by β-transducin repeat-containing homolog protein (β-TRCP), containing E3 ubiquitin ligase, and this facilitates efficient β-catenin ubiquitination and degradation by the proteosome. Thus, under basal conditions, the cellular amount of total β-catenin is maintained at a low level, and nuclear β-catenin, or the nonphosphorylated form, is essentially undetectable. Mutation of β-catenin at the key phosphorylation residues results in decreased degradation of β-catenin and subsequent nuclear accumulation. Cells expressing such mutants are not dependent on extracellular Wnt activation and can activate Wnt-responsive genes constitutively. Somatic mutation of β-catenin has been reported in human cancers, including colorectal and hepatocellular carcinomas (Giles et al. 2003). In addition, at the cell surface, β-catenin associates with α-catenin and cadherin proteins (adherens junctions) and modulates cell-cell adhesion and the actin cytoskeleton (Ozawa et al. 1989). This pool of β-catenin stays membrane-associated and is protected from the destruction complex. Furthermore, this pool of β-catenin may also act as a reservoir of β-catenin that could then be mobilized depending upon cellular needs. This suggests a complex interplay between cell-cell adhesion, cell shape and polarity, cell migration, and the canonical Wnt pathway (Nelson and Nusse 2004). Thus, in the absence of Wnt signaling, free cytosolic β-catenin is maintained at a low level as a result of sequestration in the destruction complex and in the adherens junction complex.

Upon Wnt activation, the destruction complex is disrupted, although the exact mechanism has not been identified. The composition of the destruction complex has not been fully elucidated, and it may be dependent on various cellular contexts and different tissues and their microenvironment. Accumulating evidence supports that the recruitment of Axin to the phosphorylated cytoplasmic tail of the LRP-5 or -6 co-receptor or to the Fz receptor-Dsh complex is a key initiating event (Mao et al. 2001). In addition, Dsh also interacts with GSK3 binding protein (GBP), also known as frequently rearranged in advanced T-cell lymphoma (FRAT), which is another cellular mechanism that recruits proteins in the destruction complex to the membrane upon Wnt activation (Jonkers et al. 1997; Yost et al. 1998). FRAT binds to GSK3β and Dsh, resulting in disruption of the destruction complex, and is a positive regulator of the Wnt pathway. In one report, a kinase-defective GSK mutant was not sufficient for lymphoid enhancing factor 1 (LEF1)-mediated transcriptional activation. The authors conclude that other mechanisms are required to disrupt the destruction complex (Yuan et al. 1999). Other mechanisms that regulate β-catenin phosphorylation include protein phosphatase 2A (PP2A), which removes the phosphate group on β-catenin and possibly other GSK3β substrates (Yang et al. 2003). The accumulation of nonphosphorylated β-catenin and its nuclear translocation (Tolwinski and Wieschaus 2004) result in binding to the LEF1 and T-cell factor (TCF) family of transcription factors and activation of Wnt-responsive genes such as c-myc, cyclin D1, and CD44 (Behrens et al. 1996; Molenaar et al. 1996; van de Wetering et al. 1997).

Depending on the binding partners, LEF/TCF transcription factor complexes can act as transcriptional activators or transcriptional repressors. As a family, the LEF/TCF transcription factors have a β-catenin-binding domain, a context-dependent regulatory domain (CRD), a high mobility group (HMG) DNA-binding domain, a nuclear localization signal domain, and a highly variable and not fully characterized C-terminal tail (reviewed in Arce et al. 2006). The regions of highest amino acid identity are the β-catenin-binding domain (60%) and HMG DNA-binding domain (95%–99%). The CRD is modestly conserved (at best, 15%–20%) and is thought to be responsible for much of the cell type and context-specific diversity of LEF/TCF function. This region promotes binding of corepressors such as Groucho, first discovered in *Drosophila* (Cavallo et al. 1998). In mammalians, the Groucho orthologs are amino-terminal enhancer of split (AES) and transducin-like enhancer of split (TLE). By binding to the CRD domain on LEF/TCF transcription factors, TLE blocks the access and binding of β-catenin to LEF/TCF transcriptional factors. Furthermore, Groucho binds to histone deacetylases, which recruits these enzymes to a promoter region targeted for transcriptional silencing. Histone deacetylases remove negatively charged acetyl groups of histones that neutralize positively charged lysine and arginine residues. Removal of acetyl groups enhances the binding of positively charged histones to negatively charged DNA, and allows for more tightly folded, transcriptionally silent chromatin structure. As a result of Wnt activation and subsequent nuclear β-catenin accumulation, β-catenin binds to LEF/TCF and displaces corepressor binding. Although the exact mechanism is unclear, this results in recruitment of a histone acetylase, cyclic AMP response element-binding protein (CBP) (Hecht et al. 2000; Takemaru and Moon 2000). Addition of acetyl groups on histones reduces its affinity to DNA and "loosens" the chromatin structure and enhances access to other transcriptional activators or chromatin remodeling proteins. Modulation of Wnt target gene expression by β-catenin-LEF/TCF is further achieved by nuclear proteins that (1) affect the binding affinity of β-catenin and LEF/TCF to each other and to DNA (such as pygopus and legless in *Drosophila*), (2) disrupt the β-catenin-LEF/TCF-CBP complex, and (3) alter nuclear export of some of the proteins involved in the transcriptional activator complex (Kramps et al. 2002; Parker et al. 2002; Thompson et al. 2002).

For a comprehensive listing of Wnt target genes, the readers are referred to a frequently updated website established by Roel Nusse (http://www.stanford.edu/~rnusse/wntwindow.html). These genes are regulators of survival and proliferation, apoptosis, cell fate determination and differentiation, and self-renewal. Emerging data support the hypothesis that the Wnt pathway is important for embryonic development and maintenance of the stem cell compartment in the base of the hair follicle, the crypt of colonic epithelium, and the bone marrow niche (reviewed in Reya and Clevers 2005). However, the Wnt pathway is tightly regulated and modulated under these cellular contexts. Too much or too little Wnt pathway activity can produce significant phenotypic and functional aberration. Multiple Wnt target genes encode proteins that are directly involved in the canonical Wnt pathway. Notable negative regulators of the pathway that are upregulated include Axin2 and Dkk, providing a negative feedback loop of Wnt pathway activation.

Although there is some apparent functional redundancy in the Wnt pathway, such as Axin1 and 2, GSK3α and β, LEF1, TCF1, 3, and 4, β-catenin and γ-catenin, there are also examples where there are cell type and context-specific differences. Through genetic approaches, the observed mutant phenotypes are not always rescued by expressing another isoform. For example, such is the case with GSK3α and -β in zebrafish cardiogenesis (Lee et al. 2007). This area of intense research interest continues to produce controversy in the field, but most will agree that it is highly unlikely that there is absolute functional redundancy of the various isoforms in all cell types and tissue contexts. Adding to this complexity is the presence of multiple Wnt proteins in vertebrates that can activate one or more Fz receptors and nonconventional Wnt receptors. Only some Wnt proteins can activate the canonical pathway. Furthermore, physiologically relevant binding between the different Wnts and their cognate Fz receptors remains to be determined.

Deregulation of the Wnt pathway is involved in human cancers and other human diseases (reviewed in Logan and Nusse 2004; Giles et al. 2003; Reya and Clevers 2005). Inherited truncations of APC, resulting in loss of function, are found in 85% of cases of familial adenomatous polyposis (FAP) (Groden et al. 1991; Nishisho et al. 1991). This is an autosomal-dominant condition that is characterized by the formation of thousands of adenomatous polyps throughout the colon and rectum. Invariably, these polyps accumulate additional genetic aberrations and undergo malignant transformation in early adulthood. Patients with inherited Axin2 mutations, resulting in loss of function, are predisposed to early colorectal cancer development. Somatic mutations of Axin2 have also been reported (Satoh et al. 2000). Acquired mutations in β-catenin (at key serine sites that are phosphorylated for targeted proteosome degradation) and in APC have been reported in cases of sporadic colorectal cancer (Giles et al. 2003). In hepatocellular carcinoma, 18%–34% of patient samples had exon 3 mutations in the CTNNB1 gene (encoding β-catenin) (Kondo et al. 1999; Legoix et al. 1999; Nhieu et al. 1999). Interestingly, similar mutations have not been found in hematopoietic malignancies. More relevant to this review is the role of the Wnt pathway in maintaining the stem cell compartment, and the focus of this review will be on hematopoiesis and the bone marrow niche, and the leukemia stem cell population.

Wnt Pathway in Self-Renewal and Differentiation of Hematopoietic Stem Cells

Embryonic stem cells are the most primitive stem cells. In mammals, hematopoietic development begins within the mesoderm germ layer in a region described as the aorta-gonad-mesonephros (AGM) Dzierzak 2002). Certain Wnt genes are highly expressed in this area (Takada et al. 1994). In particular, Wnt3a, is highly expressed during embryonic stem cell differentiation, and its overexpression can enhance in vitro hematopoietic commitment (Takada et al. 1994; Lako et al. 2001). Stem cells have the unique property of essentially infinite self-renewal capacity. This entails asymmetric cell division and careful partitioning of a cell to generate a copy of itself and a more differentiated progenitor, or symmetric cell division, which gives rise to two daughter stem cells or two daughter progenitor cells. Such a process could give rise to daughter cells that have maintained the stem cell phenotype by retaining certain subcellular organelles and large protein complexes yet to be characterized. Little is known about how this occurs so precisely, and as a result it is a highly active area of research. Stem cells are required for tissue regeneration at times of normal tissue turnover but also during tissue injury. The Wnt pathway has been identified as one of the key pathways involved in this process. Evidence that the canonical Wnt pathway is important in self-renewal of normal stem cells is best described in the intestinal crypt, hair follicle unit, and hematopoietic bone marrow niche. For an excellent recent review of this subject, readers are referred to the review article written by Reya and Clevers (2005). The ensuing discussion will focus on the hematopoietic bone marrow niche.

Hematopoietic stem cells (HSCs) are a rare population of cells in the bone marrow, estimated at a frequency of 1 in 100,000 cells in mice (Baum et al. 1992; Spangrude et al. 1988). The HSC exists in a stromal cell niche, which maintains the stem cell phenotype and modulates stem cell activity in tissue regeneration. The niche is created by a complex architecture of cells, extracellular matrix, and soluble factors, provided by osteoblasts (Calvi et al. 2003; Zhang et al. 2003) and endothelial cells (Heissig et al. 2002; Kiel et al. 2005; reviewed in Lichtman 1981 and more recently Kiel and Morrison 2006). Hematopoietic tissues were found to express Wnt proteins and their cognate Fzd receptors. LEF-TCF reporter activity is detected in the native bone marrow microenvironment, which provides supporting evidence that Wnt signaling is relevant in vivo (Reya et al. 2003). Evidence that the Wnt pathway is important for hematopoiesis came initially from a set of experiments looking at the effects of Wnt proteins on murine $CD34^+$ cells on hematopoietic colony formation and engraftment in lethally irradiated mice (Austin et al. 1997; Murdoch et al. 2003; Van Den Berg et al. 1998). In these experiments, $CD34^+$ cells were cocultured with stromal cells transfected to overexpress Wnt proteins. This resulted in enhanced colony formation and engraftment in mice. Using a more HSC-enriched population, expression of constitutively active β-catenin (Reya et al. 2003) and Wnt3a treatment of a single cell (Willert et al. 2003) also enhanced reconstitution of lethally irradiated mouse recipients. In the first study, Bcl-2 transgenic recipient mice were used, so this may have modulate the effects of constitutively active β-catenin in vivo. Several reports using conditional genetic approaches support the emerging theory that too much β-catenin activity can inhibit differentiation and inhibit HSC self-renewal and function (Baba et al. 2006; Kirstetter et al. 2006; Scheller et al. 2006). Mice that overexpressed β-catenin developed a bone marrow failure phenotype, suggesting exhaustion of the long-term stem cell pool. Thus, Wnt pathway activation is likely kept in fine balance to meet the demands of tissue regeneration, as physiologically appropriate. This balance may be achieved by Wnt5a, a negative regulator of canonical Wnt3a signaling.

Wnt5a treatment of cultured HSCs under serum-free conditions promoted GSK3b-independent β-catenin degradation and inhibited Wnt3-dependent canonical signaling (Nemeth et al. 2007). However, these cells had enhanced HSC engraftment and repopulation, suggesting a role for the noncanonical pathway in HSC function. Other studies have suggested a similar role for Wnt5a, as only stromal cell lines that support HSC growth and maintenance produce Wnt5a and those stromal cell lines that do not lack Wnt5a production (Hackney et al. 2002). Similar findings have been reported in human hematopoietic progenitors with Wnt5a treatment in vitro (Van Den Berg et al. 1998) and in vivo (Murdoch et al. 2003).

The evidence suggests that both canonical (Wnt3a) and noncanonical pathways (Wnt5a) play a role in expansion of HSCs and progenitors, and maintenance of an undifferentiated phenotype. Interestingly, $Wnt5a^{+/-}$ mice are predisposed to development of myeloid leukemias and lymphomas, and loss of heterozygosity was found in the mice that developed disease (Liang et al. 2003). Thus, in vivo, Wnt5a may act primarily as a tumor-suppressor gene and inhibit the canonical Wnt pathway signaling. The downstream signaling pathway of Wnt5a activation is not clearly defined, but G-protein coupling to phospholipase C and subsequent rise in intracellular calcium and activation of calcium-dependent kinases and protein kinase C have been reported as possibilities (Sheldahl et al. 1999; Kuhl et al. 2000a; Kuhl et al. 2000b).

Wnt proteins are produced by the stem cells (autocrine) and the cells supporting the bone marrow niche (paracrine) (Chiba et al. 2004). Other soluble ligands, such as Sonic hedgehog and Notch, have also been shown to be present in the bone marrow niche (Campbell et al. 2008). Soluble Wnt proteins synergize with stem cell factor (SCF, also known as steel factor) to promote growth and inhibit differentiation of murine hematopoietic progenitors (Austin et al. 1997). Forced overexpression of Axin, a negative regulator of the Wnt pathway, in murine HSC decreased their expansion and inhibited hematopoietic reconstitution in mice (Reya et al. 2003). In other genetic approaches, conditional deletion of β-catenin in mice (Cobas et al. 2004) had no observable consequences in hematopoiesis, suggesting a much needed clarification of β-catenin function in vivo and the potential capacity of other Wnt pathway components to compensate, such as γ-catenin. However, reconstitution in mice was not tested beyond the primary recipient or for more than 8 weeks, but both are better functional discriminators of the HSC versus progenitor population. Conditional deletion of other self-renewal genes such as HOXB4 had few phenotypic consequences indicative of the redundant roles of self-renewal agonists in vivo (Brun et al. 2004). In a recent paper, mice lacking β-catenin in their hematopoietic cells formed HSCs but were deficient in long-term stem cell function. Interestingly, these β-catenin-deficient mice formed BCR-ABL-induced acute lymphoblastic leukemia (ALL), but not AML, suggesting β-catenin is required for myeloid differentiation and transformation (Zhao et al. 2007). Recent in vitro data support a role for γ-catenin in HSC renewal and expansion. Although dissecting the signaling components in the in situ hematopoietic bone marrow niche is challenging, it nonetheless will provide important insights into the interplay of the Wnt, Shh, and Notch pathways in HSC self-renewal and capacity for tissue regeneration. A recent report shows that Notch signaling is required for canonical Wnt pathway-mediated maintenance of undifferentiated HSC (Duncan et al. 2005). In cancer, an emerging theme is that a subpopulation of cells harbors the necessary genetic perturbations and either has the intrinsic capacity to self-renew or acquires this phenotype by subverting elaborate regulatory signals. This subpopulation of cells is capable of propagating cancer and has been described as cancer stem cells (CSCs). It is believed that CSCs, like HSCs, have the cellular property of infinite self-renewal and are relatively quiescent. These are characteristics that impede the cure of human cancers using cytotoxic chemotherapies or cytostatic biologic agents alone. These agents predominately target more differentiated and actively dividing cells.

HSCs are found in higher abundance in the endosteal surface of trabecular bone when compared with the bone marrow. This interaction appears to be facilitated by N-cadherin–expressing osteoblasts (Zhang et al. 2003). The functional significance of this interaction was revealed in a set of experiments using an osteoblast-specific promoter for targeted ablation of osteoblasts in vivo, and there was subsequent decreased bone marrow cellularity, decreased number of c-Kit$^+$/Sca$^+$/Lin$^-$ cells, and increased extramedullary hematopoiesis (Visnjic et al. 2004). In another set of experiments, mature osteoblasts were ablated, however, the above defects were not observed (Corral et al. 1998). Thus, this suggests the HSC niche is likely dependent on immature osteoblasts, in particular osteoblast progenitors, to maintain the self-renewal cues. Increasing bone density by parathyroid hormone administration to mice increased the number of osteoblasts and HSCs in the bone marrow (Calvi et al. 2003). In vitro studies show that multiple Wnt proteins are secreted by osteoblasts, and thus, may be a potential source of Wnt proteins in vivo (Kato et al. 2002; Zhang et al. 2004). Other secreted factors, such as angiopoietin-1 (Suri et al. 1996) and jagged-1 (Calvi et al. 2003), and adhesion molecules (e.g., vascular cell-adhesion molecule-1) facilitate homing and adhesion to the bone marrow niche. The Wnt pathway is essential in normal osteoblast differentiation (Day et al. 2005; Hill et al. 2005) and maintenance of progenitors (Rodda and McMahon 2006). Wnt proteins stimulate osteoblast precursor growth and differentiation (Bradbury et al. 1994; Gong et al. 2001; Rawadi et al. 2003). Transgenic mice expressing Wnt-10b under an adipocyte-specific promoter developed increased bone density and trabecular bone mass (Bennett et al. 2003). Wnt-10b-deficient mice had the opposite phenotype. Knockout of β-catenin produced some contradicting results, in one case a decrease in osteoblasts (Holmen et al. 2005) and in another case no decrease in osteoblasts but an increase in osteoclasts (Glass et al. 2005). Both cases resulted in decrease bone density, and the difference is likely due to the specific genetic approach for generating the knockout, specifically relating to the completeness of the knockout in mature osteoblasts and the promoter used to target osteoblasts. The latter report does raise the question of a role for the canonical Wnt pathway in regulating osteoclast progenitor proliferation and survival. In a mouse model (Gong et al. 2001) and in a human disease, described as osteoporosis-pseudoglioma syndrome, loss of function of LRP5 results in decreased bone density. A dominant-acting missense mutation of LRP5 (Gly Val at codon 171) discovered in human patients abrogates physical interaction with Dkk, a negative regulator of Wnt-mediated activation of the coreceptor (Ai et al. 2005; Boyden et al. 2002). Affected patients have increased bone density, with a thickened mandible and torus palatinus (Boyden et al. 2002). These data clearly support a role for the Wnt pathway in balancing osteoblast and osteoclast activity, and ultimately, the bone marrow niche. The secreted Wnt proteins in this niche likely modulate and regulate hematopoiesis under physiologically relevant conditions.

Deregulation of the Wnt Pathway in Leukemia Stem Cells

The functional description of the leukemia stem cell (LSC) population comes from xenotransplant models what show that a subset of cells engraft and form tumors robustly in serial transplantations. The origin of the LSC is much debated, but one can envision the accumulation of genetic changes occurring in a HSC, which already intrinsically has the cellular mechanisms in place for infinite self-renewal and tissue regeneration. A second scenario is the accumulation of genetic changes in more committed progenitor cells or more differentiated cells. However, these changes must also deregulate self-renewal pathways as these cells intrinsically have only *limited* self-renewal capacity. Indeed, in human acute myelogenous leukemia (AML), as well as in mouse transgenic models, a malignant self-renewing committed myeloid progenitor population could faithfully recapitulate all aspects of acute myelogenous leukemia (Hope et al. 2004; Somervaille and Cleary 2006). In human AML (Simon et al. 2005) and chronic myeloid leukemia (CML) patient samples (Jamieson et al. 2004; AE Abrahamsson and CH Jamieson et al., manuscript submitted), the Wnt pathway is activated as detected by a LEF-TCF reporter construct transduced in the progenitor subpopulation highly enriched for LSCs. A third scenario is that the majority of mutations occur in the HSCs, which retain some ability to differentiate into progenitors that aberrantly gain the capacity to self-renew as demonstrated in human CML as a result of β-catenin activation (AE Abrahamsson and CH Jamieson et al., manuscript submitted; Jamieson et al. 2004; Eisterer et al. 2005). Furthermore, β-catenin activation appears to be important for generation of myeloid blast crisis and BCR-ABL inhibitor-resistant LSC in mouse models of CML (Zhao et al. 2007; Hu et al. 2008). Loss of β-catenin impairs the renewal of normal and CML stem cells in vivo (Zhao et al. 2007), and β-catenin is essential for survival of LSCs insensitive to kinase inhibition (Hu et al. 2008). Moreover, transgenic T-cell acute lymphoblastic leukemia (T-ALL) mouse model research demonstrated that β-catenin activation collaborates with a number of mutations in LSC generation (Guo et al. 2008). Multigenetic events collaboratively contribute to Pten-null leukemia stem cell formation (Guo et al. 2008). Thus, the LSC may originate from either an HSC or a more committed progenitor. It remains to be determined whether the exact origin of the LSC translates to biologically different diseases (Barabe et al. 2007). A key unifying feature of the LSC that is strikingly different from the HSC is the loss of response to physiologic signals, for example, to inhibit or promote tissue regeneration and to give rise to daughter cells capable of orderly differentiation.

Acute Myelogenous Leukemia

AML is a heterogeneous group of diseases characterized by numerous genetic aberrations found in human patient samples (Kelly et al. 2002a). Despite a large compendium of mutations, phenotypically, the leukemia stem cells acquire similar characteristics, including enhanced proliferative and survival potential, impaired differentiation, and infinite self-renewal capacity. This strongly suggests that these genetic aberrations disrupt common signal transduction and transcription factor pathways. Indeed, in AML, perturbations of various genes have been simplified into two functional complementation groups based on current data (Gilliland 2002). Mutations in each complementation group cooperate to promote leukemogenesis.

In the first group, collectively, activating mutations in either FLT3 or RAS occur in approximately 50% of cases of AML. Mutations in the FLT3 gene, either in the juxtamembrane domain or tyrosine kinase domain, result in a constitutively active receptor tyrosine kinase (Nakao et al. 1996; Yamamoto et al. 2001), whereas mutations in the N- and K-RAS genes result in small G-proteins that are insensitive to GTPase activating proteins (Mrozek et al. 2007). As a result, these Ras proteins are unable to hydrolyze GTP and remain in an active, GTP-bound state. Functionally, activating mutations in FLT3 or RAS genes result in constitutive activation of various signaling pathways, such as Ras/MAPK, JAK/STAT, and PI3K/Akt, that confer proliferative and survival advantage but do not seem to affect hematopoietic differentiation or endow the cells with self-renewal capacity (Chan et al. 2004; Kelly et al. 2002b). Mutations in other genes the regulate proliferation and survival are likely present in cases of AML with nonmutated FLT3 or RAS genes.

In the second group, translocations found in AML (also known as AML-associated translocation products, or AATPs) frequently involve the core binding factor (CBF), retinoic acid receptor-α (RARα), myeloid/lymphoid leukemia (MLL), and homeotic (HOX) genes (reviewed in Kelly et al. 2002a; Gilliland 2002). The gene rearrangements produce various fusion genes, for example, AML1-ETO and PML-RARα that acquire a gain-of-function. As a group, there is recruitment of transcriptional corepressors or coactivators, not ordinarily inherent in the wild-type proteins individually, at the promoters of genes important in regulating hematopoietic differentiation and self-renewal. In transgenic murine models, expression of these fusion genes does not uniformly result in AML development. A common theme is that additional mutations, acquired by spontaneous karyotypic abnormalities or induced by chemical mutagenesis, are required to induce leukemic transformation. In cases of AML with normal cytogenetics, mutation in nucleophosmin (NPM1) is a frequent finding. The mutation disrupts the normal pool of nucleophosmin available for corepressor action upon RA-responsive target genes (Liu et al. 2007a). These data support the existence of two different but cooperating complementation groups, and other experimental murine models to test genes in the two different complementation groups support this phenomenon (Gilliland 2002). However, recently, an in vivo model of MLL fusion gene leukemogenesis was established by overexpression in primitive human hematopoietic cells and transplantation into immunodeficient mice (Barabe et al. 2007). The leukemia-initiating cells recapitulated the human disease in mice and retained the capacity to express both myeloid and lymphoid lineage markers. The MLL fusion gene may act as a master regulator of multiple pathways from each complementation group.

There is increasing evidence that the Wnt pathway is frequently deregulated in AML. Unlike in human colorectal cancer, in which inactivating mutations of negative regulators of the pathway, including APC and Axin, and N-terminal activating mutations of β-catenin have been reported (reviewed in Segditsas and Tomlinson 2006), such mutations have not been found in human leukemias. This suggests that deregulation of the Wnt pathway is a consequence of genetic or epigenetic aberrations of genes that cross-talk with or regulate the Wnt pathway. Emerging data suggest overactivation of the Wnt pathway alone is unlikely to produce an LSC because of the propensity of cells with constitutively active β-catenin to undergo apoptosis. Instead, a balanced activation of the Wnt pathway may occur in the human LSCs that have acquired a survival advantage prior to Wnt activation. So which of the two complementation groups would the Wnt pathway fall under? Based on the current evidence, the answer is likely that the Wnt pathway contributes to both complementation groups.

Mutant FLT3, PML-RARα, and AML1-ETO induce downstream Wnt signaling events (reviewed in Mikesch et al. 2007). In a myeloid progenitor cell line 32D, Fzd4 was upregulated in Flt3-ITD-expressing cells. The β-catenin protein level was increased. Furthermore, in five of seven AML samples with Flt3-ITD mutations, there were higher β-catenin protein levels compared with wild-type Flt3 patient samples. Flt3-ITD induced TCF-dependent transcriptional activity and its downstream target gene, c-Myc. Thus, this provides evidence that the Flt3-ITD and Wnt signaling pathway synergize in myeloid leukemic transformation (Tickenbrock et al. 2005). AATPs modulate Wnt pathway genes. For example, AATPs directly activate a γ-catenin promoter reporter construct and *in vitro* replating efficiency is dependent on γ-catenin as tested by RNA interference assays (Zheng et al. 2004). Furthermore, in cell lines expressing either AML-ETO, PML-RARα, or PLZF-RARα, global gene expression analysis showed a group of genes similarly regulated. Multiple Wnt genes were regulated by these fusion proteins, including upregulation of γ-catenin at the mRNA and protein levels (Muller-Tidow et al. 2004). In a recent report, proliferation of MLL-transformed human leukemia cell lines or murine myeloid progenitors was dependent on GSK3β and its destabilization of the cyclin-dependent kinase inhibitor p27 (Wang et al. 2008). Pharmacologic inhibition of GSK3β may be a new therapeutic approach in this subtype of AML, however, this would have to be tested further.

Transcriptional silencing of promoters of tumor suppressor genes is a well-described phenomenon in human cancers (Issa 2007; Glozak and Seto 2007; Wang et al. 2007). This could occur by methylation of CpG islands by DNA methyltransferases that impede binding of transcriptional activators and recruit other chromatin-remodeling proteins that contain methyl-CpG-binding domains. In addition, chromatin structure is directly regulated by histones. Acetylation of histones neutralizes positive charges and reduces their affinity to DNA. As a result, the chromatin structure is looser and more amenable to transcriptional activation. Conversely, histone deacetylases reverse this process. Both of these mechanisms are important in the transcriptional silencing of tumor suppressor genes in leukemogenesis. Medications in clinical use or clinical trials include hypomethylating agents (e.g., 5-azacytidine, 5-aza-2'-deoxycytidine) and histone deacetylase inhibitors, and are variably effective in multiple myeloma, myelodysplastic syndrome, and AML.

Hypermethylation of the promoters of negative regulators of the Wnt pathway, including TLE1, sFRPs, and Dickkopf-1 (DKK1) genes, have been reported in AML patient samples (Fraga et al. 2008; Jost et al. 2008; Suzuki et al. 2007). Methylation of WIF-1 was detected in 47% of acute promyelocytic leukemia patient samples (15 of 32), but not in other subtypes of AML (0 of 50), and was determined to be an independent poor prognosis factor (Chim et al. 2006a). The same group showed less frequent methylation of WIF-1 in ALL (5 of 20) and chronic lymphocytic leukemia (CLL) (5 of 43) (Chim et al. 2006a, 2006b). In vitro studies using human $CD34^+$ bone marrow cells or cord blood cells showed that pretreatment with valproic acid, a histone deacetylase inhibitor, enhanced proliferation and increased self-renewal. This is associated with phosphorylation of GSK3β at serine 9 and upregulation of HoxB4, a downstream Wnt-responsive gene (Bug et al. 2005a). A small clinical study by the same group treating patients with valproic acid and all-transretinoic acid, however, showed only transient disease control in a subset of AML originating from a myeloproliferative disorder (Bug et al. 2005b). In another study, loss of α-catenin expression due to hypermethylation of the CTNNA1 promoter was present in the leukemia-initiating cells of AML or myelodysplastic syndrome patient samples with 5q deletion (Liu et al. 2007b). α-catenin acts as a tumor suppressor, and restoration of its expression in HL-60 cells resulted in reduced proliferation and increased apoptosis.

Increased β-catenin activity, as determined by nuclear β-catenin staining (Mai et al. 2007; Xu et al. 2008) and LEF-TCF reporter activity (Simon et al. 2005), has been reported in AML patient samples. In situ bone marrow staining of β-catenin was associated with a higher International Prognostic Scoring System (IPSS) score and a poor prognosis (Xu et al. 2008). Retrospective analysis of 82 AML patient samples revealed a correlation between detectable β-catenin expression by Western blot analysis and enhanced clonogenic capacity in vitro and worse overall survival when compared with samples with undetectable β-catenin (Ysebaert et al. 2006). A novel zinc finger transcription factor SALL4, a human homolog of *Drosophila* Spalt, is constitutively expressed in AML patient samples and binds to β-catenin and activate genes in the Wnt pathway (Ma et al. 2006). Furthermore, in transgenic SALL4-expressing mice, a myelodysplastic syndrome-like disease preceded development of AML (Ma et al. 2006). A role for γ-catenin has been demonstrated by loss of replating efficiency when γ-catenin was inhibited by RNA interference and augmentation of replating efficiency upon overexpression of γ-catenin in HSCs (Zheng et al. 2004). Furthermore, the AATPs directly activate the γ-catenin promoter (Zheng et al. 2004).

Two myeloid translocation genes, MTG1 and -2, also known as ETO1 and ETO2/CBFA2T3, respectively, are transcriptional corepressors that are disrupted upon fusion gene formation in AML. MTG-related-1 (MTGR1 or CBFA2T2) has been shown to bind to TCF4 and disrupt TCF4-β-catenin activation of Wnt target genes. In the small intestines of mice lacking MTGR-1, there was c-Myc overexpression (Moore et al. 2008). AML/CBFα runt domain transcription factors, such as AML1 and AML2, contain C-terminal motifs, including VWRPY, that recruit binding of TLE1 and convert them to transcriptional repressors of T-cell receptor

(TCR) regulation (Levanon et al. 1998). Similarly, binding of LEF1 to TLE1 changes LEF1 into a transcriptional repressor of TCR and Wnt-responsive genes. This could contribute to the preferential suppression of lymphoid differentiation and lymphoid marker expression in AML leukemogenesis. Others report that a balanced activity of LEF1 is required for normal hematopoietic development. Retroviral transduction of LEF1 or constitutively active LEF1 in murine bone marrow cells transplanted into mice resulted in severe impairment of hematopoiesis and subsequent formation of ALL and AML. Interestingly, it was noted that there was promiscuous expression of myeloid and lymphoid markers in these cells (Petropoulos et al. 2008).

Two hereditary human leukemia syndromes, severe congenital neutropenia (mutation ELA2 gene, encoding protease neutrophil elastase) and familial platelet disorder with AML (mutation AML1 gene, encoding CBFα), have a possible link through LEF1. Lef1 cooperates with CBFα to activate the ELA2 promoter (Li et al. 2004). A state of ELA2 deficiency may predispose myeloid precursors to undergo malignant transformation, although the cellular and molecular bases of this have yet to be determined. MM-1, a newly identified tumor suppressor protein, binds c-Myc and inactivates its transcriptional activity. Mutation of MM-1 has been found in human leukemias and lymphomas. Knockdown of MM-1 resulted in increased expression of Wnt4 and accumulation and translocation of β-catenin to the nucleus followed by transcriptional activation through TCF/LEF1. MM-1 negatively regulates Wnt4 expression by directly binding to its promoter region. Thus, MM-1 influences c-Myc activity by two mechanisms. First, MM-1 directly inhibits its activity, and, second, it downregulates Wnt4 and the Wnt pathway (Yoshida et al. 2008). TCF4 microsatellite instability evaluation in leukemia cell lines (HL60 and Jurkat) showed that frameshift mutations in a region of A(9) repeats gave rise to gene products that localized to the nucleus, but, interestingly, failed to activate the reporter construct when cotransfected with β-catenin (Chang et al. 2006).

Overall, these data indicate that the Wnt pathway is commonly deregulated in AML cells derived from human patients, underscoring the importance of Wnt pathway activation in leukemic transformation.

Acute Lymphoblastic Leukemia

The origin of the LSC in ALL is likely a lymphoid progenitor cell that harbors structural or numeric chromosomal abnormalities, which often involve transcription factors that control differentiation in both B- and T-cell development (Cleary 1991; Look 1997; Rabbitts 1994). However, the most frequent chromosomal aberration is the Philadelphia chromosome or t(9;22), resulting in BCR-ABL p190 or p210 (Faderl et al. 1998). The incidence of Ph-positive ALL increases in older patients and is associated with a poor prognosis. However, the clinical use of tyrosine kinase inhibitors targeting BCR-ABL in combination with chemotherapy or allogeneic stem cell transplant is expected to improve overall survival. As with

AML, deregulated growth and arrested differentiation are the overt phenotype of these cells. Approximately 80% of ALL have immunophenotypes consistent with B-cell progenitors, usually positive for CD10, CD19, and CD20. B-cell ALL cells express surface immunoglobulins, suggesting a more committed progenitor as the origin of the LSC in this rare phenotype. Pre-B ALL cells have cytoplasmic immunoglobulin-μ expression, and patients with this variant have a worse prognosis than early pre-B ALL, who do not have cytoplasmic immunoglobulin-μ expression. More recently, the aggressiveness of pre-B ALL disease has been attributed more specifically to the t(1;19) (Crist et al. 1990). T-cell ALL is less common, accounting for 10%–15% of cases of pediatric ALL and even less frequent in cases of adult ALL, and is historically associated with a worse prognosis than B-lineage ALL. Expression of T-cell surface markers includes CD1, CD2, CD4, CD5, CD8, and CD10. In addition to cytogenetic findings, clonality is established by uniform rearrangements of immunoglobulin genes for B-cell lineage or T-cell receptor genes for T-cell lineage ALL. Mixed-lineage leukemia is a subtype of ALL that has immunophenotype expression of antigens from both B-cell and T-cell lineage. The origin of the LSC has not been clearly established (Greaves et al. 1986).

Chromosomal translocations are found in 75% of ALL, and the distribution of specific subtypes differs remarkably for pediatric versus adult ALL (Pui et al. 1990; Williams et al. 1984). Frequently, the chromosomal breakpoints involve a region in the heavy chain (14q32), κ light chain (2p12), λ light chain (22q11), TCR enhancer (7q35), or TCRA/D enhancer (14q11). A unifying theme is the central role of transcription factor genes translocated to these regions of highly active transcription in lymphoid progenitors. The result is deregulated growth and arrested differentiation. These transcription factors are grouped according to their DNA-binding domains: basic region helix-loop-helix (MYC, LYL1, TAL1, TAL2, BHLHB1), cysteine-rich (LMO1, LMO2), homeodomain (HOX11, HOX11L2, E2A-PBX1), basic region leucine-zipper zinc finger (E2A-HLF), A-T hook minor groove (MLL-AF4, MLL-ENL), and ETS-like or Runt homology (TEL-AML1) (Cleary 1991; Look 1997; Rabbitts 1994). The translocation products form fusion proteins that have altered DNA or protein-binding properties. The deregulated activity of these transcription factors occurs by overexpression of nonmutated genes or chimeric genes or by formation of chimeric transcription factors that result in de novo activity or enhanced activity on promoters of target genes (Rabbitts 1991). Target genes control cell proliferation, survival, apoptosis, and differentiation.

Loss of tumor suppressor genes, by either mutation or deletion, is most notable for TP53 (Wada et al. 1993) and p16INK4A (reviewed in Drexler 1998). Gain of function of RAS genes, by point mutations that lock the GTPase in an active, GTP-bound state, are found in ALL as well, especially in the NRAS genes (Rodenhuis et al. 1986). Hyperdiploidy is a relatively common occurrence in pediatric ALL, and when greater than 50 chromosomes, this subset of patients has a very good prognosis (Ferrando and Look 2000). Although not as well defined for ALL when compared with AML, there are likely cooperating complementation groups that give rise to ALL. Similar to AML, a single genetic event is not likely sufficient to produce the full spectrum of ALL disease. The genetic and molecular events described so far in ALL have been used effectively as prognostic factors to stratify patients into risk groups and to guide therapeutic approaches in pediatric ALL. Detection of minimal residual disease and gene expression profiling are molecular modalities that might improve therapeutic approach even further, especially in adult ALL where cure rates with intensive chemotherapy and allogeneic stem cell transplantation are still very limited.

The importance of activation of self-renewal pathways in ALL has been best demonstrated in T-ALL, in the context of Notch pathway activation (reviewed in Demarest et al. 2008). In this chapter, known connections to the Wnt pathway will be discussed. Wnt signaling is required for normal B- and T-cell development, with the earliest evidence provided by knockout mouse models. LEF1-deficient mice displayed impaired growth and survival of B-cell progenitors (Ranheim et al. 2005; Reya et al. 2000). TCF1-deficient mice have reduced thymocyte numbers (van Genderen et al. 1994; Verbeek et al. 1995), and this phenotype is more severe with $TCF1^{-/-}/LEF1^{-/-}$ double knockout (Held et al. 2003; Okamura et al. 1998). Additional studies support that TCF1 is essential for maintenance of early thymocyte progenitors (Prieve and Waterman 1999; Schilham et al. 1998). In ALL leukemogenesis, a study showed that Wnt3a treatment increased proliferation and survival of ALL cells under conditions of serum deprivation (Khan et al. 2007). The remainder of the evidence in support for a role of the canonical Wnt pathway in ALL leukemogenesis is available in relation to the E2A-Pbx1 translocation product and to the high incidence of downregulation of negative regulators of the pathway.

E2A-Pbx1 is a translocation product of t(1;19) (q23;p13) present in 25% of the cases of pre-B ALL. It is present in 3% of all adult ALL and has been shown to arrest B-cell development (Ferrando and Look 2000). The chimeric transcription factor retains the N-terminal transactivation domain of E2A, while replacing the bHLH DNA- and protein-binding domain with the homeobox DNA-binding domain of Pbx1. Wnt16b is induced by E2A-Pbx1 expression in pre-B ALL cells (McWhirter et al. 1999). Inhibition of Wnt16b induced apoptosis in ALL cell lines (Mazieres et al. 2005). Similar results were obtained in a study using siRNA knockdown of E2A-Pbx1 in 697, an ALL t(1;19) cell line, resulting in decreased EB1 and Wnt16b mRNA transcripts. Interestingly, a t(1;19)-negative cell line also had similar findings, suggesting that the knockdown could cause other cellular effects independent of the E2A-Pbx1 translocation (Casagrande et al. 2006).

A series of papers reported on the hypermethylation of the promoters of negative regulators (Wnt5a, Dkk-3, sFRP1, sFRP2, sFRP4, sFRP5, WIF1, and Hdpr1) of the Wnt pathway in ALL patient samples (Roman-Gomez et al. 2007a; Roman-Gomez et al. 2004; Roman-Gomez et al. 2007b). These findings had downstream effects in the canonical Wnt pathway, including increased nuclear β-catenin staining and upregulation of Wnt-responsive genes WNT16, FZ3, TCF1, LEF1, and cyclin D1 (Roman-Gomez et al. 2007a). Hypermethylation of these genes in newly diagnosed, ALL patient samples is associated with worse disease-free survival and overall survival (Roman-Gomez et al. 2007a). In a large series of 75 Ph^+ ALL patient samples, the methylated group (defined as methylation of at least one of the following genes: sFRP1, sFRP2, sFRP4, sFRP5, Wif1, Dkk3, or Hdpr1), which consisted of 49 patients, had a significantly worse disease-free survival and overall survival when compared with the unmethylated group (Martin et al. 2008).

The importance of the canonical Wnt pathway in ALL development has been best demonstrated thus far in relation to the E2A-Pbx1 translocation product and hypermethylation of genes, which act as negative regulators of the pathway. Thus, canonical Wnt pathway activation, by upstream signaling events yet to be fully characterized, may be the final common thread connecting these molecular events in ALL.

Chronic Myeloid Leukemia

CML is a clonal myeloproliferative disorder initiated at the level of HSC and is characterized by BCR-ABL translocation gene (t(9;22)). The clinical disease is divided into chronic, accelerated, and blast crisis phases. However, based on recent gene expression profiling, the latter two phases are nearly indistinguishable (Radich et al. 2006). Without treatment, virtually all patients in the chronic phase typically progress to blast crisis phase in a median of 5–6 years depending on when the disease was first detected. In the chronic phase, there is typically an elevated leukocyte count, and the cells in the peripheral blood consist of the full spectrum of immature and mature myeloid cells. In other words, there is no obvious maturation defect. However, in the blast crisis phase, characterized by increased myeloblasts or lymphoblasts, there is arrested maturation, and these cells have all the features of malignant transformation including high nuclear:cytoplasmic ratio and prominent nucleoli, and the capacity to metastasize. BCR-ABL induces aberrant proliferation and resistance to apoptosis, while simultaneously also promoting abnormal, error-prone DNA repair mechanisms (reviewed in Melo and Barnes 2007; Skorski 2002). This permissive cellular environment created by BCR-ABL has been described as endowing cells with a "mutator phenotype." Thus, consistent with this, the blast crisis phase is usually accompanied by additional genetic aberrations, including structural or numeric changes in chromosomes, BCR-ABL amplification or duplication, and mutation of oncogenes or tumor suppressor genes (Radich 2007). Patients with this disease have a poor prognosis and are typically refractory to intensive chemotherapy. They have a high rate of death from transplant-related mortality and from relapsed disease after a brief period of remission, following allogeneic stem cell transplant (Radich 2007). Cure rates are only 5% for blast crisis phase CML when compared with at least 60% for de novo AML.

There is emerging evidence that the Wnt pathway is activated in CML, and more specifically in the blast crisis phase of CML. Recently, the granulocyte-macrophage progenitor (GMP) population has been shown to be expanded in the blast crisis phase in patient blood and bone marrow samples (Jamieson et al. 2004). The blast crisis GMP population of cells behaves like the LSC population, as supported by increased nuclear β-catenin staining and LEF-TCF reporter construct activity.

Furthermore, this population serially transplants the disease in recipient mice more readily than the HSC or blast population, in some cases, because of GSK3β deregulation, leading to β-catenin activation. In K562 cells, Fzd5 and Frat2 were highly expressed, suggesting other possible mechanisms of Wnt pathway activation (Saitoh et al. 2001a; Saitoh et al. 2001b). Differential expression of Wnt13 isoforms were detected in leukemic cell lines (U937 and K562) undergoing differentiation, suggesting a role in lineage-specific differentiation (Bunaciu et al. 2008). Alternative promoter usage is thought to give rise to the alternative splice Wnt13-A, -B, and -C isoforms. Wild-type BCR acts as a negative regulator of the canonical Wnt pathway by direct binding to TCF1 (Ress and Moelling 2005; Ress and Moelling 2006) and displacing β-catenin from TCF1-β-catenin complexes. Phosphorylation of BCR leads to the dissociation of the transcriptionally inactive BCR-TCF1 complex. As a fusion protein, BCR-ABL alters the phosphorylation status of BCR and subcellular localization in a manner that enhances TCF1 activity. DNA microarray gene expression analyses were carried out in CML patient samples, which were separated into chronic, accelerated, and blast crisis phases of the disease. Based on signature patterns, a two-phase gene expression chronic versus accelerated/blast crisis phase was observed. Genes involved in the WNT/β-catenin pathway were among those found to be deregulated in the advanced phases of disease (Radich et al. 2006). Deregulation of Sonic hedgehog, Wnt, Hox, and Notch pathway genes is detectable in CML blast crisis patient samples, and signal transducer and activator of transcription 3 (STAT3) may be integral to connect these pathways and modulate their gene expression (Sengupta et al. 2007). Treatment of the K562 cell line with imatinib or etoposide induced apoptosis and the mRNA expression of FZD4, FZD5, FZD7, and WNT5b, suggesting a role for noncanonical Wnt pathway in programmed cell death (Sercan et al. 2007).

Taken together, these data support an important role of the canonical Wnt pathway in the blast crisis phase of CML. During the transformation from chronic phase to blast crisis phase of CML, there are additional genetic aberrations that impart the cells with additional characteristics, including differentiation arrest and the capacity to metastasize. We believe that activation of the canonical Wnt pathway is a key cellular event leading to blast crisis transformation.

Chronic Lymphocytic Leukemia

CLL is a lymphoproliferative disorder characterized by progressive accumulation of functionally defective but mature-appearing lymphocytes with decreased apoptosis or prolonged survival in the peripheral blood. In approximately 95% of cases, CLL is a clonal disorder of more mature B-cell progenitors, coexpressing CD5, CD19, CD20, and CD23. Mutated versus unmutated immunoglobulin VH (IgVH) gene has emerged as two distinct subtypes of CLL disease (Francis et al. 2006). Mutated IgVH, a product of somatic hypermutation during affinity maturation, is associated with a clinically more indolent disease than unmutated IgVH.

Subsequently, expression of zeta chain-associated protein (ZAP-70), a tyrosine kinase associated with the CD3 receptor, has been identified through gene array studies to differentiate accurately between the mutated versus unmutated IgVH groups (Rassenti et al. 2004). The authors conclude that ZAP-70 expression is a clinically useful prognostic marker and identifies a subset of patients that require treatment. Hereditary predisposition for CLL development is based on monozygotic twin analyses and the higher incidence of CLL and other B-cell malignancies in first-degree relatives of CLL patients (Cuttner 1992; Rawstron et al. 2002). Approximately 90% of CLL patient samples have Bcl2 overexpression, and hypomethylation of the promoter is one possible mechanism (Hanada et al. 1993). No specific genetic lesions cause this upregulation. Thus, Bcl2 overexpression may be the final common downstream pathway of yet to be fully characterized upstream events. Some of these events may result from chromosomal abnormalities frequently reported in CLL, including deletion of 13q in 55%, deletion of 11q in 18%, trisomy of 12q in 16%, and deletion of 17p in 7% of CLL (Grever et al. 2007; Dohner et al. 2000), the latter having the overall worse median survival of only 32 months (Dohner et al. 2000).

At present, the role of the Wnt pathway in CLL has been demonstrated based on expression analysis of Wnt pathway members and methylation status of promoters of sFRPs. Subgroup analysis of CLL patient samples based on cytogenetic findings, deletion of common regions, mutation status of IgVH, or ZAP-70 expression is lacking. In a survey of Wnt ligand and Fzd receptor expression in primary CLL patient samples compared with normal samples, Wnt3, Wnt5b, Wnt6, Wnt10a, Wnt14, Wnt16, and Fzd3 were highly expressed in CLL. Downstream gene targets of activated Wnt pathway, LEF1 and cyclinD1, were correspondingly overexpressed. Survival and antiapoptosis cues could be modulated by altering the Wnt pathway (Lu et al. 2004). A cell line engineered to express constitutively active, phosphorylated CD5 had upregulation of genes involved in antiapoptosis, survival, growth and proliferation, and self-renewal (Wnt genes) (Gary-Gouy et al. 2007). Methylation of CpG islands of sFRP promoters has also been reported in CLL patient samples. 5-aza-2'-deoxycytidine treatment resulted in reinduction of sFRP expression. Of the five sFRP family members tested, sFRP1 was hypermethylated in all patient samples tested (Liu et al. 2006). Recently, autologous infusion of leukemic CLL cells transduced ex vivo with adenovirus-CD154 to illicit an immune response and break immune tolerance and induced formation of antibodies to ROR1, an oncofetal receptor for Wnt5a (Fukuda et al. 2008). In vitro studies show Wnt5a can bind to ROR1 and activate NF-κB and enhance survival of CLL cells. These results underscore a role of noncanonical Wnt pathway in leukemia cell survival.

These results also support that the Wnt pathway is commonly deregulated in CLL. It would be important, however, to further characterize this finding in relation to clinical behavior and ZAP-70 expression or mutation status of IgVH. A unifying upstream mechanism for Wnt pathway deregulation that results in overexpression of Wnt proteins or Fzd receptors and hypermethylation of sFRPs has not been determined.

Common Themes and Future Directions

In this chapter, we described the canonical Wnt pathway in some detail, providing a summary of the evidence for its role in self-renewal of HSC and LSC, and gave examples of how Wnt pathway activation is seen in AML, ALL, CML, and CLL. Due to space constraints, we did not discuss the importance of the Wnt pathway in multiple myeloma and how this area of research has increased our understanding of the complex interplay between the bone marrow niche and the CSC. Regrettably we could not cite all the key papers in the field. Several common themes can be ascertained from the topics discussed in this chapter.

First, the Wnt pathway is clearly important in hematopoiesis and self-renewal of HSC and LSC based on the available evidence. However, dissecting the in vivo and in situ role of the Wnt pathway, canonical and noncanonical, remains a major challenge in normal and leukemic human patient samples. Hematopoietic stem and progenitor cells constitute only a small fraction of patient samples and are costly to isolate and study. The limited cell number is a major hindrance in carrying out studies that have been easily done historically in cell lines, murine cells, or bulk tumor cells. Furthermore, the cells have to be isolated away from the bone marrow niche, which plays an important role for maintenance and development of the HSC and LSC. We believe the role of the Wnt pathway in normal development and in human disease is exquisitely dependent on the cell type and specific context.

Second, multiple, sequential acquisition of genetic aberrations is responsible for the development of the acute leukemias, AML and ALL. The LSCs invariably produce progeny cells with deregulated growth and arrested differentiation. Recurring translocation products found in AML and ALL often involve transcription factors that are important in normal hematopoiesis. In addition, the specific translocation products found in AML and ALL patient samples provide clues to clinical outcomes, underscoring that the molecular and cellular events driven by these translocation products are important in defining subgroups in very heterogeneous diseases. The chronic leukemias, CML and CLL, have the common characteristic of retaining some capacity to differentiate into mature-appearing cells. In the case of CML blast crisis transformation, Wnt pathway activation may be the defining event that transforms CML into a rapidly fatal disease if left untreated. Correspondingly, these cells lose the capacity to differentiate, as the name blast crisis suggests.

Third, no mutations have been found in genes involved in the Wnt pathway in human leukemias, yet activation of the Wnt pathway is a common finding in leukemia patient samples. Studies investigating the role of the canonical Wnt pathway in normal hematopoiesis and HSC function have clearly demonstrated that too much or too little Wnt pathway activity can have profound functional consequences in cellular proliferation and self-renewal capacity. Mutations in APC, Axin, and β-catenin, such as that found in colorectal cancer, may activate the canonical Wnt pathway excessively in hematopoietic stem or progenitor cells such that the cells undergo apoptosis before they could acquire additional genetic changes to become LSCs. In contrast, activation of the canonical Wnt pathway by upstream events or coupled with other pathways may provide a balance of Wnt pathway activation such that the cells could survive and acquire additional genetic changes to become LSCs. The noncanonical Wnt pathway may provide some of this balance. As discussed, some of the AML- and ALL-associated translocation products upregulate genes involved in the canonical Wnt pathway. There is also increasing evidence that hypermethylation is a key mechanism for gene silencing of negative regulators of the Wnt pathway, including sFzds, WIF-1, and Dkk. This could provide the mechanistic basis of the relatively effective clinical use of hypomethylating agents, 5-azacytidine and 5-aza-2'-deoxycytidine, in MDS and AML.

Finally, hematologic malignancies are generally not curable using cytotoxic chemotherapies or cytostatic biologic agents alone. The relatively quiescent LSC is not eliminated by these agents. Clearly, a better understanding of the biology of the LSC, and not just the bulk tumor or progeny cells, would give rise to novel molecular targets that could then be exploited to treat specifically the LSC and to assist with detection of minimal residual disease. One could envision that the future of treating hematologic malignancies will rely on combination therapies, cytotoxic agents that effectively eliminate the progenitor population and bulk tumor burden, and molecularly targeted agents that could specifically kill the LSC or induce cycling of the LSC. At present, the only chance for cure is allogeneic stem cell transplantation. The LSC is targeted by a graft-versus-leukemia effect, mediated by a T-cell response to the LSC via minor histocompatibility antigens (Bonnet et al. 1999) or to other LSC-specific antigens. However, this procedure is associated with significant infection-related mortality, graft-versus-host disease–related mortality, and long-term posttransplant complications. Understanding the graft T-cell response to various antigens presented by the host LSC could give rise to alternative cellular therapies that involve ex vivo manipulation of the patient's own T cells to stimulate or augment this response.

References

Ai M, Holmen SL, Van Hul W, Williams BO, Warman ML (2005) Reduced affinity to and inhibition by DKK1 form a common mechanism by which high bone mass-associated missense mutations in LRP5 affect canonical Wnt signaling. Mol Cell Biol 25:4946–4955

Amit S, Hatzubai A, Birman Y, Andersen JS, Ben-Shushan E, Mann M, Ben-Neriah Y, Alkalay I (2002) Axin-mediated CKI phosphorylation of beta-catenin at Ser 45: A molecular switch for the Wnt pathway. Genes Dev 16:1066–1076

Arce L, Yokoyama NN, Waterman ML (2006) Diversity of LEF/TCF action in development and disease. Oncogene 25:7492–7504

Austin TW, Solar GP, Ziegler FC, Liem L, Matthews W (1997) A role for the Wnt gene family in hematopoiesis: Expansion of multilineage progenitor cells. Blood 89:3624–3625

Baba Y, Yokota T, Spits H, Garrett KP, Hayashi S, Kincade PW (2006) Constitutively active beta-catenin promotes expansion of multipotent hematopoietic progenitors in culture. J Immunol 177:2294–2303

Barabe F, Kennedy JA, Hope KJ, Dick JE (2007) Modeling the initiation and progression of human acute leukemia in mice. Science 316:600–604

Baum CM, Weissman IL, Tsukamoto AS, Buckle AM, Peault B (1992) Isolation of a candidate human hematopoietic stem-cell population. Proc Natl Acad Sci USA 89:2804–2808

Behrens J, von Kries JP, Kuhl M, Bruhn L, Wedlich D, Grosschedl R, Birchmeier W (1996) Functional interaction of beta-catenin with the transcription factor LEF-1. Nature 382:638–642

Bennett CN, Hodge CL, MacDougald OA, Schwartz J (2003) Role of Wnt10b and C/EBPalpha in spontaneous adipogenesis of 243 cells. Biochem Biophys Res Commun 302:12–16

Bhanot P, Brink M, Samos CH, Hsieh JC, Wang Y, Macke JP, Andrew D, Nathans J, Nusse R (1996) A new member of the frizzled family from *Drosophila* functions as a Wingless receptor. Nature 382:225–230

Bonnet D, Warren EH, Greenberg PD, Dick JE, Riddell SR (1999) CD8(+) minor histocompatibility antigen-specific cytotoxic T lymphocyte clones eliminate human acute myeloid leukemia stem cells. Proc Natl Acad Sci USA 96:8639–8644

Boyden LM, Mao J, Belsky J, Mitzner L, Farhi A, Mitnick MA, Wu D, Insogna K, Lifton RP (2002) High bone density due to a mutation in LDL-receptor-related protein 5. N Engl J Med 346:1513–1521

Bradbury JM, Niemeyer CC, Dale TC, Edwards PA (1994) Alterations of the growth characteristics of the fibroblast cell line C3H 10T1/2 by members of the Wnt gene family. Oncogene 9:2597–2603

Brun AC, Bjornsson JM, Magnusson M, Larsson N, Leveen P, Ehinger M, Nilsson E, Karlsson S (2004) Hoxb4-deficient mice undergo normal hematopoietic development but exhibit a mild proliferation defect in hematopoietic stem cells. Blood 103:4126–4133

Bug G, Gul H, Schwarz K, Pfeifer H, Kampfmann M, Zheng X, Beissert T, Boehrer S, Hoelzer D, Ottmann OG, Ruthardt M (2005a) Valproic acid stimulates proliferation and self-renewal of hematopoietic stem cells. Cancer Res 65:2537–2541

Bug G, Ritter M, Wassmann B, Schoch C, Heinzel T, Schwarz K, Romanski A, Kramer OH, Kampfmann M, Hoelzer D, Neubauer A, Ruthardt M, Ottmann OG (2005b) Clinical trial of valproic acid and all-trans retinoic acid in patients with poor-risk acute myeloid leukemia. Cancer 104:2717–2725

Bunaciu RP, Tang T, Mao CD (2008) Differential expression of Wnt13 isoforms during leukemic cell differentiation. Oncol Rep 20:195–201

Cadigan KM, Fish MP, Rulifson EJ, Nusse R (1998) Wingless repression of Drosophila frizzled 2 expression shapes the Wingless morphogen gradient in the wing. Cell 93:767–777

Calvi LM, Adams GB, Weibrecht KW, Weber JM, Olson DP, Knight MC, Martin RP, Schipani E, Divieti P, Bringhurst FR, Milner LA, Kronenberg HM, Scadden DT (2003) Osteoblastic cells regulate the haematopoietic stem cell niche. Nature 425:841–846

Campbell C, Risueno RM, Salati S, Guezguez B, Bhatia M (2008) Signal control of hematopoietic stem cell fate: Wnt, Notch, and Hedgehog as the usual suspects. Curr Opin Hematol 15:319–325

Casagrande G, te Kronnie G, Basso G (2006) The effects of siRNA-mediated inhibition of E2A-PBX1 on EB-1 and Wnt16b expression in the 697 pre-B leukemia cell line. Haematologica 91:765–771

Cavallo RA, Cox RT, Moline MM, Roose J, Polevoy GA, Clevers H, Peifer M, Bejsovec A (1998) *Drosophila* Tcf and Groucho interact to repress Wingless signalling activity. Nature 395:604–608

Chan IT, Kutok JL, Williams IR, Cohen S, Kelly L, Shigematsu H, Johnson L, Akashi K, Tuveson DA, Jacks T, Gilliland DG (2004) Conditional expression of oncogenic K-ras from its endogenous promoter induces a myeloproliferative disease. J Clin Invest 113:528–538

Chang HR, Cheng TL, Liu TZ, Hu HS, Hsu LS, Tseng WC, Chen CH, Tsao DA (2006) Genetic and cellular characterizations of human TCF4 with microsatellite instability in colon cancer and leukemia cell lines. Cancer Lett 233:165–171

Chiba H, Kobune M, Kato J, Kawano Y, Ito Y, Nakamura K, Asakura S, Hamada H, Niitsu Y (2004) Wnt3 modulates the characteristics and cobblestone area-supporting activity of human stromal cells. Exp Hematol 32:1194–11203

Chim CS, Chan WW, Pang A, Kwong YL (2006a) Preferential methylation of Wnt inhibitory factor-1 in acute promyelocytic leukemia: An independent poor prognostic factor. Leukemia 20:907–909

Chim CS, Fung TK, Wong KF, Lau JS, Liang R (2006b) Infrequent Wnt inhibitory factor-1 (Wif-1) methylation in chronic lymphocytic leukemia. Leuk Res 30:1135–1139

Cleary ML (1991) Oncogenic conversion of transcription factors by chromosomal translocations. Cell 66:619–622

Cliffe A, Hamada F, Bienz M (2003) A role of dishevelled in relocating Axin to the plasma membrane during wingless signaling. Curr Biol 13:960–966

Cobas M, Wilson A, Ernst B, Mancini SJ, MacDonald HR, Kemler R, Radtke F (2004) Beta-catenin is dispensable for hematopoiesis and lymphopoiesis. J Exp Med 199:221–229

Corral DA, Amling M, Priemel M, Loyer E, Fuchs S, Ducy P, Baron R, Karsenty G (1998) Dissociation between bone resorption and bone formation in osteopenic transgenic mice. Proc Natl Acad Sci USA 95:13835–13840

Crist WM, Carroll AJ, Shuster JJ, Behm FG, Whitehead M, Vietti TJ, Look AT, Mahoney D, Ragab A, Pullen DJ et al (1990) Poor prognosis of children with pre-B acute lymphoblastic leukemia is associated with the t(1;19)(q23;p13): A Pediatric Oncology Group study. Blood 76:117–122

Cuttner J (1992) Increased incidence of hematologic malignancies in first-degree relatives of patients with chronic lymphocytic leukemia. Cancer Invest 10:103–109

Day TF, Guo X, Garrett-Beal L, Yang Y (2005) Wnt/beta-catenin signaling in mesenchymal progenitors controls osteoblast and chondrocyte differentiation during vertebrate skeletogenesis. Dev Cell 8:739–750

Demarest RM, Ratti F, Capobianco AJ (2008) It's T-ALL about Notch. Oncogene 27:5082–5091

Dohner H, Stilgenbauer S, Benner A, Leupolt E, Krober A, Bullinger L, Dohner K, Bentz M, Lichter P (2000) Genomic aberrations and survival in chronic lymphocytic leukemia. N Engl J Med 343:1910–1916

Drexler HG (1998) Review of alterations of the cyclin-dependent kinase inhibitor INK4 family genes p15, p16, p18 and p19 in human leukemia-lymphoma cells. Leukemia 12:845–859

Duncan AW, Rattis FM, DiMascio LN, Congdon KL, Pazianos G, Zhao C, Yoon K, Cook JM, Willert K, Gaiano N, Reya T (2005) Integration of Notch and Wnt signaling in hematopoietic stem cell maintenance. Nat Immunol 6:314–322

Dzierzak E (2002) Hematopoietic stem cells and their precursors: developmental diversity and lineage relationships. Immunol Rev 187:126–138

Eisterer W, Jiang X, Christ O, Glimm H, Lee KH, Pang E, Lambie K, Shaw G, Holyoake TL, Petzer AL, Auewarakul C, Barnett MJ, Eaves CJ, Eaves AC (2005) Different subsets of primary chronic myeloid leukemia stem cells engraft immunodeficient mice and produce a model of the human disease. Leukemia 19:435–441

Faderl S, Kantarjian HM, Talpaz M, Estrov Z (1998) Clinical significance of cytogenetic abnormalities in adult acute lymphoblastic leukemia. Blood 91:3995–4019

Ferrando AA, Look AT (2000) Clinical implications of recurring chromosomal and associated molecular abnormalities in acute lymphoblastic leukemia. Semin Hematol 37:381–395

Fraga MF, Berdasco M, Ballestar E, Ropero S, Lopez-Nieva P, Lopez-Serra L, Martin-Subero JI, Calasanz MJ, de Silanes IL, Setien F, Casado S, Fernandez AF, Siebert R, Stifani S, Esteller M (2008) Epigenetic inactivation of the Groucho homologue gene TLE1 in hematologic malignancies. Cancer Res 68:4116–4122

Francis S, Karanth M, Pratt G, Starczynski J, Hooper L, Fegan C, Pepper C, Valcarcel D, Milligan DW, Delgado J (2006) The effect of immunoglobulin VH gene mutation status and other prognostic factors on the incidence of major infections in patients with chronic lymphocytic leukemia. Cancer 107:1023–1033

Fukuda T, Chen L, Endo T, Tang L, Lu D, Castro JE, Widhopf GF 2nd, Rassenti LZ, Cantwell MJ, Prussak CE, Carson DA, Kipps TJ (2008) Antisera induced by infusions of autologous Ad-CD154-leukemia B cells identify ROR1 as an oncofetal antigen and receptor for Wnt5a. Proc Natl Acad Sci USA 105:3047–3052

Gao ZH, Seeling JM, Hill V, Yochum A, Virshup DM (2002) Casein kinase I phosphorylates and destabilizes the beta-catenin degradation complex. Proc Natl Acad Sci USA 99:1182–1187

Gary-Gouy H, Sainz-Perez A, Marteau JB, Marfaing-Koka A, Delic J, Merle-Beral H, Galanaud P, Dalloul A (2007) Natural phosphorylation of CD5 in chronic lymphocytic leukemia B cells and analysis of CD5-regulated genes in a B cell line suggest a role for CD5 in malignant phenotype. J Immunol 179:4335–4344

Giles RH, van Es JH, Clevers H (2003) Caught up in a Wnt storm: Wnt signaling in cancer. Biochim Biophys Acta 1653:1–24

Gilliland DG (2002) Molecular genetics of human leukemias: New insights into therapy. Semin Hematol 39:6–11

Glass DA 2nd, Bialek P, Ahn JD, Starbuck M, Patel MS, Clevers H, Taketo MM, Long F, McMahon AP, Lang RA, Karsenty G (2005) Canonical Wnt signaling in differentiated osteoblasts controls osteoclast differentiation. Dev Cell 8:751–764

Glozak MA, Seto E (2007) Histone deacetylases and cancer. Oncogene 26:5420–5432

Gong Y, Slee RB, Fukai N, Rawadi G, Roman-Roman S, Reginato AM, Wang H, Cundy T, Glorieux FH, Lev D, Zacharin M, Oexle K, Marcelino J, Suwairi W, Heeger S, Sabatakos G, Apte S, Adkins WN, Allgrove J, Arslan-Kirchner M, Batch JA, Beighton P, Black GC, Boles RG, Boon LM, Borrone C, Brunner HG, Carle GF, Dallapiccola B, De Paepe A, Floege B, Halfhide ML, Hall B, Hennekam RC, Hirose T, Jans A, Juppner H, Kim CA, Keppler-Noreuil K, Kohlschuetter A, LaCombe D, Lambert M, Lemyre E, Letteboer T, Peltonen L, Ramesar RS, Romanengo M, Somer H, Steichen-Gersdorf E, Steinmann B, Sullivan B, Superti-Furga A, Swoboda W, van den Boogaard MJ, Van Hul W, Vikkula M, Votruba M, Zabel B, Garcia T, Baron R, Olsen BR, Warman ML (2001) LDL receptor-related protein 5 (LRP5) affects bone accrual and eye development. Cell 107:513–523

Greaves MF, Chan LC, Furley AJ, Watt SM, Molgaard HV (1986) Lineage promiscuity in hemopoietic differentiation and leukemia. Blood 67:1–11

Greco V, Hannus M, Eaton S (2001) Argosomes: A potential vehicle for the spread of morphogens through epithelia. Cell 106:633–645

Grever MR, Lucas DM, Dewald GW, Neuberg DS, Reed JC, Kitada S, Flinn IW, Tallman MS, Appelbaum FR, Larson RA, Paietta E, Jelinek DF, Gribben JG, Byrd JC (2007) Comprehensive assessment of genetic and molecular features predicting outcome in patients with chronic lymphocytic leukemia: Results from the US Intergroup Phase III Trial E2997. J Clin Oncol 25:799–804

Groden J, Thliveris A, Samowitz W, Carlson M, Gelbert L, Albertsen H, Joslyn G, Stevens J, Spirio L, Robertson M et al (1991) Identification and characterization of the familial adenomatous polyposis coli gene. Cell 66:589–600

Guo W, Lasky JL, Chang CJ, Mosessian S, Lewis X, Xiao Y, Yeh JE, Chen JY, Iruela-Arispe ML, Varella-Garcia M, Wu H (2008) Multi-genetic events collaboratively contribute to Pten-null leukaemia stem-cell formation. Nature 453:529–533

Hackney JA, Charbord P, Brunk BP, Stoeckert CJ, Lemischka IR, Moore KA (2002) A molecular profile of a hematopoietic stem cell niche. Proc Natl Acad Sci USA 99:13061–130666

Hanada M, Delia D, Aiello A, Stadtmauer E, Reed JC (1993) bcl-2 gene hypomethylation and high-level expression in B-cell chronic lymphocytic leukemia. Blood 82:1820–1828

Hart MJ, de los Santos R, Albert IN, Rubinfeld B, Polakis P. 1998. Downregulation of beta-catenin by human Axin and its association with the APC tumor suppressor, beta-catenin and GSK3 beta. Curr Biol 8:573–81

Hecht A, Vleminckx K, Stemmler MP, van Roy F, Kemler R (2000) The p300/CBP acetyltransferases function as transcriptional coactivators of beta-catenin in vertebrates. EMBO J 19:1839–1850

Hedgepeth CM, Deardorff MA, Rankin K, Klein PS (1999) Regulation of glycogen synthase kinase 3beta and downstream Wnt signaling by axin. Mol Cell Biol 19:7147–7157

Heissig B, Hattori K, Dias S, Friedrich M, Ferris B, Hackett NR, Crystal RG, Besmer P, Lyden D, Moore MA, Werb Z, Rafii S (2002) Recruitment of stem and progenitor cells from the bone marrow niche requires MMP-9 mediated release of kit-ligand. Cell 109:625–637

Held W, Clevers H, Grosschedl R (2003) Redundant functions of TCF-1 and LEF-1 during T and NK cell development, but unique role of TCF-1 for Ly49 NK cell receptor acquisition. Eur J Immunol 33:1393–1398

Hendrickx M, Leyns L (2008) Non-conventional Frizzled ligands and Wnt receptors. Dev Growth Differ 50:229–243

Hill TP, Spater D, Taketo MM, Birchmeier W, Hartmann C (2005) Canonical Wnt/beta-catenin signaling prevents osteoblasts from differentiating into chondrocytes. Dev Cell 8:727–738

Hoang B, Moos M Jr, Vukicevic S, Luyten FP (1996) Primary structure and tissue distribution of FRZB, a novel protein related to *Drosophila* frizzled, suggest a role in skeletal morphogenesis. J Biol Chem 271:26131–26137

Holmen SL, Zylstra CR, Mukherjee A, Sigler RE, Faugere MC, Bouxsein ML, Deng L, Clemens TL, Williams BO (2005) Essential role of beta-catenin in postnatal bone acquisition. J Biol Chem 280:21162–21168

Hope KJ, Jin L, Dick JE (2004) Acute myeloid leukemia originates from a hierarchy of leukemic stem cell classes that differ in self-renewal capacity. Nat Immunol 5:738–743

Hsieh JC, Kodjabachian L, Rebbert ML, Rattner A, Smallwood PM, Samos CH, Nusse R, Dawid IB, Nathans J (1999) A new secreted protein that binds to Wnt proteins and inhibits their activities. Nature 398:431–436

Hsu W, Zeng L, Costantini F (1999) Identification of a domain of Axin that binds to the serine/threonine protein phosphatase 2A and a self-binding domain. J Biol Chem 274:3439–3445

Hu Y, Chen Y, Douglas L, Li S. (2008). Beta-catenin is essential for survival of leukemic stem cells insensitive to kinase inhibition in mice with BCR-ABL-induced chronic myeloid leukemia. Leukemia

Issa JP (2007) DNA methylation as a therapeutic target in cancer. Clin Cancer Res 13:1634–1637

Itasaki N, Jones CM, Mercurio S, Rowe A, Domingos PM, Smith JC, Krumlauf R (2003) Wise, a context-dependent activator and inhibitor of Wnt signalling. Development 130:4295–4305

Itoh K, Antipova A, Ratcliffe MJ, Sokol S (2000) Interaction of dishevelled and Xenopus axin-related protein is required for wnt signal transduction. Mol Cell Biol 20:2228–2238

Jamieson CH, Ailles LE, Dylla SJ, Muijtjens M, Jones C, Zehnder JL, Gotlib J, Li K, Manz MG, Keating A, Sawyers CL, Weissman IL (2004) Granulocyte-macrophage progenitors as candidate leukemic stem cells in blast-crisis CML. N Engl J Med 351:657–667

Jonkers J, Korswagen HC, Acton D, Breuer M, Berns A (1997) Activation of a novel proto-oncogene, Frat1, contributes to progression of mouse T-cell lymphomas. EMBO J 16:441–450

Jost E, Schmid J, Wilop S, Schubert C, Suzuki H, Herman JG, Osieka R, Galm O (2008) Epigenetic inactivation of secreted Frizzled-related proteins in acute myeloid leukaemia. Br J Haematol

Katanaev VL, Ponzielli R, Semeriva M, Tomlinson A (2005) Trimeric G protein-dependent frizzled signaling in *Drosophila*. Cell 120:111–222

Kato M, Patel MS, Levasseur R, Lobov I, Chang BH, Glass DA 2nd, Hartmann C, Li L, Hwang TH, Brayton CF, Lang RA, Karsenty G, Chan L (2002) Cbfa1-independent decrease in osteoblast proliferation, osteopenia, and persistent embryonic eye vascularization in mice deficient in Lrp5, a Wnt coreceptor. J Cell Biol 157:303–314

Kaykas A, Yang-Snyder J, Heroux M, Shah KV, Bouvier M, Moon RT (2004) Mutant Frizzled 4 associated with vitreoretinopathy traps wild-type Frizzled in the endoplasmic reticulum by oligomerization. Nat Cell Biol 6:52–58

Kelly L, Clark J, Gilliland DG (2002a) Comprehensive genotypic analysis of leukemia: Clinical and therapeutic implications. Curr Opin Oncol 14:10–18

Kelly LM, Liu Q, Kutok JL, Williams IR, Boulton CL, Gilliland DG (2002b) FLT3 internal tandem duplication mutations associated with human acute myeloid leukemias induce myeloproliferative disease in a murine bone marrow transplant model. Blood 99:310–318

Khan NI, Bradstock KF, Bendall LJ (2007) Activation of Wnt/beta-catenin pathway mediates growth and survival in B-cell progenitor acute lymphoblastic leukaemia. Br J Haematol 138:338–348

Kiel MJ, Morrison SJ (2006) Maintaining hematopoietic stem cells in the vascular niche. Immunity 25:862–864

Kiel MJ, Yilmaz OH, Iwashita T, Terhorst C, Morrison SJ (2005) SLAM family receptors distinguish hematopoietic stem and progenitor cells and reveal endothelial niches for stem cells. Cell 121:1109–1121

Kirstetter P, Anderson K, Porse BT, Jacobsen SE, Nerlov C (2006) Activation of the canonical Wnt pathway leads to loss of hematopoietic stem cell repopulation and multilineage differentiation block. Nat Immunol 7:1048–1056

Kishida S, Yamamoto H, Ikeda S, Kishida M, Sakamoto I, Koyama S, Kikuchi A (1998) Axin, a negative regulator of the wnt signaling pathway, directly interacts with adenomatous polyposis coli and regulates the stabilization of beta-catenin. J Biol Chem 273:10823–10826

Kondo Y, Kanai Y, Sakamoto M, Genda T, Mizokami M, Ueda R, Hirohashi S (1999) Beta-catenin accumulation and mutation of exon 3 of the beta-catenin gene in hepatocellular carcinoma. Jpn J Cancer Res 90:1301–1309

Kramps T, Peter O, Brunner E, Nellen D, Froesch B, Chatterjee S, Murone M, Zullig S, Basler K (2002) Wnt/wingless signaling requires BCL9/legless-mediated recruitment of pygopus to the nuclear beta-catenin-TCF complex. Cell 109:47–60

Kuhl M, Sheldahl LC, Malbon CC, Moon RT (2000a) Ca(2+)/calmodulin-dependent protein kinase II is stimulated by Wnt and Frizzled homologs and promotes ventral cell fates in Xenopus. J Biol Chem 275:12701–12711

Kuhl M, Sheldahl LC, Park M, Miller JR, Moon RT (2000b) The Wnt/Ca2+ pathway: A new vertebrate Wnt signaling pathway takes shape. Trends Genet 16:279–283

Lagna G, Carnevali F, Marchioni M, Hemmati-Brivanlou A (1999) Negative regulation of axis formation and Wnt signaling in Xenopus embryos by the F-box/WD40 protein beta TrCP. Mech Dev 80:101–106

Lako M, Lindsay S, Lincoln J, Cairns PM, Armstrong L, Hole N (2001) Characterization of Wnt gene expression during the differentiation of murine embryonic stem cells in vitro: Role of Wnt3 in enhancing haematopoietic differentiation. Mech Dev 103:49–59

Lee HC, Tsai JN, Liao PY, Tsai WY, Lin KY, Chuang CC, Sun CK, Chang WC, Tsai HJ (2007) Glycogen synthase kinase 3 alpha and 3 beta have distinct functions during cardiogenesis of zebrafish embryo. BMC Dev Biol 7:93

Legoix P, Bluteau O, Bayer J, Perret C, Balabaud C, Belghiti J, Franco D, Thomas G, Laurent-Puig P, Zucman-Rossi J (1999) Beta-catenin mutations in hepatocellular carcinoma correlate with a low rate of loss of heterozygosity. Oncogene 18:4044–4046

Levanon D, Goldstein RE, Bernstein Y, Tang H, Goldenberg D, Stifani S, Paroush Z, Groner Y (1998) Transcriptional repression by AML1 and LEF-1 is mediated by the TLE/Groucho corepressors. Proc Natl Acad Sci USA 95:11590–11595

Li FQ, Person RE, Takemaru K, Williams K, Meade-White K, Ozsahin AH, Gungor T, Moon RT, Horwitz M (2004) Lymphoid enhancer factor-1 links two hereditary leukemia syndromes through core-binding factor alpha regulation of ELA2. J Biol Chem 279:2873–2884

Liang H, Chen Q, Coles AH, Anderson SJ, Pihan G, Bradley A, Gerstein R, Jurecic R, Jones SN (2003) Wnt5a inhibits B cell proliferation and functions as a tumor suppressor in hematopoietic tissue. Cancer Cell 4:349–360

Lichtman MA (1981) The ultrastructure of the hemopoietic environment of the marrow: A review. Exp Hematol 9:391–410

Lin X, Perrimon N (2000) Role of heparan sulfate proteoglycans in cell-cell signaling in *Drosophila*. Matrix Biol 19:303–307

Liu C, Li Y, Semenov M, Han C, Baeg GH, Tan Y, Zhang Z, Lin X, He X (2002) Control of beta-catenin phosphorylation/degradation by a dual-kinase mechanism. Cell 108:837–847

Liu TH, Raval A, Chen SS, Matkovic JJ, Byrd JC, Plass C (2006) CpG island methylation and expression of the secreted frizzled-related protein gene family in chronic lymphocytic leukemia. Cancer Res 66:653–658

Liu H, Tan BC, Tseng KH, Chuang CP, Yeh CW, Chen KD, Lee SC, Yung BY (2007a) Nucleophosmin acts as a novel AP2alpha-binding transcriptional corepressor during cell differentiation. EMBO Rep 8:394–400

Liu TX, Becker MW, Jelinek J, Wu WS, Deng M, Mikhalkevich N, Hsu K, Bloomfield CD, Stone RM, DeAngelo DJ, Galinsky IA, Issa JP, Clarke MF, Look AT (2007b) Chromosome 5q deletion and epigenetic suppression of the gene encoding alpha-catenin (CTNNA1) in myeloid cell transformation. Nat Med 13:78–83

Logan CY, Nusse R (2004) The Wnt signaling pathway in development and disease. Annu Rev Cell Dev Biol 20:781–810

Look AT (1997) Oncogenic transcription factors in the human acute leukemias. Science 278:1059–1064

Lu D, Zhao Y, Tawatao R, Cottam HB, Sen M, Leoni LM, Kipps TJ, Corr M, Carson DA (2004) Activation of the Wnt signaling pathway in chronic lymphocytic leukemia. Proc Natl Acad Sci USA 101:3118–3123

Ma Y, Cui W, Yang J, Qu J, Di C, Amin HM, Lai R, Ritz J, Krause DS, Chai L (2006) SALL4, a novel oncogene, is constitutively expressed in human acute myeloid leukemia (AML) and induces AML in transgenic mice. Blood 108:2726–2735

Mai YJ, Qiu LG, Li ZJ, Yu Z, Li CH, Wang YF, Wang GR, Li Q (2007) The expression of beta-catenin and its significance in leukemia cells. Zhonghua Xue Ye Xue Za Zhi 28:541–544

Mao J, Wang J, Liu B, Pan W, Farr GH 3rd, Flynn C, Yuan H, Takada S, Kimelman D, Li L, Wu D (2001) Low-density lipoprotein receptor-related protein-5 binds to Axin and regulates the canonical Wnt signaling pathway. Mol Cell 7:801–809

Martin V, Agirre X, Jimenez-Velasco A, Jose-Eneriz ES, Cordeu L, Garate L, Vilas-Zornoza A, Castillejo JA, Heiniger A, Prosper F, Torres A, Roman-Gomez J (2008) Methylation status of Wnt signaling pathway genes affects the clinical outcome of Philadelphia-positive acute lymphoblastic leukemia. Cancer Sci

Mazieres J, You L, He B, Xu Z, Lee AY, Mikami I, McCormick F, Jablons DM (2005) Inhibition of Wnt16 in human acute lymphoblastoid leukemia cells containing the t(1;19) translocation induces apoptosis. Oncogene 24:5396–5400

McWhirter JR, Neuteboom ST, Wancewicz EV, Monia BP, Downing JR, Murre C (1999) Oncogenic homeodomain transcription factor E2A-Pbx1 activates a novel WNT gene in pre-B acute lymphoblastoid leukemia. Proc Natl Acad Sci USA 96:11464–11469

Melo JV, Barnes DJ (2007) Chronic myeloid leukaemia as a model of disease evolution in human cancer. Nat Rev Cancer 7:441–453

Mikesch JH, Steffen B, Berdel WE, Serve H, Muller-Tidow C (2007) The emerging role of Wnt signaling in the pathogenesis of acute myeloid leukemia. Leukemia 21:1638–1647

Miller JR (2002) The Wnts. Genome Biol 3: REVIEWS3001

Molenaar M, van de Wetering M, Oosterwegel M, Peterson-Maduro J, Godsave S, Korinek V, Roose J, Destree O, Clevers H (1996) XTcf-3 transcription factor mediates beta-catenin-induced axis formation in Xenopus embryos. Cell 86:391–399

Moore AC, Amann JM, Williams CS, Tahinci E, Farmer TE, Martinez JA, Yang G, Luce KS, Lee E, Hiebert SW (2008) Myeloid translocation gene family members associate with T-cell factors (TCFs) and influence TCF-dependent transcription. Mol Cell Biol 28:977–987

Mrozek K, Dohner H, Bloomfield CD (2007) Influence of new molecular prognostic markers in patients with karyotypically normal acute myeloid leukemia: Recent advances. Curr Opin Hematol 14:106–114

Muller-Tidow C, Steffen B, Cauvet T, Tickenbrock L, Ji P, Diederichs S, Sargin B, Kohler G, Stelljes M, Puccetti E, Ruthardt M, deVos S, Hiebert SW, Koeffler HP, Berdel WE, Serve H (2004) Translocation products in acute myeloid leukemia activate the Wnt signaling pathway in hematopoietic cells. Mol Cell Biol 24:2890–2904

Murdoch B, Chadwick K, Martin M, Shojaei F, Shah KV, Gallacher L, Moon RT, Bhatia M (2003) Wnt-5A augments repopulating capacity and primitive hematopoietic development of human blood stem cells in vivo. Proc Natl Acad Sci USA 100:3422–3427

Nakao M, Yokota S, Iwai T, Kaneko H, Horiike S, Kashima K, Sonoda Y, Fujimoto T, Misawa S (1996) Internal tandem duplication of the flt3 gene found in acute myeloid leukemia. Leukemia 10:1911–1918

Nelson WJ, Nusse R (2004) Convergence of Wnt, beta-catenin, and cadherin pathways. Science 303:1483–1487

Nemeth MJ, Topol L, Anderson SM, Yang Y, Bodine DM (2007) Wnt5a inhibits canonical Wnt signaling in hematopoietic stem cells and enhances repopulation. Proc Natl Acad Sci USA 104:15436–15441

Nhieu JT, Renard CA, Wei Y, Cherqui D, Zafrani ES, Buendia MA (1999) Nuclear accumulation of mutated beta-catenin in hepatocellular carcinoma is associated with increased cell proliferation. Am J Pathol 155:703–710

Nishisho I, Nakamura Y, Miyoshi Y, Miki Y, Ando H, Horii A, Koyama K, Utsunomiya J, Baba S, Hedge P (1991) Mutations of chromosome 5q21 genes in FAP and colorectal cancer patients. Science 253:665–669

Nusse R, Brown A, Papkoff J, Scambler P, Shackleford G, McMahon A, Moon R, Varmus H (1991) A new nomenclature for int-1 and related genes: The Wnt gene family. Cell 64:231

Nusslein-Volhard C, Wieschaus E (1980) Mutations affecting segment number and polarity in Drosophila. Nature 287:795–801

Okamura RM, Sigvardsson M, Galceran J, Verbeek S, Clevers H, Grosschedl R (1998) Redundant regulation of T cell differentiation and TCRalpha gene expression by the transcription factors LEF-1 and TCF-1. Immunity 8:11–20

Ozawa M, Baribault H, Kemler R (1989) The cytoplasmic domain of the cell adhesion molecule uvomorulin associates with three independent proteins structurally related in different species. EMBO J 8:1711–1717

Parker DS, Jemison J, Cadigan KM (2002) Pygopus, a nuclear PHD-finger protein required for Wingless signaling in *Drosophila*. Development 129:2565–2576

Petropoulos K, Arseni N, Schessl C, Stadler CR, Rawat VP, Deshpande AJ, Heilmeier B, Hiddemann W, Quintanilla-Martinez L, Bohlander SK, Feuring-Buske M, Buske C (2008) A novel role for Lef-1, a central transcription mediator of Wnt signaling, in leukemogenesis. J Exp Med 205:515–522

Prieve MG, Waterman ML (1999) Nuclear localization and formation of beta-catenin-lymphoid enhancer factor 1 complexes are not sufficient for activation of gene expression. Mol Cell Biol 19:4503–4015

Pui CH, Crist WM, Look AT (1990) Biology and clinical significance of cytogenetic abnormalities in childhood acute lymphoblastic leukemia. Blood 76:1449–1463

Rabbitts TH (1991) Translocations, master genes, and differences between the origins of acute and chronic leukemias. Cell 67:641–644

Rabbitts TH (1994) Chromosomal translocations in human cancer. Nature 372:143–149

Radich JP (2007) The biology of CML blast crisis. Hematology Am Soc Hematol Educ Program 2007:384–391

Radich JP, Dai H, Mao M, Oehler V, Schelter J, Druker B, Sawyers C, Shah N, Stock W, Willman CL, Friend S, Linsley PS (2006) Gene expression changes associated with progression and response in chronic myeloid leukemia. Proc Natl Acad Sci USA 103:2794–2799

Ranheim EA, Kwan HC, Reya T, Wang YK, Weissman IL, Francke U (2005) Frizzled 9 knock-out mice have abnormal B-cell development. Blood 105:2487–2494

Rassenti LZ, Huynh L, Toy TL, Chen L, Keating MJ, Gribben JG, Neuberg DS, Flinn IW, Rai KR, Byrd JC, Kay NE, Greaves A, Weiss A, Kipps TJ (2004) ZAP-70 compared with immunoglobulin heavy-chain gene mutation status as a predictor of disease progression in chronic lymphocytic leukemia. N Engl J Med 351:893–901

Rattner A, Hsieh JC, Smallwood PM, Gilbert DJ, Copeland NG, Jenkins NA, Nathans J (1997) A family of secreted proteins contains homology to the cysteine-rich ligand-binding domain of frizzled receptors. Proc Natl Acad Sci USA 94:2859–2863

Rawadi G, Vayssiere B, Dunn F, Baron R, Roman-Roman S (2003) BMP-2 controls alkaline phosphatase expression and osteoblast mineralization by a Wnt autocrine loop. J Bone Miner Res 18:1842–1853

Rawstron AC, Yuille MR, Fuller J, Cullen M, Kennedy B, Richards SJ, Jack AS, Matutes E, Catovsky D, Hillmen P, Houlston RS (2002) Inherited predisposition to CLL is detectable as subclinical monoclonal B-lymphocyte expansion. Blood 100:2289–2290

Ress A, Moelling K (2005) Bcr is a negative regulator of the Wnt signalling pathway. EMBO Rep 6:1095–1100

Ress A, Moelling K (2006) Bcr interferes with beta-catenin-Tcf1 interaction. FEBS Lett 580:1227–1230

Reya T, Clevers H (2005) Wnt signalling in stem cells and cancer. Nature 434:843–850
Reya T, O'Riordan M, Okamura R, Devaney E, Willert K, Nusse R, Grosschedl R (2000) Wnt signaling regulates B lymphocyte proliferation through a LEF-1 dependent mechanism. Immunity 13:15–24
Reya T, Duncan AW, Ailles L, Domen J, Scherer DC, Willert K, Hintz L, Nusse R, Weissman IL (2003) A role for Wnt signalling in self-renewal of haematopoietic stem cells. Nature 423:409–414
Rijsewijk F, Schuermann M, Wagenaar E, Parren P, Weigel D, Nusse R (1987) The *Drosophila* homolog of the mouse mammary oncogene INT-1 is identical to the segment polarity gene wingless. Cell 50:649–657
Rodda SJ, McMahon AP (2006) Distinct roles for Hedgehog and canonical Wnt signaling in specification, differentiation and maintenance of osteoblast progenitors. Development 133:3231–3244
Rodenhuis S, Bos JL, Slater RM, Behrendt H, van 't Veer M, Smets LA (1986) Absence of oncogene amplifications and occasional activation of N-ras in lymphoblastic leukemia of childhood. Blood 67:1698–1704
Roman-Gomez J, Jimenez-Velasco A, Agirre X, Castillejo JA, Navarro G, Barrios M, Andreu EJ, Prosper F, Heiniger A, Torres A (2004) Transcriptional silencing of the Dickkopfs-3 (Dkk-3) gene by CpG hypermethylation in acute lymphoblastic leukaemia. Br J Cancer 91:707–713
Roman-Gomez J, Cordeu L, Agirre X, Jimenez-Velasco A, San Jose-Eneriz E, Garate L, Calasanz MJ, Heiniger A, Torres A, Prosper F (2007a) Epigenetic regulation of Wnt-signaling pathway in acute lymphoblastic leukemia. Blood 109:3462–3469
Roman-Gomez J, Jimenez-Velasco A, Cordeu L, Vilas-Zornoza A, San Jose-Eneriz E, Garate L, Castillejo JA, Martin V, Prosper F, Heiniger A, Torres A, Agirre X (2007b) WNT5A, a putative tumour suppressor of lymphoid malignancies, is inactivated by aberrant methylation in acute lymphoblastic leukaemia. Eur J Cancer 43:2736–2746
Saitoh T, Hirai M, Katoh M (2001a) Molecular cloning and characterization of human Frizzled-5 gene on chromosome 2q33.3–q34 region. Int J Oncol 19:105–110
Saitoh T, Moriwaki J, Koike J, Takagi A, Miwa T, Shiokawa K, Katoh M (2001b) Molecular cloning and characterization of FRAT2, encoding a positive regulator of the WNT signaling pathway. Biochem Biophys Res Commun 281:815–820
Sakanaka C, Williams LT (1999) Functional domains of axin. Importance of the C terminus as an oligomerization domain. J Biol Chem 274:14090–14093
Satoh S, Daigo Y, Furukawa Y, Kato T, Miwa N, Nishiwaki T, Kawasoe T, Ishiguro H, Fujita M, Tokino T, Sasaki Y, Imaoka S, Murata M, Shimano T, Yamaoka Y, Nakamura Y (2000) AXIN1 mutations in hepatocellular carcinomas, and growth suppression in cancer cells by virus-mediated transfer of AXIN1. Nat Genet 24:245–250
Scheller M, Huelsken J, Rosenbauer F, Taketo MM, Birchmeier W, Tenen DG, Leutz A (2006) Hematopoietic stem cell and multilineage defects generated by constitutive beta-catenin activation. Nat Immunol 7:1037–1047
Schilham MW, Wilson A, Moerer P, Benaissa-Trouw BJ, Cumano A, Clevers HC (1998) Critical involvement of Tcf-1 in expansion of thymocytes. J Immunol 161:3984–3991
Segditsas S, Tomlinson I (2006) Colorectal cancer and genetic alterations in the Wnt pathway. Oncogene 25:7531–7537
Sengupta A, Banerjee D, Chandra S, Banerji SK, Ghosh R, Roy R, Banerjee S (2007) Deregulation and cross talk among Sonic hedgehog, Wnt, Hox and Notch signaling in chronic myeloid leukemia progression. Leukemia 21:949–955
Sercan HO, Pehlivan M, Simsek O, Ates H, Sercan Z (2007) Induction of apoptosis increases expression of non-canonical WNT genes in myeloid leukemia cell lines. Oncol Rep 18:1563–1569
Sheldahl LC, Park M, Malbon CC, Moon RT (1999) Protein kinase C is differentially stimulated by Wnt and Frizzled homologs in a G-protein-dependent manner. Curr Biol 9:695–698
Simon M, Grandage VL, Linch DC, Khwaja A (2005) Constitutive activation of the Wnt/beta-catenin signalling pathway in acute myeloid leukaemia. Oncogene 24:2410–2420
Skorski T (2002) Oncogenic tyrosine kinases and the DNA-damage response. Nat Rev Cancer 2:351–360
Slusarski DC, Corces VG, Moon RT (1997) Interaction of Wnt and a Frizzled homologue triggers G-protein-linked phosphatidylinositol signalling. Nature 390:410–413

Somervaille TC, Cleary ML (2006) Identification and characterization of leukemia stem cells in murine MLL-AF9 acute myeloid leukemia. Cancer Cell 10:257–268

Spangrude GJ, Heimfeld S, Weissman IL (1988) Purification and characterization of mouse hematopoietic stem cells. Science 241:58–62

Staal FJ, Clevers HC (2005) WNT signalling and haematopoiesis: A WNT-WNT situation. Nat Rev Immunol 5:21–30

Suri C, Jones PF, Patan S, Bartunkova S, Maisonpierre PC, Davis S, Sato TN, Yancopoulos GD (1996) Requisite role of angiopoietin-1, a ligand for the TIE2 receptor, during embryonic angiogenesis. Cell 87:1171–1180

Suzuki R, Onizuka M, Kojima M, Shimada M, Fukagawa S, Tsuboi K, Kobayashi H, Shintani A, Ogawa Y, Kawada H, Hotta T, Ando K (2007) Preferential hypermethylation of the Dickkopf-1 promoter in core-binding factor leukaemia. Br J Haematol 138:624–631

Takada S, Stark KL, Shea MJ, Vassileva G, McMahon JA, McMahon AP (1994) Wnt-3a regulates somite and tailbud formation in the mouse embryo. Genes Dev 8:174–189

Takemaru KI, Moon RT (2000) The transcriptional coactivator CBP interacts with beta-catenin to activate gene expression. J Cell Biol 149:249–254

Tamai K, Semenov M, Kato Y, Spokony R, Liu C, Katsuyama Y, Hess F, Saint-Jeannet JP, He X (2000) LDL-receptor-related proteins in Wnt signal transduction. Nature 407:530–535

Tamai K, Zeng X, Liu C, Zhang X, Harada Y, Chang Z, He X (2004) A mechanism for Wnt coreceptor activation. Mol Cell 13:149–156

Thompson B, Townsley F, Rosin-Arbesfeld R, Musisi H, Bienz M (2002) A new nuclear component of the Wnt signalling pathway. Nat Cell Biol 4:367–373

Tickenbrock L, Schwable J, Wiedehage M, Steffen B, Sargin B, Choudhary C, Brandts C, Berdel WE, Muller-Tidow C, Serve H (2005) Flt3 tandem duplication mutations cooperate with Wnt signaling in leukemic signal transduction. Blood 105:3699–3706

Tolwinski NS, Wieschaus E (2004) A nuclear function for armadillo/beta-catenin. PLoS Biol 2:E95

van de Wetering M, Cavallo R, Dooijes D, van Beest M, van Es J, Loureiro J, Ypma A, Hursh D, Jones T, Bejsovec A, Peifer M, Mortin M, Clevers H (1997) Armadillo coactivates transcription driven by the product of the *Drosophila* segment polarity gene dTCF. Cell 88:789–799

Van Den Berg DJ, Sharma AK, Bruno E, Hoffman R (1998) Role of members of the Wnt gene family in human hematopoiesis. Blood 92:3189–3202

van Genderen C, Okamura RM, Farinas I, Quo RG, Parslow TG, Bruhn L, Grosschedl R (1994) Development of several organs that require inductive epithelial-mesenchymal interactions is impaired in LEF-1-deficient mice. Genes Dev 8:2691–2703

Verbeek S, Izon D, Hofhuis F, Robanus-Maandag E, te Riele H, van de Wetering M, Oosterwegel M, Wilson A, MacDonald HR, Clevers H (1995) An HMG-box-containing T-cell factor required for thymocyte differentiation. Nature 374:70–74

Visnjic D, Kalajzic Z, Rowe DW, Katavic V, Lorenzo J, Aguila HL (2004) Hematopoiesis is severely altered in mice with an induced osteoblast deficiency. Blood 103:3258–3264

Wada M, Bartram CR, Nakamura H, Hachiya M, Chen DL, Borenstein J, Miller CW, Ludwig L, Hansen-Hagge TE, Ludwig WD et al (1993) Analysis of p53 mutations in a large series of lymphoid hematologic malignancies of childhood. Blood 82:3163–3169

Wang S, Krinks M, Lin K, Luyten FP, Moos M Jr (1997) Frzb, a secreted protein expressed in the Spemann organizer, binds and inhibits Wnt-8. Cell 88:757–766

Wang GG, Allis CD, Chi P (2007) Chromatin remodeling and cancer, Part I: Covalent histone modifications. Trends Mol Med 13:363–372

Wang Z, Smith KS, Murphy M, Piloto O, Somervaille TC, Cleary ML (2008) Glycogen synthase kinase 3 in MLL leukaemia maintenance and targeted therapy. Nature

Wehrli M, Dougan ST, Caldwell K, O'Keefe L, Schwartz S, Vaizel-Ohayon D, Schejter E, Tomlinson A, DiNardo S (2000) Arrow encodes an LDL-receptor-related protein essential for Wingless signalling. Nature 407:527–530

Willert K, Brown JD, Danenberg E, Duncan AW, Weissman IL, Reya T, Yates JR 3rd, Nusse R (2003) Wnt proteins are lipid-modified and can act as stem cell growth factors. Nature 423:448–452

Williams DL, Look AT, Melvin SL, Roberson PK, Dahl G, Flake T, Stass S (1984) New chromosomal translocations correlate with specific immunophenotypes of childhood acute lymphoblastic leukemia. Cell 36:101–109

Winston JT, Strack P, Beer-Romero P, Chu CY, Elledge SJ, Harper JW (1999) The SCFbeta-TRCP-ubiquitin ligase complex associates specifically with phosphorylated destruction motifs in IkappaBalpha and beta-catenin and stimulates IkappaBalpha ubiquitination in vitro. Genes Dev 13:270–283

Wu J, Cohen SM (2002) Repression of teashirt marks the initiation of wing development. Development 129:2411–2418

Xu J, Suzuki M, Niwa Y, Hiraga J, Nagasaka T, Ito M, Nakamura S, Tomita A, Abe A, Kiyoi H, Kinoshita T, Naoe T (2008) Clinical significance of nuclear non-phosphorylated beta-catenin in acute myeloid leukaemia and myelodysplastic syndrome. Br J Haematol 140:394–401

Yamamoto Y, Kiyoi H, Nakano Y, Suzuki R, Kodera Y, Miyawaki S, Asou N, Kuriyama K, Yagasaki F, Shimazaki C, Akiyama H, Saito K, Nishimura M, Motoji T, Shinagawa K, Takeshita A, Saito H, Ueda R, Ohno R, Naoe T (2001) Activating mutation of D835 within the activation loop of FLT3 in human hematologic malignancies. Blood 97:2434–2439

Yanagawa S, van Leeuwen F, Wodarz A, Klingensmith J, Nusse R (1995) The dishevelled protein is modified by wingless signaling in *Drosophila*. Genes Dev 9:1087–1097

Yanagawa S, Matsuda Y, Lee JS, Matsubayashi H, Sese S, Kadowaki T, Ishimoto A (2002) Casein kinase I phosphorylates the Armadillo protein and induces its degradation in *Drosophila*. EMBO J 21:1733–1742

Yang J, Wu J, Tan C, Klein PS (2003) PP2A:B56epsilon is required for Wnt/beta-catenin signaling during embryonic development. Development 130:5569–5578

Yoshida T, Kitaura H, Hagio Y, Sato T, Iguchi-Ariga SM, Ariga H (2008) Negative regulation of the Wnt signal by MM-1 through inhibiting expression of the wnt4 gene. Exp Cell Res 314:1217–1228

Yoshikawa S, McKinnon RD, Kokel M, Thomas JB (2003) Wnt-mediated axon guidance via the *Drosophila* derailed receptor. Nature 422:583–588

Yost C, Torres M, Miller JR, Huang E, Kimelman D, Moon RT (1996) The axis-inducing activity, stability, and subcellular distribution of beta-catenin is regulated in Xenopus embryos by glycogen synthase kinase 3. Genes Dev 10:1443–1454

Yost C, Farr GH 3rd, Pierce SB, Ferkey DM, Chen MM, Kimelman D (1998) GBP, an inhibitor of GSK-3, is implicated in Xenopus development and oncogenesis. Cell 93:1031–1041

Ysebaert L, Chicanne G, Demur C, De Toni F, Prade-Houdellier N, Ruidavets JB, Mansat-De Mas V, Rigal-Huguet F, Laurent G, Payrastre B, Manenti S, Racaud-Sultan C (2006) Expression of beta-catenin by acute myeloid leukemia cells predicts enhanced clonogenic capacities and poor prognosis. Leukemia 20:1211–1216

Yuan H, Mao J, Li L, Wu D (1999) Suppression of glycogen synthase kinase activity is not sufficient for leukemia enhancer factor-1 activation. J Biol Chem 274:30419–30423

Zecca M, Basler K, Struhl G (1996) Direct and long-range action of a wingless morphogen gradient. Cell 87:833–844

Zhang J, Niu C, Ye L, Huang H, He X, Tong WG, Ross J, Haug J, Johnson T, Feng JQ, Harris S, Wiedemann LM, Mishina Y, Li L (2003) Identification of the haematopoietic stem cell niche and control of the niche size. Nature 425:836–841

Zhang Y, Wang Y, Li X, Zhang J, Mao J, Li Z, Zheng J, Li L, Harris S, Wu D (2004) The LRP5 high-bone-mass G171V mutation disrupts LRP5 interaction with Mesd. Mol Cell Biol 24:4677–4684

Zhao C, Blum J, Chen A, Kwon HY, Jung SH, Cook JM, Lagoo A, Reya T (2007) Loss of beta-catenin impairs the renewal of normal and CML stem cells in vivo. Cancer Cell 12:528–541

Zheng X, Beissert T, Kukoc-Zivojnov N, Puccetti E, Altschmied J, Strolz C, Boehrer S, Gul H, Schneider O, Ottmann OG, Hoelzer D, Henschler R, Ruthardt M (2004) Gamma-catenin contributes to leukemogenesis induced by AML-associated translocation products by increasing the self-renewal of very primitive progenitor cells. Blood 103:3535–3543

Index